Industry Guide to Polymer Nanocomposites

Edited by Dr Günter Beyer
Kabelwerk Eupen AG

First published by Plastics Information Direct – a division
of Applied Market Information Ltd – in 2009
Republished by Elsevier

AMSTERDAM • BOSTON • HEIDELBERG • LONDON
NEW YORK • OXFORD • PARIS • SAN DIEGO
SAN FRANCISCO • SINGAPORE • SYDNEY • TOKYO

ELSEVIER

Industry Guide to Polymer Nanocomposites

First published by Plastics Information Direct – a division of Applied Market
Information Ltd – in 2009
Republished by Elsevier

Elsevier
Radarweg 29, PO Box 211, 1000 AE Amsterdam, Netherlands
The Boulevard, Langford Lane, Kidlington, Oxford OX5 1GB, UK
50 Hampshire St, 5th Floor, Cambridge, MA 02139, USA

ISBN: 978-1-906479-04-6

Contents

Industry Guide to Nanocomposites

Developments in Commercial Polymer Nanocomposite Materials

Working with Nanocomposite Materials

Unique Properties of Polymer Nanocomposites

Polymer Nanocomposites in Demanding Industrial Applications

Chapter 1

Synthesis, structure, properties, and characterization of organically modified clay minerals

Hendrik Heinz, Department of Polymer Engineering, University of Akron, Ohio, USA

Clay minerals, and particularly organically modified clay minerals, have been extensively used as filler materials in polymer/layered silicate nanocomposites. Already at low single digit volume percentages, the nanometer-thick layers reduce the gas permeability, increase mechanical stability, and improve electrical insulating properties. A remaining challenge is the achievement of control over the morphology at the nanoscale in which the modification of the surface of the clay mineral layers using surfactants plays a critical role.

1.1 Overview of clay minerals

Clay minerals are abundant in soil and form a variety of complex chemical species such as montmorillonite and mica [1,2]. Clay minerals can be considered as annealing products of mixtures of oxides such as SiO_2, Al_2O_3, K_2O, Na_2O, MgO, CaO, Fe_3O_4, as well as H_2O, in variable molar ratio under high temperature and high pressure. The variation in stoichiometry and reaction conditions, including weathering and hydrothermal conditions, accounts for the large number of known clay minerals, and provides an opportunity to manufacture synthetic clay minerals of defined composition such as laponite using similar reaction conditions.

Clay minerals are typically layered silicates and aluminates containing alkali or earth alkali cations between the layers, as shown in Figure 1.1. The cations are electrostatically bound to the interlayer space via charge defects in the layers such as $SiO_2 \rightarrow AlO_2^- \cdots K^+$ or $AlO(OH) \rightarrow MgO(OH)^- \cdots Na^+$ defects. The deficiency in one valence electron of Al versus Si as well as of Mg versus Al introduces a negative charge on the surrounding oxygen atoms at the defect sites in the lattice to maintain an isoelectronic framework.[3] The cations between the layers compensate for the negative charge, render the mineral surface hydrophilic, and contribute to the swelling of soil. Industrial interest in clay minerals originates from their role as waste products in mining operations, and first reported uses include mica paper for electrical insulation and montmorillonite-containing slurries for drilling liquids [4,5].

The structure of the plate-like minerals is a sequence of rigid, bonded layers and non-bonded, more flexible interlayers containing alkali or earth alkali cations. In 2:1 layered silicates, each layer typically consists of three covalently interconnected sheets: a tetrahedral SiO_2 sheet, an octahedral $AlO(OH)$ sheet, and a tetrahedral SiO_2 sheet. The octahedral sheet can also consist of other oxides such as $Mg_3O_2(OH)_2$. Each of the sheets can include charge defects, which determine the amount of cations in the interlayer space. The 2:1 layered silicate structures are called dioctahedral or trioctahedral depending on amount of metal cations in the middle sheet, and form the majority of clay minerals currently used in nanocomposites. There are also 1:1 layered silicates with a two-sheet structure, e.g., including a tetrahedral SiO_2 sheet and an octahedral $AlO(OH)$ sheet. Charge defects and interlayer cations are rarely found in 1:1 layered silicates. Examples include allophane and halloysite, for which further details can be found in the literature [1,6].

The 2:1 layered silicates are mechanically more stable than 1:1 layered silicates, and we distinguish smectites, vermiculites, and micas according to an increasing amount of exchangeable cations per surface area, Table 1.1. The cation density and physico-chemical properties are best characterized by the chemical formula, however, the cation exchange capacity (CEC) is an efficient and widespread value used to express the molar amount of exchangeable cations of charge $+1e$ per unit mass of clay mineral in meq/100g. For example, if 100 g clay mineral contains 0.1 mol (= 100 mmol) of exchangeable Na^+ ions or 0.05 mol (= 50 mmol) of exchangeable Ca^{2+} ions, the CEC equals 100 meq/100g, Table 1.1. The exact value of the CEC depends somewhat on the amount of isomorphous substitution in the clay mineral, e.g., Fe^{3+} against Al^{3+}, even when the amount of exchangeable cations per unit of surface area is constant. Muscovite mica possesses the highest possible cation density per surface area in clay minerals (2.14 cations per nm^2) because each cavity in the surface is occupied with exactly one alkali cation in the layered structure. Therefore, muscovite mica can also serve as a reference (1.0 or 100 %) to express the number of exchangeable cations per formula unit. Due to essentially identical structure of the tetrahedral SiO_2 sheets, including some $SiO_2 \rightarrow AlO_2^- \cdots K^+$ defects, the lattice parameters in the plane of the sheets for pyrophyllite, montmorillonites, and mica are almost identical. However, variable amounts of defects, differences in cations, and isomorphous substitution cause variations in the typical gallery height (perpendicular to the plane of the sheets) of 1.0 nm by about ±10 %. The synthetic clay mineral laponite, see Table 1.1, differs from montmorillonite in the replacement of Al^{3+}/Mg^{2+} in the octahedral layer by Mg^{2+}/Li^+, so that the octahedral layer adopts a modified trioctahedral structure to accommodate the excess cations while the amount of surface area for the Na^+ cations is the same as in montmorillonite for the same number of defects per formula unit.

The utility of clay minerals lies in the wide range of possible CECs from 0 meq/100g to 251 meq/100g. Suitable materials can be obtained according to geographic origin (supplier) or by choice of synthetic alternatives. The specific utility of layered silicates as filler materials in nanocomposites is associated with the ability to exchange the alkali or earth alkali ions for cationic surfactants such as alkylammonium or alkylphosphonium halides, as shown in Figure 1.2. Ion exchange procedures have been developed over the past century by Weiss et al.[7], Gaines [8], Lagaly et al. [9,10], Vaia et al. [11], and Osman et al. [12–15], which make it possible to change the polarity of the clay mineral surface from hydrophilic to hydrophobic in a controlled way. This helps to match the surface properties of the organically modified clay mineral with the surface properties of polymer matrices.

Table 1.1

Chemical composition and CEC of representative clay minerals. The bar at the right indicates a broader classification according to CEC

Mineral name	Simplified chemical formula (Impurities neglected)	CEC (meq/100g)
Pyrophyllite	$[Si_2O_4][Al_2O_2(OH)_2][Si_2O_4]$	0
Laponite (synthetic)	$Na_{0.25}[Si_2O_4][Mg_{2.75}Li_{0.15}Na_{0.1}O_2(OH)_2][Si_2O_4]$	60–65
Montmorillonite (Southern Clay)	$Na_{0.33}[Si_2O_4][Al_{1.67}Mg_{0.33}O_2(OH)_2][Si_2O_4]$	90
Montmorillonite (Nanocor)	$Na_{0.53}[Si_2O_4][Al_{1.47}Mg_{0.53}O_2(OH)_2][Si_2O_4]$	145
Muscovite Mica	$K_{1.0}[Si_{1.5}Al_{0.5}O_4][Al_2O_2(OH)_2][Si_{1.5}Al_{0.5}O_4]$	251

Reproduced with permission from reference 16, Heinz, H.; Koerner, H.; Vaia, R. A.; Anderson, K. L.; Farmer, B. L. Force field for mica-type silicates and dynamics of octadecylammonium chains grafted to montmorillonite. *Chemistry of Materials,* **2005,** 17, 5658. Copyright 2005 American Chemical Society, and from reference 17 for laponite

Smectites

Vermiculites

Micas

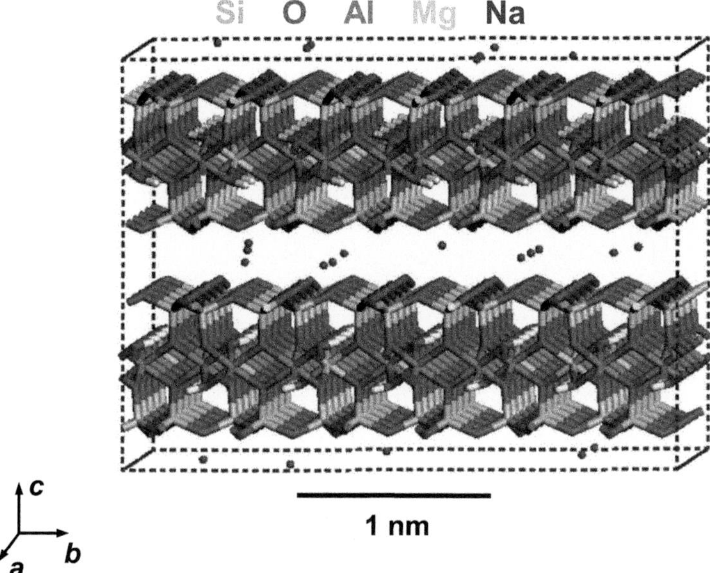

Figure 1.1

Montmorillonite model (CEC = 90 meq/100g) showing the three-dimensional structure and repeat. Two layers and sodium cations between the layers are shown. Each layer is composed of interconnected silica-aluminum oxide hydroxide-silica sheets. $AlO(OH) \rightarrow MgO(OH)^- \cdots Na^+$ charge defects are found in the octahedral aluminum oxide hydroxide sheet, and the associated sodium ions are located in the interlayer space (see also Fig. 12.1, Chapter 12)

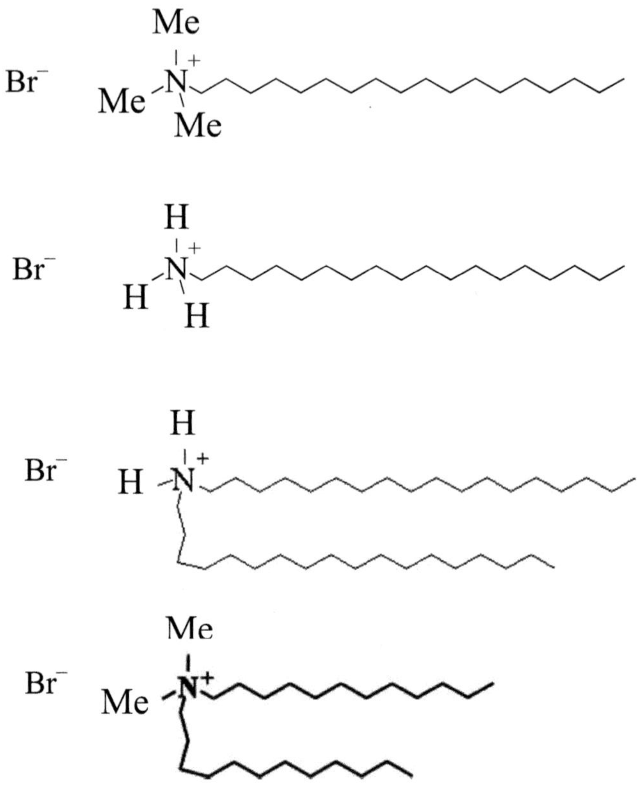

Figure 1.2

Alkylammonium salts to render the clay mineral surface hydrophobic upon exchange of alkali and

earth alkali metal cations

1.2 Synthesis of organically modified clay minerals

Organically modified clay minerals can be obtained by cation exchange reactions, for example:

$$\text{mont-Na}_{0.4} + 0.4\,(n\text{-}C_{18}H_{37})_2N(CH_3)_2^+ + 0.4\,Br^- \rightleftharpoons$$

$$\text{mont-}[(CH_3)_2N(n\text{-}C_{18}H_{37})_2]_{0.4} + 0.4\,Na^+ + 0.4\,Br^- \tag{1.1}$$

Due to the limited solubility of ammonium ions with long alkyl tails in water, ion exchange reactions are often carried out in ethanol-water mixtures as a solvent. In addition, ultrasound is applied to support the delamination of the layered silicate for quantitative ion exchange. Control of the reaction is essential to obtain well-characterized organically modified layered silicates. This usually requires (1) knowledge of the amount of exchangeable ions in the clay mineral (CEC), (2) an appropriate solvent to drive the ion exchange reaction, (3) a suitable stoichiometry in the reaction mixture to obtain the desired amount of ion exchange, and (4) monitoring of the yield (percentage of ion exchange) during the ion exchange reaction.

Quantitative ion exchange is easy to achieve for minerals with low CEC, while the progress of the reaction depends on the nature of the interlayer cations at high CEC. The rate of ion exchange decreases in the order Li^+ (easiest) > Na^+ > K^+ > Rb^+ > Cs^+ (most difficult). In the same order, the hydration shells of the cations become smaller so that the tendency of lamination of the layered silicates increases [18]. For example, it is challenging to exchange Cs^+ ions for alkylammonium surfactants. At high CEC such as in mica, delamination of the mineral is required prior to the efficient grafting of amphiphilic surfactants. This can be achieved through replacement of surface and interlayer cations with Li^+ ions by treatment with a hot $LiNO_3$ solution under application of ultrasound. The Li-mica can then be subjected to the ion exchange reaction with alkylammonium surfactants [12-14].

A major challenge is the measurement of the CEC which is commonly associated with uncertainties in the order of ± 5 % [12-14]. A convenient procedure to determine the CEC is based on the exchange reaction of the alkali ions with copper triethylenetetraamine Cu(trien)$^{2+}$ and the photometric detection of the decrease in Cu(trien)$^{2+}$ concentration [14]:

$$\text{mont-Na}_{0.4}\,(aq) + 0.2\,\text{Cu(trien)}^{2+}\,(aq) + 0.2\,\text{SO}_4^{2-}\,(aq) \rightleftharpoons$$

$$\text{mont-Cu}_{0.2}\,(aq) + 0.2\,\text{trien}\,(aq) + 0.4\,\text{Na}^{+}\,(aq) + 0.2\,\text{SO}_4^{2-}\,(aq) \qquad (1.2)$$

An excess of the Cu(trien)$^{2+}$ sulfate solution is typically applied so that a decrease in extinction at $\lambda = 255$ nm upon addition of the layered silicate indicates the molar amount of exchangeable ions per unit mass due to adsorption of the Cu^{2+} ions onto the clay mineral. Details of the procedure have been described by Osman et al. [14] however, the accuracy is limited because the exchange reaction may not be quantitative and some alkali cations substitute for trace amounts of ammonium ions or H$^+$ at lower pH which are not replaced by Cu^{2+}. Therefore, the method may underestimate the CEC by about 10 % [18].

An alternative approach to determine the CEC relies on the preparation of the clay mineral with Li$^+$ counter ions through substitution of all other interlayer cations using hot LiNO$_3$ solution, followed by a controlled ion exchange with an excess of Cs$^+$ ions. The Li$^+$ ions are most readily exchanged and Cs$^+$ ions have a high lamination tendency within the silicate layers so that the detection of the amount of released Li$^+$ in this procedure provides a very accurate measure of the CEC due to quantitative ion exchange and a fast reaction time [14]. The detection of released cations can be carried out by atomic absorption spectrometry (AAS) or inductively coupled plasma−mass spectrometry (ICP-MS). Experimental protocols for the ion exchange reaction have been described, for example, by Osman et al. for montmorillonites [19] and micas [15]. As an alternative to the synthesis of custom-made organically modified clay minerals, a number of organically modified clay minerals can also be purchased directly from suppliers such as Southern Clay Products Inc. or Nanocor Inc.

1.3   Structure of organically modified clay minerals

The nanoscale structures of surfactant-modified layered silicates can be grouped into three categories, Figure 1.3 [20]. Extensive results from experiment and simulation have shown that the molecular order of the surfactants is determined by the packing density λ_0 of the alkyl chains on the surface (section 1.4). The packing density is given by the ratio of the cross-sectional area of the all-anti configured hydrocarbon chains in a given surfactant ($A_{C,0}$) to the available surface area per cationic site (A_S):

$$\lambda_0 = \frac{A_C}{A_S} . \tag{1.3}$$

A single all-anti configured alkyl chain C_n such as in n-$C_{18}H_{37}$–NH_3^+ surfactants has a cross-sectional area of 0.188 nm^2, a two-arm alkyl chain $2C_n$ such as in (n-$C_{18}H_{37})_2$–NH_2^+ is of 0.376 nm^2 cross-sectional area, and so forth. Similarly, the available area per cationic site on the layered silicate surface ranges from 0.468 nm^2 in mica (CEC = 25 1meq/100g) to much larger values such as 1.40 nm^2 per cationic site in montmorillonite (CEC = 90 meq/100g). In this manner, a packing density $\lambda_0 = A_C / A_S$ between 0.0 and 1.0 can be achieved according to the choice of a surfactant-clay mineral combination. The average structure of the organic modifiers on the mineral surface can be described in these three categories, Figure 1.3. At a packing density less than 0.2, the alkyl chains are rather disordered and exploit the available space on the surface by adopting an essentially parallel orientation to the surface. Between a packing density of 0.20 and 0.75, the alkyl chains are forced into some chain tilt upwards from the surface as a consequence of volume packing restraints. In a qualitative way, this behaviour has been explained by Weiss et al [7]. The intermediate tilt angle leads to intermediate order of the hydrocarbon backbones so that reversible order-disorder transitions upon heating are common. If the packing density exceeds approximately 0.75, volume packing constraints lead to well ordered, crystal-like alkyl monolayers. The average segmental tilt angle relative to the surface normal is then < 40°, and no order-disorder transitions are found on heating.

In the following sections, the structures are discussed in detail, including the structure of the inorganic-organic interface, the different packing densities, and the effect of the structures upon non-quantitative cation exchange.

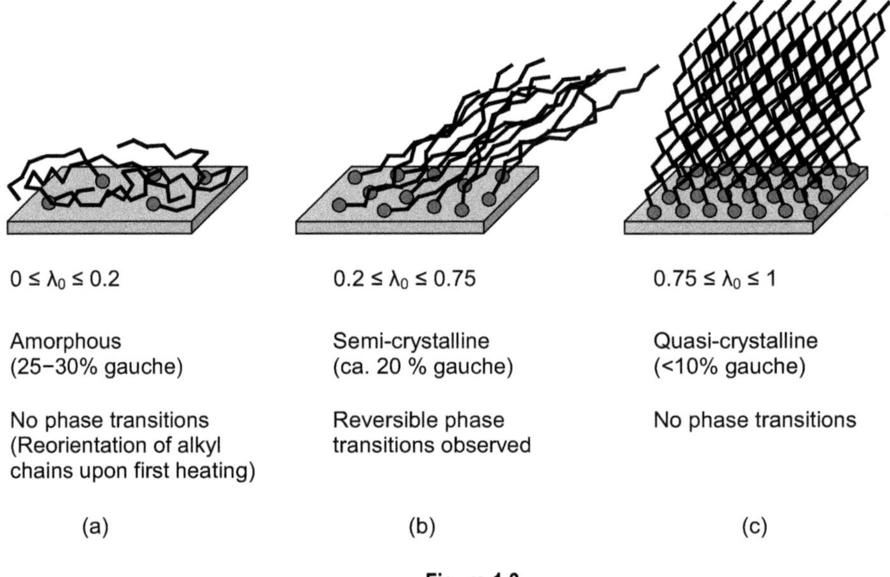

Figure 1.3

Range of homogeneous alkyl layers (chain length ≥ C_{10}) on even layered silicate surfaces

(a) Disordered chains oriented parallel to the surface

(b) Intermediately ordered chains with an intermediate collective tilt angle

(c) Nearly vertically oriented, all-anti configured chains

Reproduced with permission from reference 20, Heinz, H.; Vaia, R. A.; Farmer, B. L., Relation between packing density and thermal transitions of alkyl chains on layered silicate and metal surfaces. *Langmuir* **2008**, 24, 3727−3733. Copyright 2008 American Chemical Society.

1.3.1   Effect of cation density on the surface and the inorganic-organic interface

In the assembly of cationic surfactants, the density of cations on the layered silicate surface plays a central role. Figure 1.4 serves as a visual guide for the density of the cations for typical values of the CEC. The montmorillonite surface consists of fused (–Si–O–) 12-rings, and the alkali cations occupy positions above the cavities of these ring structures. The horizontal coordinates of the cations are centered in the cavities while the vertical position is well above the plane of the superficial Si and O atoms (see Figure 1.1). The highest cation density is found in mica (CEC = 251 meq/100g) where cations occupy every 2nd cavity on a single surface, i.e., all cavities are occupied in the layered structure. In typical montmorillonites, cations are found on every 3.75th cavity (CEC = 145 meq/100g) or every 6th cavity (CEC = 90 meq/100g) on a single surface. The distribution of the cations is determined by the distribution of charge defects in the octahedral and tetrahedral sheets [16,21].

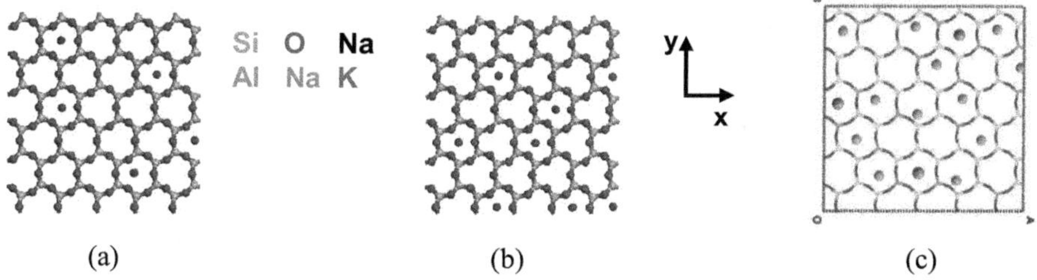

Figure 1.4

Model of the cation density on the layered silicate surface. The size of the model surfaces is 2.6 nm × 2.7 nm.

(a) Montmorillonite of CEC = 90 meq/100g

(b) Montmorillonite of CEC = 145 meq/100g

(c) Mica of CEC = 251 meq/100g.

Upon ion exchange, the positively charged head groups of the surfactants occupy similar positions to those previously occupied by the alkali ions. Depending on the chemical nature of the head group, differences in the location of the head group and in the structure of the inorganic-organic interface can be seen, Figure 1.5.

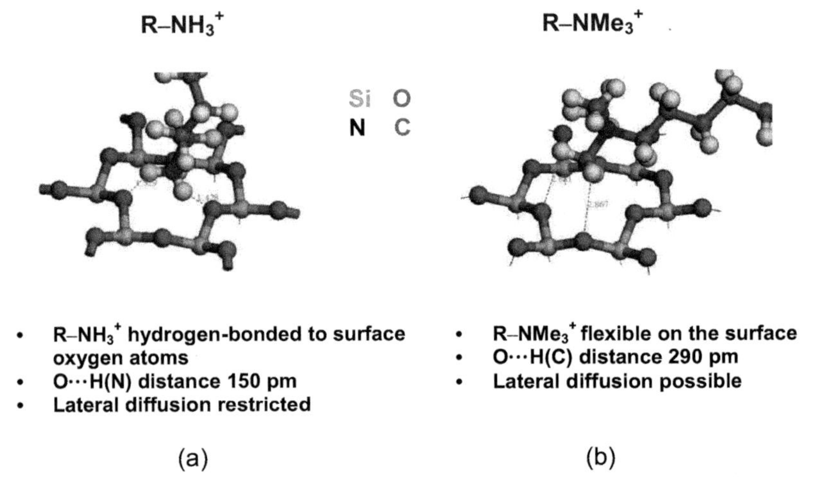

- **R–NH₃⁺ hydrogen-bonded to surface oxygen atoms**
- **O···H(N) distance 150 pm**
- **Lateral diffusion restricted**

- **R–NMe₃⁺ flexible on the surface**
- **O···H(C) distance 290 pm**
- **Lateral diffusion possible**

(a)   (b)

Figure 1.5

Comparison of the inorganic-organic interface with primary ammonium surfactants and with quaternary ammonium surfactants.

(a) Primary ammonium surfactants. The electrostatic bond of R–NH$_3$+ to the silicate surface is supported by up to 3 hydrogen bonds which reduce the translational freedom.

(b) Quaternary ammonium surfactants. The electrostatic bond of R–NMe$_3$+ surfactants is not further supported and the surfactants possess a higher probability for lateral diffusion. Reproduced with permission from reference 22, Heinz, H.; Vaia, R. A.; Krishnamoorti, R.; Farmer, B. L. Self-assembly of alkylammonium chains on montmorillonite: Effect of chain length, headgroup structure,

In the following discussion of the structures, we rely on accurate models for layered silicates and

organic surfactants which were obtained from molecular dynamics and Monte Carlo simulation by

Teppen et al.[23], Hackett et al. [24], Zeng et al. [25], Heinz et al. [3,16,18,20−22,26] and He et al. [31]. Such

models have achieved very good agreement with experimental measurements, including X-ray

diffraction (XRD), transmission electron microscopy (TEM), atomic force microscopy (AFM),

differential scanning calorimetry (DSC), nuclear magnetic resonance spectroscopy (NMR), infrared

spectroscopy (IR), near edge X-Ray absorption fine structure (NEXAFS), and surface tensions [16,26].

Since evidence of the molecular structure from experiments is still mostly indirect, visualization and

explanation of data on the basis of reliable models is a useful tool. To illustrate this view, a high

resolution TEM picture of an organically modified montmorillonite layer embedded in an epoxy

matrix is compared to a molecular model, Figure 1.6.

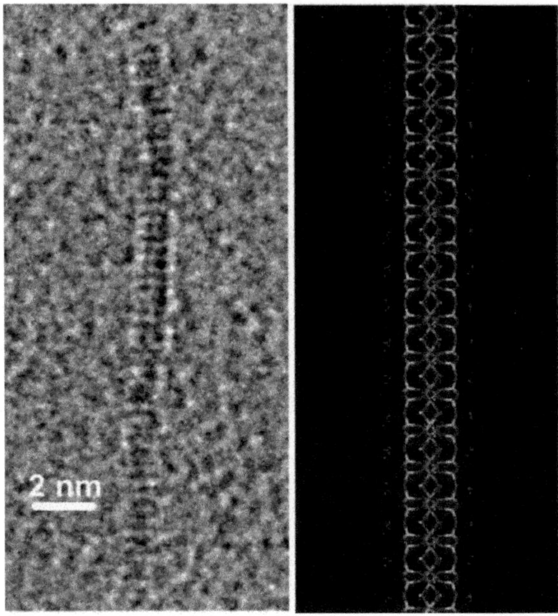

Figure 1.6

A high resolution TEM image of a single organically modified montmorillonite sheet in an epoxy

matrix in comparison with a model

Reproduced with permission from reference 28, , L. F.; Koerner, H.; Farmer, K.; Tan, A.; Farmer, B.

L.; Vaia, R. A. High-resolution electron microscopy of montmorillonite and montmorillonite/epoxy

nanocomposites. *Journal of Physical Chemistry B,* **2005**, 109, 17868−17878. Copyright 2005

American Chemical Society.

1.3.2   Low packing density

The structure of organically modified montmorillonite at low packing density is illustrated in Figure

1.7 for alkylammonium surfactants of various chain lengths and two different head groups [22]. For

short chains, an alkyl monolayer is formed parallel to the layered silicate sheets. When the chain length increases, the packing within the monolayer increases, accompanied by a small increase in gallery height. A significant increase in gallery height occurs when the lateral alkyl monolayer is densely packed so that the alkyl chains begin to build a partial second layer. Once a frustrated bilayer is formed, the gallery spacing increases again only marginally up to a chain length at which the formation of a dense bilayer is completed. For longer chains, the surfactants do not fit into the bilayer and a partial third layer begins to form, which is accompanied by a marked increase in gallery spacing again. In this series, the interlayer density and the average chain conformation undergo similar fluctuations as the basal plane spacing. The interlayer density reaches maxima when densely packed layers are formed, and minima when frustrated new layers are formed. The percentage of gauche conformations in the alkyl layers varies between 15 % and 40 %, and a dependence on the presence of the primary ammonium head group can be seen as well [22]. With a primary ammonium head group $R–NH_3^+$, lower gauche percentages are found for partial layers because the chain backbones have more conformational freedom to achieve energetically favorable anti-conformations. Higher gauche percentages are found for densely packed layers because the alkyl chains need to fit into the dense layer at the expense of more gauche arrangements. Quaternary ammonium groups such as $R–NMe_3^+$ lead to less pronounced fluctuation in the percentage of gauche conformations as a function of chain length because quaternary ammonium head groups are more flexibly bound to the surface, Figure 1.5. In summary, the alkyl chains are in a disordered, liquid-like state.

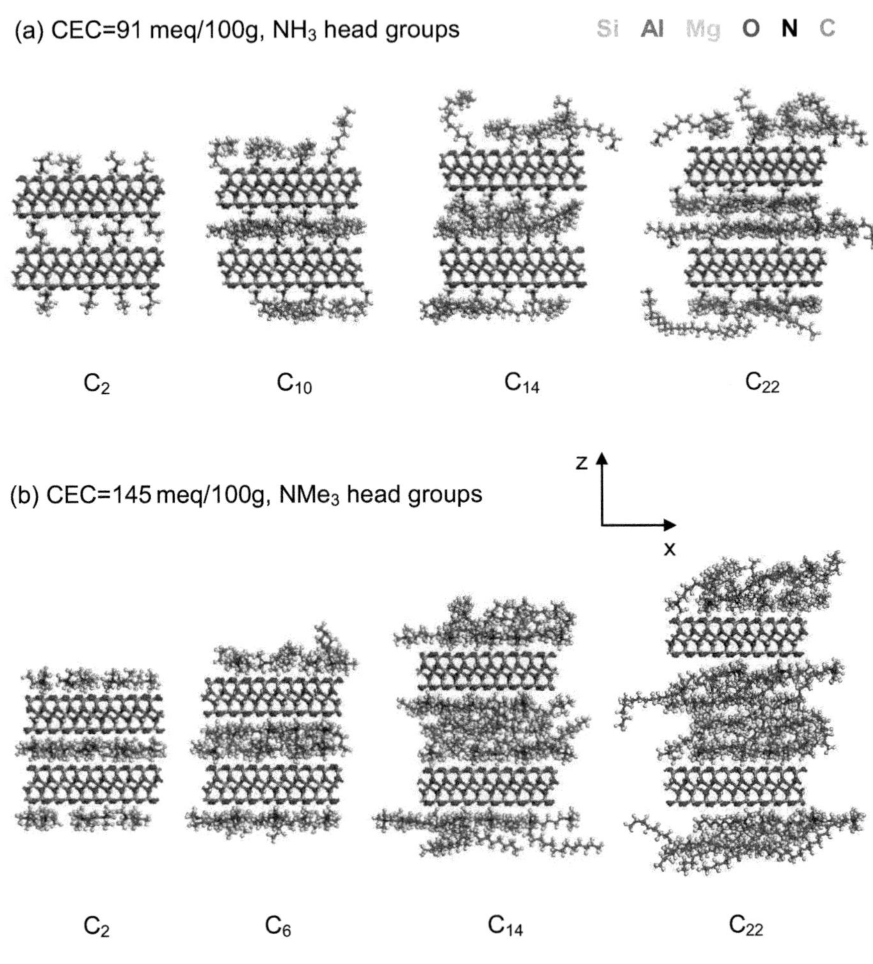

Figure 1.7

Visualization of alkylammonium montmorillonites, viewed along the *y* direction.

(a) CEC=91 meq/100g and NH_3^+–C_n chains (packing density $\lambda_0 = 0.13$). The difference between

partially formed layers (C_2, C_{14}) and completely formed layers (C_{10}, C_{22}) can be seen

(b) CEC=145 meq/100g and NMe_3^+–C_n chains (packing density $\lambda_0 = 0.21$). The successive

formation of layers with decreasing order can be seen.

Reproduced with permission from reference 22, Heinz, H.; Vaia, R. A.; Krishnamoorti, R.; Farmer, B. L. Self-assembly of alkylammonium chains on montmorillonite: Effect of chain length, headgroup structure, and cation exchange capacity. *Chemistry of Materials,* **2007**, 19, 59–68. Copyright 2007 American Chemical Society.

1.3.3   Medium packing density

A higher packing density on the surface can be achieved using montmorillonite of medium CEC and surfactants with multiple hydrocarbon backbones such as $NH_2(n\text{-}C_{18}H_{37})_2^+$ ("$2C_{18}$"), $NH(n\text{-}C_{18}H_{37})_3^+$ ("$3C_{18}$"), or "ditallow" surfactants which are mixtures of $2C_n$ surfactants with variable length n of the alkyl chains between $n = 14$ and $n = 18$. X-Ray, DSC, NMR, and IR data by Osman et al.[19] show that the structure and properties of these organically modified montmorillonites are similar to those of organically modified mica (high CEC) with a single hydrocarbon backbone such as $N(CH_3)_3(n\text{-}C_{18}H_{37})^+$ ("C_{18}") which have the same packing density [15]. Mica with C_{18} surfactants can thus serve as a representation for layered silicates with a packing density between 0.2 and 0.75. Representative structures for two different temperatures are shown in Figure 1.8 [18].

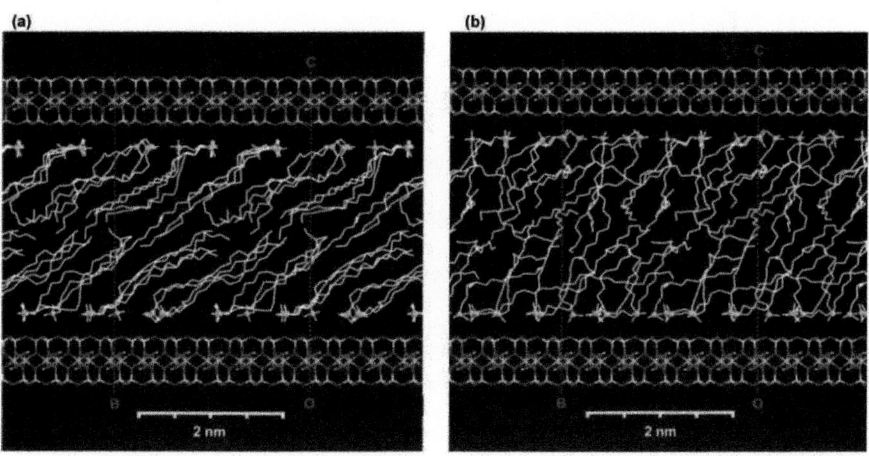

Figure 1.8

Structure of C_{18}-mica at 20 °C and at 100 °C viewed along the crystallographic *a* direction (packing

density $\lambda_0 = 0.4$). A melting transition, an increase in gallery spacing, and repositioning of the

quaternary ammonium head groups on the surface upon heating can be seen

(a) C_{18}-mica at 20 °C

(b) C_{18}-mica at 100 °C

Reproduced with permission from reference 18, Heinz, H.; Castelijns, H. J.; Suter, U. W. Structure

and phase transitions of alkyl chains on mica. *Journal of the American Chemical Society,* **2003,**

125, 9500−9510. Copyright 2003 American Chemical Society.

The increased density of alkyl chains per surface area leads to a tilted orientation of the alkyl

chains on the surface, as opposed to the parallel orientation to the surface at lower packing density

(Figure 1.7). The alkyl chains are in a semi-ordered state at lower temperature and the order of the

chain backbones is lost upon heating, which causes one or two phase transitions upon heating

(see section 1.4.2) [15,19]. The first phase transition is associated with the order-disorder transitions

of the tethered alkyl backbones and is immediately reversible upon cooling. A second transition is

seen for quaternary alkylammonium ions which further increase the melting enthalpy by changing

the location of the ammonium head groups (spheres), Figure 1.9. The second transition is not

immediately reversible upon cooling due to significant energy barriers for moving the quaternary

ammonium head groups back to the original position. The recovery of the second transition, which

does not occur for primary ammonium surfactants with hydrogen bonding (Figure 1.5), is complete

after several hours. Transition temperatures are in a range between 0°C and 100°C and enthalpies

up to 12 kJ per mol tethered C_{18} chain have been measured in DSC [15,19]. The percentage of gauche conformations in the alkyl chains is about 20 % to 25 % in the semi-crystalline state and increases to approximately 30 % in the molten state [18].

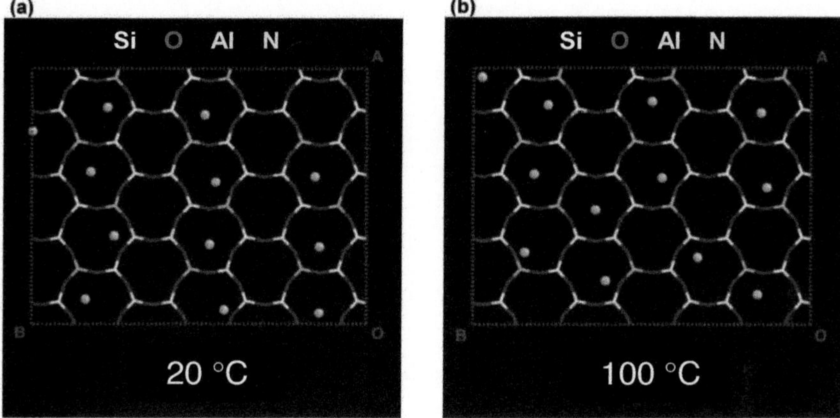

Figure 1.9

The location of the ammonium head groups in (n-$C_{18}H_{37}$)$N(CH_3)_3{}^+$ ammonium surfactants

(spheres) on the mica surface as a function of temperature. With increasing temperature,

quaternary C_{18} ions can change their position on the surface by lateral movement

Reproduced with permission from reference 18, Heinz, H.; Castelijns, H. J.; Suter, U. W. Structure

and phase transitions of alkyl chains on mica. *Journal of the American Chemical Society,* **2003**,

125, 9500–9510. Copyright 2003 American Chemical Society.

1.3.4   High packing density

High packing density of the alkyl chains can be achieved with ammonium surfactants with three or

four alkyl backbones such as $3C_{18}$ or $4C_{18}$ in combination with clay minerals of medium CEC.

Alternatively, ammonium surfactants with two or three alkyl chains such as $2C_{18}$ or $3C_{18}$ may be

used in combination with a high CEC layered silicate, e.g., mica. In these systems, significantly ordered alkyl layers are formed (Figure 1.10). Tilt angles relative to the surface normal are in a range between 0° and 40° and the structures resemble molecular crystals. The amount of gauche conformations is only between 5 % and 10 % [18,19,21,29].

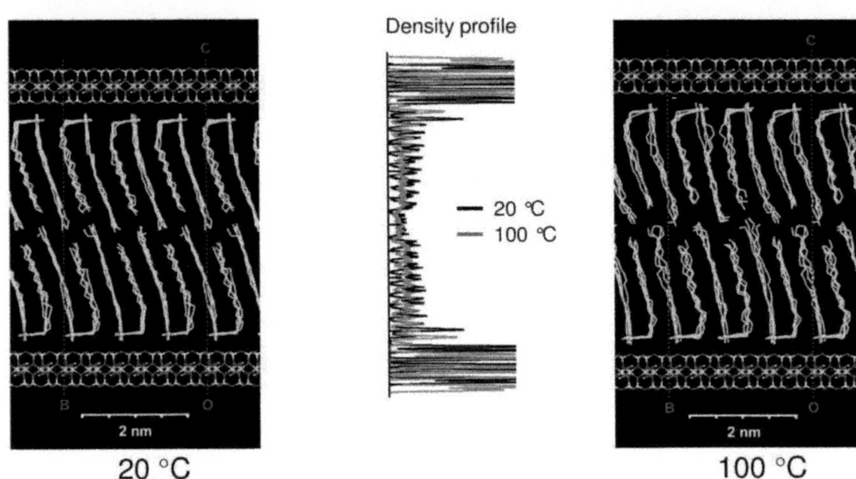

Figure 1.10

Structure and density profile of $2C_{18}$-mica at 20 °C and at 100 °C viewed along the *a* direction (packing density $\lambda_0 = 0.8$). Upon heating, some more gauche-conformations are present at 100 °C and no order-disorder transitions occur.

Reproduced with permission from reference 18, Heinz, H.; Castelijns, H. J.; Suter, U. W. Structure and phase transitions of alkyl chains on mica. *Journal of the American Chemical Society,* **2003**, 125, 9500−9510. Copyright 2003 American Chemical Society.

1.3.5   Non-quantitative ion exchange

Organically modified clay minerals can also be prepared by non-quantitative ion exchange. It is then possible to obtain homogeneous structures in which the alkali cations are interspersed between the organic surfactants, or spatially phase-separated structures in which the organic surfactants accumulate in islands and the remaining alkali cations form separate domains on the surface, Figure 1.11. The preferred structure depends on the cation density, the packing density (λ_0) adjusted for the fraction of ion exchange (f), $\lambda_{av} = \lambda_0 f$, and the chain length [21,29]. It is important to note that a different CEC may lead to different spatial distribution of the surfactants even when the average packing density λ_{av} and the chain length are the same.

(a)

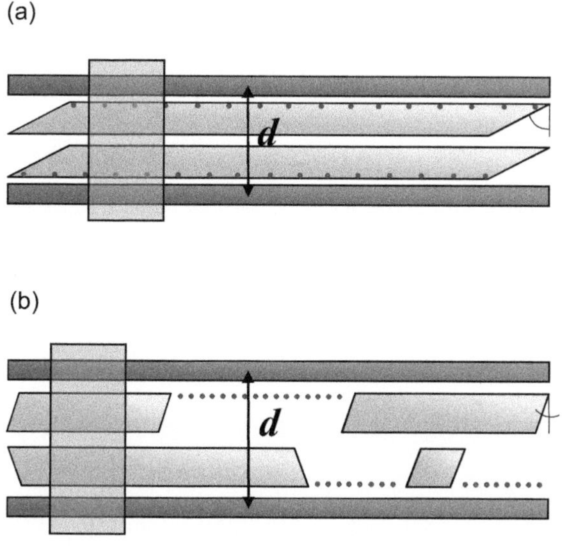

(b)

Figure 1.11

The formation of (a) one spatially homogeneous phase and (b) two spatially separated phases of alkali ions and alkylammonium ions on the surface of layered silicates. The basal plane spacing d of the two-phase structure is higher than that of the corresponding one-phase structure.

At low to medium cation density (low to medium CEC), quantitative cation exchange is routinely possible [19] and simply the supply of sub-stoichiometric amounts of alkylammonium ions during the ion exchange reaction introduces non-quantitative ion exchange [30,31]. The distribution of the alkyl chains on the surfaces for non-quantitative ion exchange is yet unknown, however, it is expected that the larger distance between neighbouring surfactants on the surface and the disorder of the alkyl chains compared to high cation density give more preference toward uniform structures, Figure 1.11a. At high cation density, spatial phase separation is more likely to occur. Nucleation and growth of islands of octadecyltrimethylammonium ions was observed on the mica surface by AFM by Hayes et al [32]. The recorded surface patterns for non-quantitative ion exchange show clearly the formation and growth of spatially separated domains, Figure 1.12, and it can be seen that quantitative ion exchange on mica surfaces is difficult to achieve [32]. Osman et al [15,29] found that exchange of alkali ions for octadecyltrimethylammonium bromide and dioctadecyldimethylammonium bromide on the mica surface does not surpass 70 % to 80 % even if a large excess of surfactant is provided. The excess surfactant remains rather loosely attached to the modified surface without substituting more alkali ions.

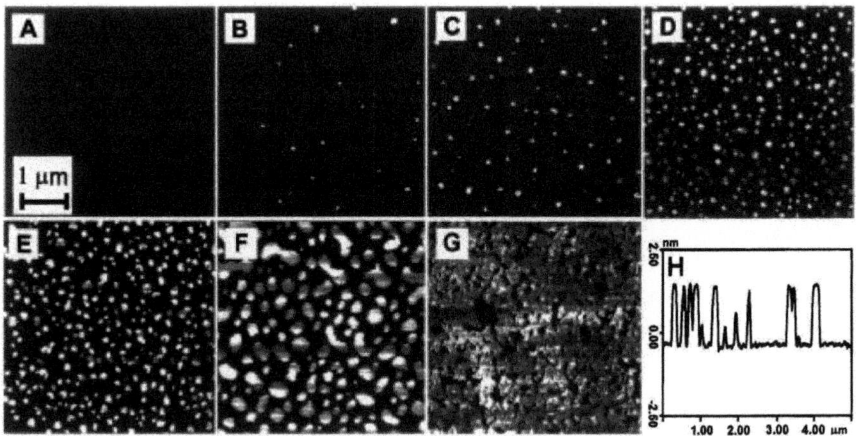

Figure 1.12

Series of ex-situ, contact-mode AFM images showing the surface topography of the coated mica surface as a function of the duration of exposure to a 0.10 mM solution of $C_{18} H_{37} N(CH_3)_3^+$ Br^-. (A) 1 h, (B) 24 h, (C) 48 h, (D) 72 h, (E) 120 h, (F) 168 h, (G) 336 h, (H) Representative line scan through image (E).

Reproduced with permission from reference 32, Hayes, W. A.; Schwartz, D. K. Two-stage growth of octadecyltrimethyl-ammonium bromide monolayers at mica from aqueous solution below the Krafft point. *Langmuir* **1998**, 14, 5913−5917. Copyright 1998 American Chemical Society.

The spatial distribution of alkylammonium ions has been analyzed on the mica surface (cation density $1 / A_S$ = 2.14 cations per nm^2) as a function of average packing density λ_{av} and chain length, Figure 1.13. A qualitative explanation can be given using the free energy difference between two spatially separated surfaces and a spatially uniform surface, Heinz et al. [21]:

$$\Delta A_{1 \to 2} = \Delta E_{1 \to 2} - T \Delta S_{1 \to 2} \tag{1.4}$$

Hereby, the free energy difference $\Delta A_{1 \to 2}$ between the two possible surface structures 2 and 1 is given by the difference in energy $\Delta E_{1 \to 2}$, the temperature T, and the difference in entropy $\Delta S_{1 \to 2}$. If the average packing density λ_{av} is small the van-der-Waals energy between the chains ($\Delta E_{1 \to 2}$ < 0) leads to island formation (negative free energy of the two-phase structure 2, $\Delta A_{1 \to 2}$ < 0). When the chains are longer, possible conformational freedom ($\Delta S_{1 \to 2}$ < 0) favors a uniform phase (positive free energy of the two-phase structure 2, $\Delta A_{1 \to 2}$ > 0). When the chains are very short, energetic and entropic factors become smaller and a remaining entropic tendency for head group disorder ($\Delta E_{1 \to 2} \approx 0$, $\Delta S_{1 \to 2}$ < 0) leads to a uniform phase(positive free energy of the two-phase structure 2, $\Delta A_{1 \to 2}$ > 0).

While the concept in equation (4) can be applied to montmorillonite or other clay minerals with significantly lower cation density, the phase diagram in Figure 1.13 is likely to change. The understanding of thermodynamically preferred structures as a function of average packing density and chain length for smectites and vermiculites (clay minerals of low and medium CEC) remains as a future challenge. Understanding the assembly of the surfactants on the surface in such systems for non-quantitative ion exchange may open new avenues to tailor interactions in the interfacial volume between the clay mineral, surfactants, and polymer matrices in nanocomposites.

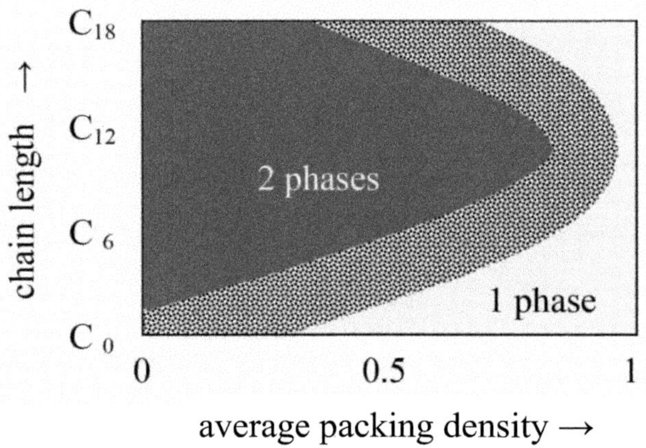

Figure 1.13

Spatial distribution of alkylammonium surfactants on mica surfaces as a function of average packing density and chain length.

Reproduced with permission from reference 21, Heinz, H.; Suter, U. W. Surface structure of organoclays. *Angewandte Chemie, International Edition,* **2004**, 43, 2239−2243. Copyright Wiley-VCH Verlag GmbH & Co. KGaA, 2004.

1.4   Characterization and properties of organically modified clay minerals

As introduced in the previous section, many experimental techniques have been employed to characterize organically modified clay minerals and understand their properties. These include XRD, TEM, scanning electron microscopy (SEM), AFM, DSC, differential thermogravimetry (DTG), IR/Raman, NMR, dielectric, elastic, NEXAFS, and surface tension measurements, as well as complementary simulation studies using molecular dynamics and Monte Carlo techniques. In the following sections, we will discuss the experimental techniques and the results for organically modified clays.

1.4.1   X-ray diffraction, microscopy, and structural properties

Wide angle X-Ray diffraction predominantly yields information on the gallery spacing. The gallery height of a variety of layered silicates with surfactants of different composition and chain length has been recorded in this manner by Lagaly et al. [10,33], Vaia et al. [11], Osman et al. [15,19,29] and Heinz et al. [22]. An example is shown in Figure 1.14. XRD can also be applied to trace small changes in gallery height as a function of temperature to detect thermal phase transitions of the organic surfactants upon heating, Figure 1.15. An increase in the basal plane spacing by 2 % to 4 % is common during thermal transitions and corresponds to a decrease in interlayer density. The thermal behaviour has often been characterized by a combination of differential scanning calorimetry (DSC), XRD, and spectroscopic techniques such as IR and NMR to obtain detailed information.

Electron microscopy and imaging techniques are useful to determine the degree of exfoliation of organically modified clay minerals in polymer matrices (TEM), to characterize the surface structure of organically modified clay minerals (AFM, Figure 1.12), and to examine the macroscopic shape of the mineral platelets including the aspect ratio (SEM, optical microscopy). The aspect ratio is the average ratio between length and thickness for a batch of organically modified layered silicates.

Typical aspect ratios are 10 to 1000 and depend on the chemical composition, the degree of delamination, and processing conditions. High aspect ratios lead to reduced gas diffusion as well as mechanical reinforcement in nanocomposites.

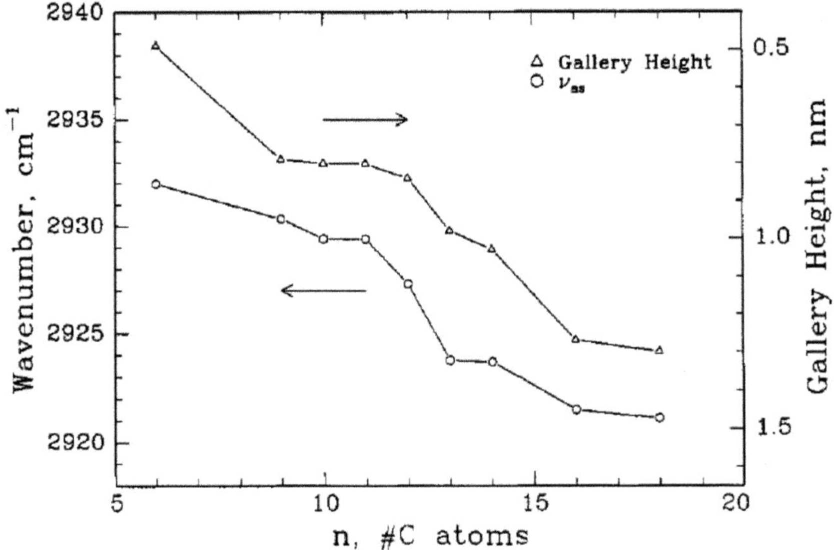

Figure 1.14

The gallery height (1.0 nm to be added for the clay mineral) and the IR frequency of the asymmetric CH_2 stretching vibration for various NH_3^+–C_n chains on fluorohectorite (CEC = 150 meq/100g). The two graphs illustrate the assembly of alkyl layers of various thickness on the layered silicate surface as well as possible conformational disorder of the surfactants, reflected in increasing IR wavenumbers

Reproduced with permission from reference 11, Vaia, R. A.; Teukolsky, R. K.; Giannelis, E. P. Interlayer structure and molecular environment of alkylammonium layered silicates. *Chemistry of Materials.* **1994**, 6, 1017−1022. Copyright 1994 American Chemical Society.

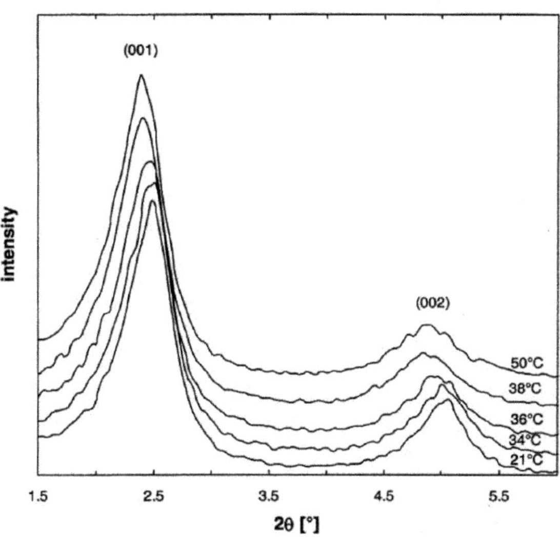

Figure 1.15

Wide Angle X-Ray diffraction patterns for $4C_{18}$ chains on montmorillonite (CEC = 72 meq/100g) at various temperatures. At 37 °C, a melting transition takes place which shifts the (001) and (002) reflections to the left, equal to an increase in the gallery spacing from 3.54 nm to 3.67 nm. Reproduced with permission from reference 19, Osman, M. A.; Ploetze, M.; Skrabal, P. Structure and properties of alkylammonium monolayers self-assembled on montmorillonite platelets. *Journal of Physical Chemistry B,* **2004**, 108, 2580−2588. Copyright 2004 American Chemical Society.

1.4.2 DSC, DTG, thermal transitions, and thermal decomposition

DSC is a powerful tool to analyze the thermal behaviour of organically modified layered silicates. Complex patterns of thermal transitions upon heating have been identified in numerous studies by Vaia et al. [11], Osman et al. [15,19,29], He et al. 2005 [31], and Jacobs et al. [34] and representative examples are shown in Figure 1.16. Transition enthalpies per C_{18} backbone range from 0 kJ/mol to 12 kJ/mol and typical transition temperatures are between 0°C and 100°C. The thermal behaviour is mainly determined by the packing density and also by the anchoring of surfactant head groups to the surface (Figure 1.5).

At low packing densities between 0.0 and 0.2, the chains are disordered (Figure 1.3), and no reversible melting transitions are seen. Often a broad peak occurs around 100 °C in the DSC trace upon first heating which is associated with the evaporation of residual water adsorbed in the interlayer (C_{18} in Figure 1.16). Annealing at 150 °C is used to eliminate water impurities, and the alkyl chains in the interlayer space typically undergo a reorientation. Subsequent heating cycles show no reversible melting transition. When the packing density approaches 0.2, the onset of a minor, reversible phase transition on heating is observed ($2C_{18}$ in Figure 1.16). Reversible phase transitions with significant melting enthalpy are seen for packing densities from 0.2 to 0.75 due to the intermediate degree of order in the chain backbones (Figure 1.3, Figure 1.8). Examples are $3C_{18}$ and $4C_{18}$ chains on montmorillonite (CEC = 72 meq/100g), Figure 1.16. Upon repeated heating, the alkyl chains need recovery time to yield identical melting enthalpies, nevertheless, the repeated occurrence of reversible transitions is not affected. $N(CH_3)_3{}^+-C_{18}H_{37}$ surfactants on mica (CEC = 251 meq/100g) have shown two transitions, Figure 1.17, which are due to a reversible melting transition (Figure 1.8) and a second transition that involves rearrangements of the ammonium head groups on the surface (Figure 1.9). The reverse rearrangement of ammonium head groups to return to the original positions on the surface is associated with energy barriers on the order of 10 kcal/mol and requires recovery times of several hours [15,18]. At high packing density between 0.75 and 1.0, the alkyl chains assume a highly ordered structure (Figure 1.3, Figure 1.10) and no peaks can be seen in the DSC curve.[29] Due to the near-crystalline order, thermal transitions are difficult to achieve unless the surfactant length increases such that grafting ceases to plays a significant role ($>>C_{20}$).

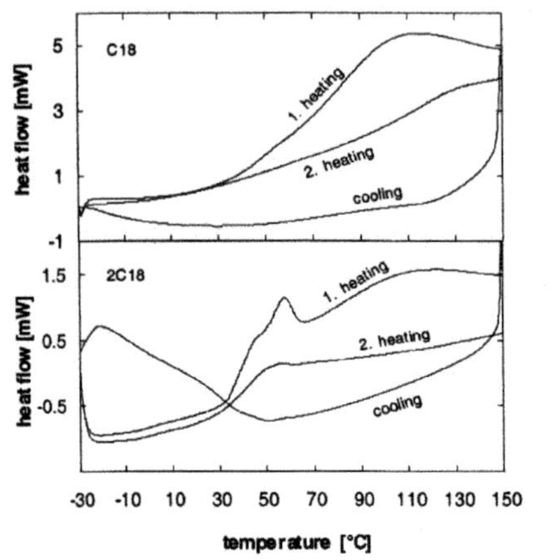
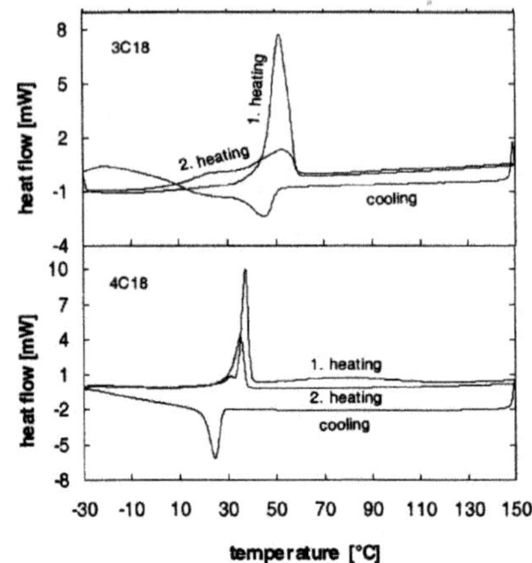

Figure 1.16

DSC traces for quaternary C_{18}, $2C_{18}$, $3C_{18}$, and $4C_{18}$ alkylammonium ions on montmorillonite (CEC

= 72 meq/100g).

Reproduced with permission from reference 19, Osman, M. A.; Ploetze, M.; Skrabal, P. Structure

and properties of alkylammonium monolayers self-assembled on montmorillonite platelets. *Journal*

of Physical Chemistry B, **2004**, 108, 2580–2588. Copyright 2004 American Chemical Society.

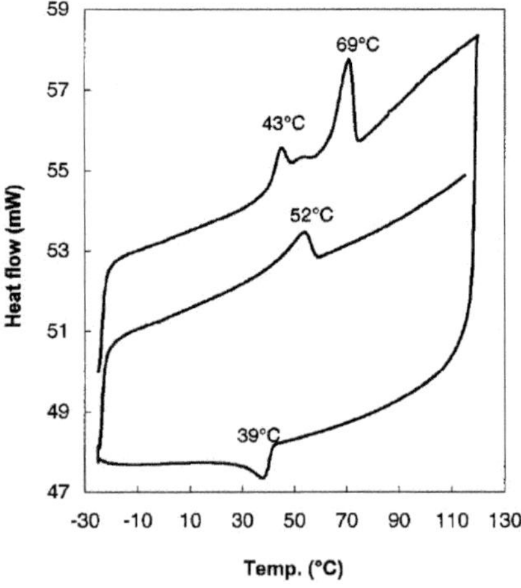

Figure 1.17

DSC trace for $N(CH_3)_3^+$–$C_{18}H_{37}$ surfactants on mica (80 % ion exchange). Reproduced with permission from reference 15, Osman, M. A.; Seyfang, G.; Suter, U. W. Two-dimensional melting of alkane monolayers ionically bonded to mica. *Journal of Physical Chemistry B*, **2000**, 104, 4433−4439. Copyright 2000 American Chemical Society.

An important aspect of thermal behaviour is also the decomposition which occurs between 250 °C and 400 °C. The likely mechanism is a Hofmann elimination. A neutral amine is formed by elimination from the N-terminal end of one of the alkyl chains and the layered silicate acts as a base to abstract a proton from the β carbon atom next to the N-terminal end of the alkyl chain [18] :

$$mont^- \cdots NR_3^+-CH_2-CH_2-R' \longrightarrow mont^- \cdots H^+ + NR_3 + CH_2{=}CH-R' \qquad (1.5)$$

Examples for differential thermogravimetric (DTG) measurements are shown in Figure 1.18 for montmorillonite (CEC = 72 meq/100g) with C_{18}, $2C_{18}$, $3C_{18}$, and $4C_{18}$ cations.

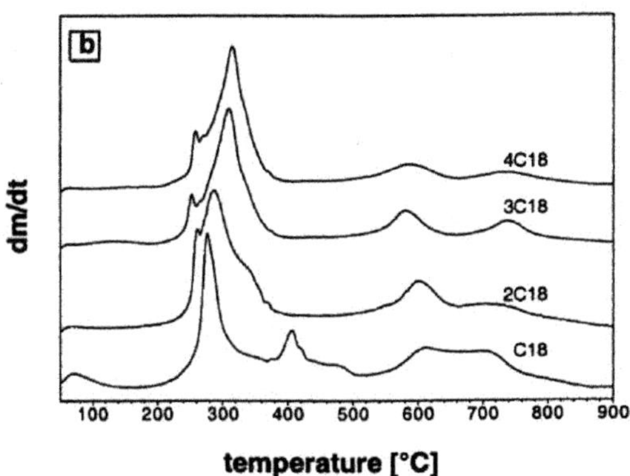

Figure 1.18

Differential thermogravimetric plot for NMe_3^+–$C_{18}H_{37}$, NMe_2^+–$(C_{18}H_{37})_2$, NMe^+–$(C_{18}H_{37})_3$, and N^+–$(C_{18}H_{37})_4$ surfactants on montmorillonite (CEC = 72 meq/100g).

Reproduced with permission from reference 19, Osman, M. A.; Ploetze, M.; Skrabal, P. Structure and properties of alkylammonium monolayers self-assembled on montmorillonite platelets. *Journal of Physical Chemistry B,* **2004**, 108, 2580–2588. Copyright 2004 American Chemical Society.

1.4.3   IR/Raman spectroscopy, NMR spectroscopy, and chain conformation

Infrared and Raman spectroscopy as well as NMR chemical shifts provide useful information on the conformational state of the alkyl chains. These methods have been employed to monitor conformations of the alkyl chains as a function of CEC, chemical nature of the head group, chain length, and temperature. In IR spectroscopy, an increase in energy (wavenumbers) of the symmetric ($v_{S,CH2}$) and of the antisymmetric CH_2 stretching vibration ($v_{AS,CH2}$) stretching vibration by a few cm^{-1} indicates an increase in gauche conformations. The values of $v_{S,CH2}$ and $v_{AS,CH2}$ typically change from 2848 cm^{-1} to 2854 cm^{-1} and from 2917 cm^{-1} to 2928 cm^{-1}, respectively, during a solid-liquid transition; the absolute values depend somewhat on the system [11]. Similarly, ^{13}C NMR spectral shifts for backbone CH_2 groups indicate an all-anti conformation with a single peak at 33 ppm and the presence of gauche conformations by a weaker single peak at 33 ppm as well as a second peak at 30 ppm [29]. In contrast to IR spectroscopy, NMR is more quantitative, and the absolute values are reproducible in different systems. Relaxation processes on larger time scales (ms and s) can also be analyzed by NMR [35].

For example, at high CEC with NH_3 head groups, IR data for the series C_6 to C_{18} have been measured, Figure 1.14. The clay mineral, fluorohectorite, is structurally similar to montmorillonite with a CEC of 150 meq/100g. A decrease in gauche conformations was observed by decreasing wavenumbers from 2932 cm^{-1} (C_6) to 2921 cm^{-1} (C_{18}) for the antisymmetric CH_2 stretching vibration. Reference frequencies for the neat alkylammonium salt in the liquid and in the crystalline state were determined as 2929 cm⁻¹ and 2918 cm⁻¹. These numbers suggest very high fraction of gauche conformations for C_6 (higher than in a liquid alkylammonium salt) and a remaining amount of gauche conformations for C_{18}. This trend could be closely reproduced in molecular dynamics simulation, Figure 1.19, where 45 % gauche conformation were found in C_6 versus 39 % in liquid C_{19}, and 25% gauche conformation for longer chains on montmorillonite. The high percentage of gauche conformations for short chains agrees with proposed hydrogen bonds between the NH_3 head group and the surface of the clay mineral (Figure 1.5) as suggested by Lagaly et al. [9]. As a further test for IR data, an increase in wavenumbers for C_{18} was observed upon heating (2921 $\rightarrow$ 2928 cm⁻¹) while only a minimal reversible peak in the DSC trace was seen [11]. Therefore, IR spectroscopy can be very sensitive to detect conformational changes even if no significant reversible phase transitions occur.

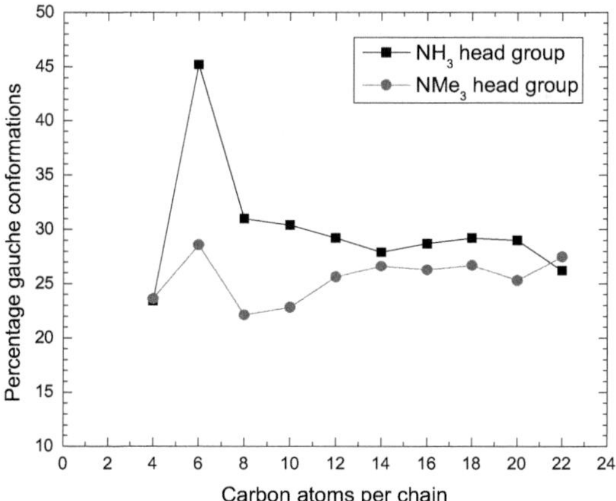

Figure 1.19

Average conformation of the alkyl chains between the silicate layers for alkylammonium-modified montmorillonite with CEC=145 meq/100g according to molecular dynamics simulation. The series $NH_3^+-C_nH_{2n+1}$ and $N(CH_3)_3^+-C_nH_{2n+1}$ with n = 2, 4, ..., 22 are shown. Reproduced with permission from reference 22, Heinz, H.; Vaia, R. A.; Krishnamoorti, R.; Farmer, B. L. Self-assembly of alkylammonium chains on montmorillonite: Effect of chain length, headgroup structure, and cation exchange capacity. *Chemistry of Materials,* **2007**, 19, 59–68. Copyright 2007 American Chemical Society.

As an example for ^{13}C-NMR measurements, we consider mica with 80 % alkali exchange with dioctadecyldimethylammonium ions $2C_{18}$, Figure 1.20. At low temperature, the ^{13}C chemical shift of 33 ppm for backbone CH_2 groups indicates nearly all-anti conformations. Heating above the melting temperature of the alkyl chains at 59°C results in a mixture of gauche and anti conformations resulting in a chemical shift of 30 ppm and 33 ppm for the backbone CH_2 groups. NMR spectroscopy is thus a sensitive tool to analyze chain conformations, although the quantitative interpretation in terms of molecular-level structures is not yet directly possible. In another example, a series of C_{18}, $2C_{18}$, $3C_{18}$, and $4C_{18}$ alkylammonium montmorillonites was investigated, Figure 1.21. C_{18} did not show any shifts in NMR or IR signals, and no reversible phase transition in DSC. A weak reversible melting transition occurs at 57 °C for $2C_{18}$ montmorillonite according to DSC (Fig. 16), which cannot be confirmed by NMR or by IR data (Figure 1.21). However, substantial melting transitions can be consistently seen for $3C_{18}$ montmorillonite at 51 °C and for $4C_{18}$ montmorillonite at 37 °C in DSC (Figure 1.16), NMR, and IR measurements (Figure 1.21).

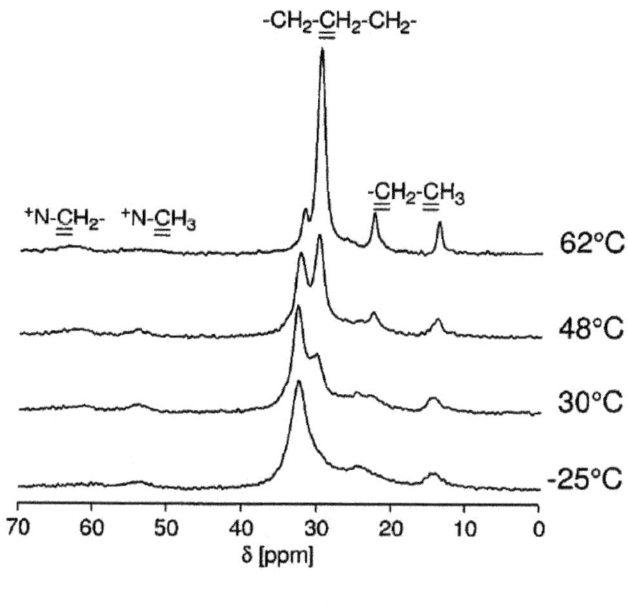

Figure 1.20

^{13}C-NMR on 2C$_{18}$ mica (CEC = 251 meq/100g) at various temperatures.

Reproduced with permission from reference 29, Osman, M. A.; Ernst, M.; Meier, B. H.; Suter, U. W. Structure and molecular dynamics of alkane monolayers self-assembled on mica platelets. *Journal of Physical Chemistry B*, **2002**, 106, 653–662. Copyright 2002 American Chemical Society.

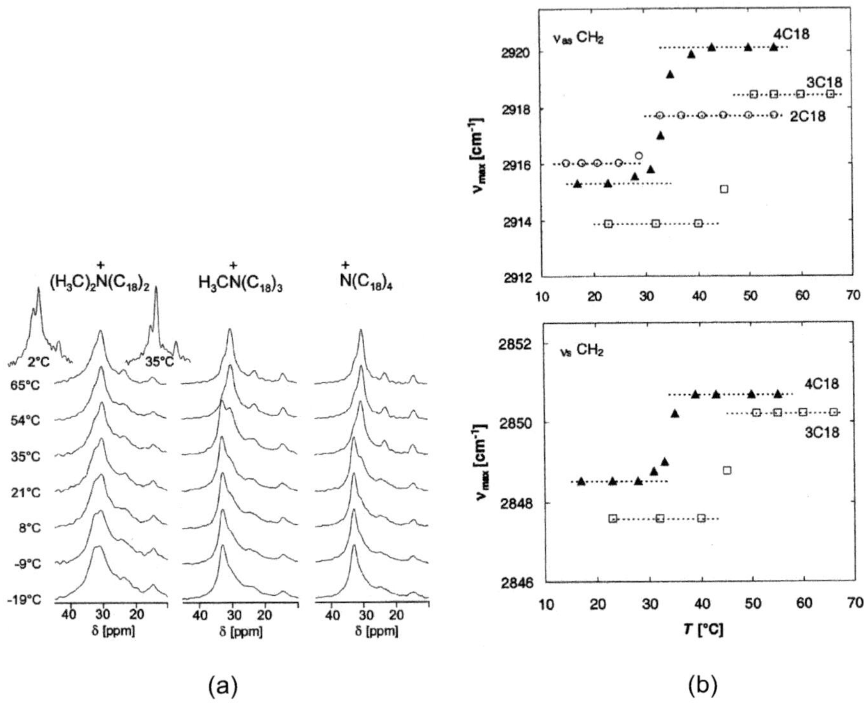

Figure 1.21

(a) Temperature dependence of the ^{13}C NMR resonances (C2-C16) of $2C_{18}$, $3C_{18}$, and $4C_{18}$ montmorillonite (CEC = 72 meq/100g). Insets for $2C_{18}$: Resolution enhancement at 2 °C and at 35 °C.

(b) Corresponding IR absorptions ν_{AS} and ν_{S} of the CH_2 groups.

Reproduced with permission from reference 19, Osman, M. A.; Ploetze, M.; Skrabal, P. Structure and properties of alkylammonium monolayers self-assembled on montmorillonite platelets. *Journal of Physical Chemistry B,* **2004**, 108, 2580–2588. Copyright 2004 American Chemical Society.

1.4.4   Dielectric, elastic, and tilt angle measurements

Dielectric and elastic properties, as well as tilt angles have been measured for specific organically modified layered silicates, although they are not yet well known for a broad variety of structures. Jacobs et al. [34] have measured the impedance on ditallowdimethylammonium montmorillonite (CEC = 92 meq/100g) which confirm the expected insulating behaviour. A hierarchical dynamics was observed which may be correlated with the particle morphology. Macroscopic behaviour was dominated by long-range charge transport (i.e., DC conductivity) along a percolation path defined by the surfaces of agglomerates and crystallites of the organically modified montmorillonite. Bound and unbound surfactant molecules (in the case of super-stoichiometric ion exchange) facilitate transport and act as the mobile charge-carrying species. At the microscopic level, charge separation occurs along this percolation path, at the heterogeneous layer surfaces and between neighbouring crystallites. The nanoscopic length scale is characterized by short-range interlayer dynamic behaviour of the surfactant molecules, which are governed by local viscosity, electrostatic interactions, and adjacent heterogeneous environments. Dipolar relaxation at high frequencies and the temperature dependence suggest a correlation between the dynamics of the electrostatic coupling between the cationic surfactant, the anionic aluminosilicate, and the dynamics of the alkyl tails within the interlayer. The organo-montmorillonite pressed powders displayed universal conductivity features, including time-temperature superposition, common in other types of disordered ion-conducting media. A significant dependence of the dielectric properties on small amounts of impurities was noticed as well [34].

The amount of data on mechanical properties of organically modified montmorillonites is relatively limited. In a review of elastic moduli of unmodified silicates by Chen et al. [36], Young's moduli in the range of 178 to 265 GPa are reported for montmorillonite and mica. Tensile and shear moduli along the stacking direction are expected to be much smaller than moduli perpendicular to the stacking direction. This is associated with the relative ease of compression and expansion perpendicular to the interlayer plane as well as shear in the interlayer plane, particularly in the presence of organic surfactants (Figures 7, 8, 10). Perpendicular to the stacking direction, the rigid covalent bonding framework leads to higher moduli. The maximum radius of bending was analyzed on the basis of TEM images by Drummy et al. [28]. Single sheets of C_{18} modified montmorillonite have been shown to be flexible up to a bending radius of about 15 nm while further deformation leads to failure by kink formation, Figure 1.22. However, comprehensive knowledge about the elastic properties of organically modified clays still remains as a future challenge.

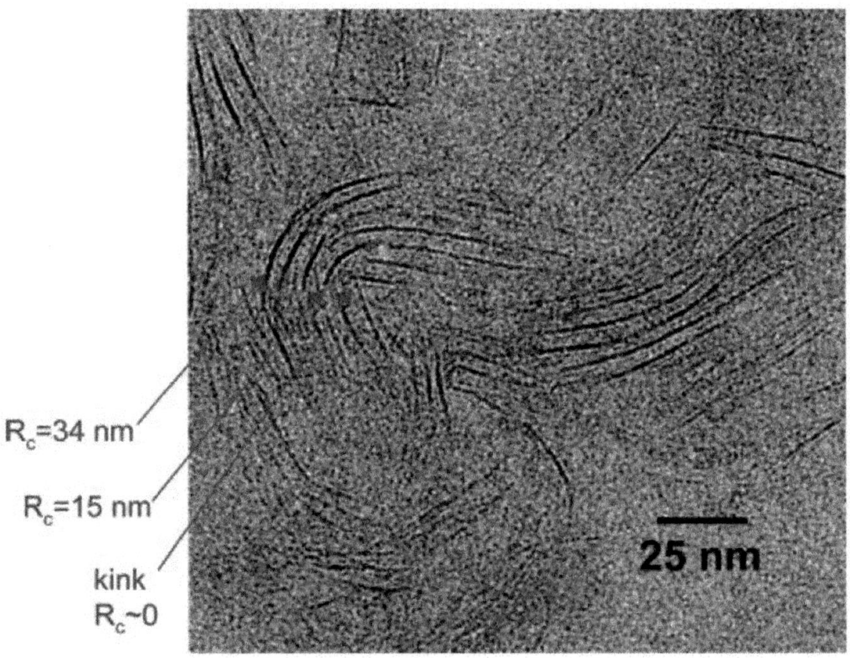

Figure 1.22

Bright-field TEM image of a region of $C_{18}H_{37}NH_3^+$ montmorillonite (CEC = 145 meq/100g, I30E) layers in an epoxy nanocomposite showing localized bending. The radius of curvature R_C is indicated.

Reproduced with permission from reference 28, Drummy, L. F.; Koerner, H.; Farmer, K.; Tan, A.; Farmer, B. L.; Vaia, R. A. High-resolution electron microscopy of montmorillonite and montmorillonite/epoxy nanocomposites. *Journal of Physical Chemistry B,* **2005**, 109, 17868−17878. Copyright 2005 American Chemical Society.

Average tilt angles of the alkyl chains with respect to the surface area a useful criterion to estimate the structure and thermal behaviour, Figure 1.3. The average tilt angle θ of the hydrocarbon backbone from end to end has been defined relative to the surface normal by Nuzzo et al. [37] and the measurement of this angle has been carried out by NEXAFS [38], IR spectroscopy, or X-Ray diffraction with subsequent calculation of the tilt angle. This definition and measurement of the tilt angle of the alkyl chains usually assumes end-to-end vectors of the alkyl chains and is unique when the alkyl chains are all-anti configured. When the chains contain gauche conformations the average tilt angle decreases and even converges to 0° in the limit of the melting transition so that the concept of tilt angles based on end-to-end vectors is useful as an order parameter rather than as a material constant. An alternative definition of the tilt angle can be made on the basis of the average segmental tilt angle θ_0 over all carbon-to-carbon vectors which is consistent with the total volume of the chain and does not depend on the chain conformation [20]. The segmental tilt angle θ_0 equals the backbone end-to-end tilt angle θ for all-anti configured chains and is directly related to the packing density λ_0 by $\lambda_0 = \cos\theta_0$. This relationship is helpful to relate the packing density, average segmental tilt angle, and occurrence of thermal transitions for a wide range of organically modified clay minerals, Figure 1.23. In this diagram, the combination of surfactant and clay mineral directly yields the packing density, the segmental tilt angle, and the expected thermal behaviour.

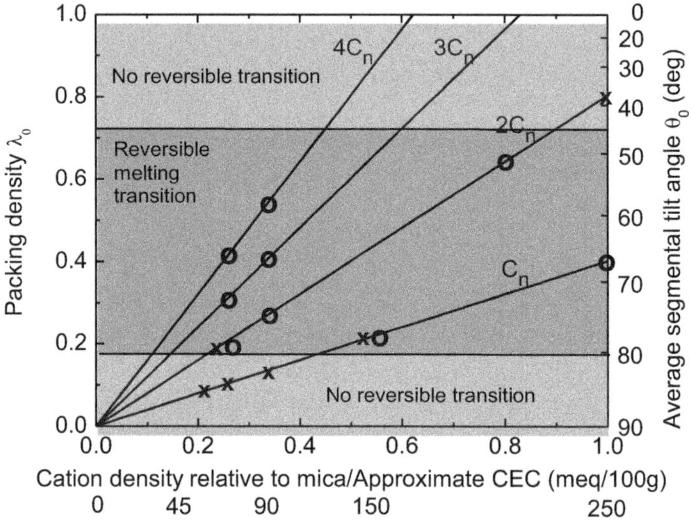

Figure 1.23

Packing density λ_0 and average segmental tilt angle θ_0 of alkyl chains (chain length ≥ C_{10}) as a function of the cation density and Cation Exchange Capacity. The symbol **O** indicates the occurrence and **x** the absence of reversible phase transitions on heating.

Reproduced with permission from reference 20, Heinz, H.; Vaia, R. A.; Farmer, B. L. Relation between packing density and thermal transitions of alkyl chains on layered silicate and metal surfaces. *Langmuir* **2008**, 24, 3727−3733. Copyright 2008 American Chemical Society.

1.4.5   Surface tension measurements and cleavage energies

Surface tensions of organically modified clay minerals have been determined by contact angle measurements with different liquids or thin layer wicking, including polar and van-der-Waals components. Examples are given by Giese et al. [39], Yariv et al. [40] and Lewin et al. [41].

Contact angle measurements involve placing a small drop (3–5 mm) of the liquid on the solid surface and using a video camera, a computer, and appropriate software for the angle measurement. Application of the Young equation and the Dupre equation yields the surface tension and the components when multiple liquids of different polarity are employed. The method works well if the surface does not reconstruct significantly in the presence of these liquids. If the liquids interact with the surface so that the surface structure changes during the measurement, e.g., by penetration of an organic solvent into the alkyl ammonium layer on the clay mineral surface, the values of the surface tension may only be meaningful in relation to the set of reference liquids employed.

Thin layer wicking is an alternative to contact angle measurements. A thin layer of the modified clay mineral is placed on a glass slide and dipped into the liquid. The rate of capillary rise of the liquid through the film is recorded and yields the contact angle using the Washburn equation [40]. For Wyoming montmorillonite (CEC = 68 meq/100g), surface tensions between 51.8 mJ/m^2 and 38.0 mJ/m^2 have been obtained in a series of primary ammonium ions NH_3^+–C_nH_{2n+1} with increasing chain length from C_0 to C_{15}. The van-der-Waals (Lifshitz-van-der-Waals) component to the surface tension is almost constant in the series from 41.7 mJ/m^2 to 37.7 mJ/m^2, and the polar component decreases from 10.1 mJ/m^2 to 0.3 mJ/m^2. In comparison, the cation $HN(CH_3)_3^+$ leads to a surface tension of 53.9 mJ/m^2 and the cation $N(C_2H_5)_4^+$ to 52.1 mJ/m^2.

Due to the interaction of the reference liquids with the surfactants or with the alkali ions on the modified clay surface, surface tension measurements are not necessarily equal to the cleavage energies when organically modified silicate sheets are separated as in the process of exfoliation. Nevertheless, very similar results have been obtained for cleavage energies using molecular dynamics simulation with accurate models for the layered silicate and the alkylammonium ions, Figure 1.24 [16,26]. The stepwise separation of the organically modified silicate layers leads to significant surface reconstruction due to conformational flexibility of the alkyl chains at short separation (<1 nm), even though the final separated structure and the original uniform structure possess a similar conformation of the alkyl chains. The energy of cleavage for C_{18}-montmorillonite (CEC = 90 meq/100g) was calculated as 40 mJ/m^2 and the free energy of cleavage as 38 mJ/m^2 by Heinz et al. [26] which provides an independent path to estimate the surface tension and is more closely related to exfoliation. The total is due to van-der-Waals interactions, with no significant electrostatic contribution, and agreement with the surface tension measured by contact angle methods is quite good. Therefore, Figure 1.24 can be seen as a suitable molecular-level representation of the cleavage process. When there are no other media present such as solvents or polymers, the interaction between the silicate layers becomes negligible after 3 nm separation.

(a)
Energy per surface area (mJ/m^2)
10
0
-10
-20
-30
-40
-50
0
1
2
3
Sheet separation (nm)
E tot
E Coul
E vdW
E int
(b) N C Si Al Mg O
d = 0.0 nm
d = 1.5 nm
d = 3.0 nm

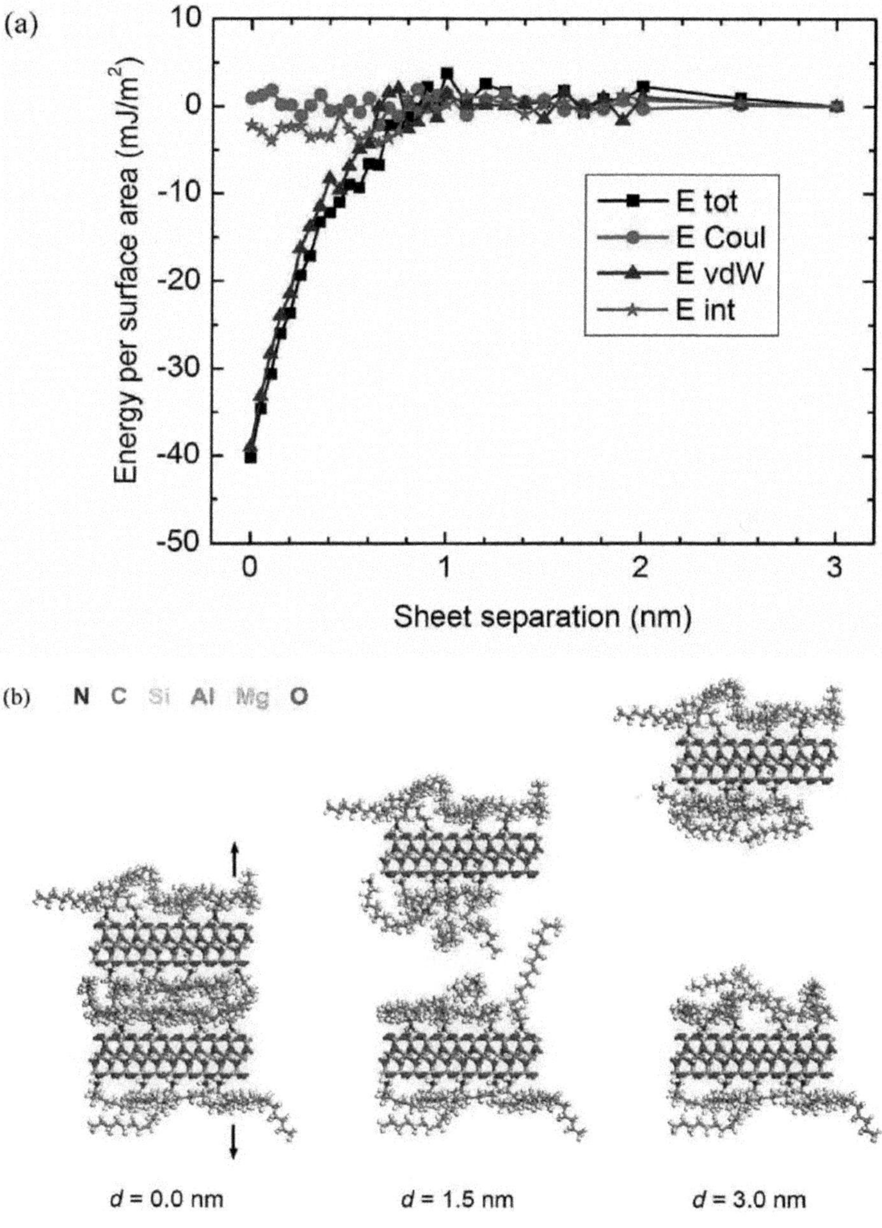

Figure 1.24

(a) Energy of interaction between two sheets of C_{18}-montmorillonite (CEC = 91 meq/100g) as a function of sheet separation. E tot = Total Energy, E Coul = Coulomb energy, E vdW = van-der-Waals energy, E int = internal energy.

(b) Snapshots of typical conformations, projected in the xz plane.

Reproduced with permission from reference 26, Heinz, H.; Vaia, R. A.; Farmer, B. L. Interaction energy and surface reconstruction between sheets of layered silicates. *Journal of Chemical Physics*, **2006**, 124, 224713. Copyright 2006 American Institute of Physics.

These considerations have implications on the interactions with polymers. Exfoliation is thermodynamically preferred when the surface tensions of the polymer and of the organically modified clay mineral are equal. Then, the free energy of exfoliation approaches zero and the two components become miscible. Differences in surface tension of the organically modified clay mineral and the polymer lead to a positive free energy of exfoliation and the two components are not miscible. Typical surface tensions of hydrophobic polymers are 30 to 35 mJ/m^2 [41,42]. Therefore polymers such as polyisoprene, polypropylene, and polyethylene are less likely to produce exfoliated nanocomposites with conventional organoclays. Somewhat more polar polymers such as epoxy resins, polyesters, and polyimides with surface tensions in the 35 mJ/m^2 to 45 mJ/m^2 range are usually more miscible More common polymers are listed in Table 1.2.

It is therefore desirable to eliminate the thermodynamic driving force for phase separation and agglomeration of the organically modified clay mineral by matching the surface tensions of the components. To achieve exfoliation and support matrix-filler interactions, additional specific interactions such as hydrogen bonds, electrostatic interactions, or even chemical crosslinks between the organically modified clay mineral and the polymer may help to introduce a negative free energy of exfoliation. From a thermodynamic viewpoint, the challenge is thus lowering the interfacial tension between the organically modified layered silicate and the polymer to negative values, or at least to zero, for the creation of a strong interface. Alternatively, the input of nanoscale and microscale work into the system by specifically designed engineering processes can help to overcome less positive interfacial free energies (see Chapter 4).

Table 1.2

Surface tensions of selected polymers under ambient conditions. Exact values depend on crystallinity, branching, tacticity, and molecular weight.

Polymer	Surface tension (mJ/m^2)
Poly(tetrafluoroethylene)	18.6–23.9
Poly(dimethylsiloxane)	19.4–22.8
Poly(propylene)	29.4 (atactic or isotactic)
cis-Poly(isoprene)	30
Poly(ethylene)	33.7–35.3 (branched), 35.7–36.8 (linear)
Poly(vinyl acetate)	36.5
Poly(styrene)	39.3–40.7
Poly(vinyl chloride)	39 (rigid), 41.9 (amorphous), 33–38 (plasticized)
Poly(methyl methacrylate)	41.0–42.7
Polycarbonate of bisphenol A	42.9 (rigid), 45 (amorphous)
Nylon-6	43–47 (crystalline), 38.4 (amorphous)
Poly(acrylamide)	52.3

From references 41 and 42.

Acknowledgment

The author acknowledges support by the University of Akron and helpful discussions with Daniel Schmidt, University of Massachusetts Lowell.

References

1   *Hydrous Phyllosilicates*, Bailey, S. W., Editor, *Reviews in Mineralogy*, Vol. 19, Mineralogical Society of America, Chelsea, MI, 1988

2   *Tonminerale und Tone*, Jasmund, K., Lagaly, G., Editors, Steinkopf: Darmstadt, 1992

3   Heinz, H.; Suter U. W. Atomic charges for classical simulations of polar systems. *Journal of Physical Chemistry. B,* **2004**, 108, 18341–18352

4   Gardiner, R. F.; Shorey, E. C. The action of solutions of ammonium sulfate on muscovite. *Industrial and Engineering Chemistry,* **1917**, 9, 589–590

5   Van Olphen, H. *An Introduction to Clay Colloidal Chemistry*; Wiley, New York, 1977

6   Webmineral: http://webmineral.com

7   Weiss, A.; Mehler, A.; Hofmann, U. Zur Kenntnis von organophilem Vermikulit. *Zeitschrift für Naturforschung,* **1956**, 11b, 431–434

8   Gaines, G. L. The ion-exchange properties of muscovite mica. *Journal of Physical Chemistry,* **1957**, 61, 1408–1413

9   Lagaly, G.; Weiss, A. Anordnung und Orientierung Kationischer Tenside auf Silicatoberflaechen. II. Paraffinähnliche Strukturen bei den n-Alkylammonium-Schichtsilicaten mit hoher Schichtladung (Glimmer). *Kolloid Zeitschrift und Zeitschrift für Polymere,* **1970**, 237, 364–368

10   Lagaly, G.; Dekany, I. Adsorption on hydrophobized surfaces: clusters and self-organization. *Advances in Colloid and Interface Science,* **2005**, 114, 189–204

11   Vaia, R. A.; Teukolsky, R. K.; Giannelis, E. P. Interlayer structure and molecular environment of alkylammonium layered silicates. *Chemistry of Materials.* **1994**, 6, 1017–1022

12   Osman, M. A.; Moor, C.; Caseri, W. R.; Suter, U. W. Alkali metals ion exchange on muscovite mica. *Journal of Colloid and Interface Science,* **1999**, 209, 232–239

13   Osman, M. A.; Suter, U. W. Dodecyl pyridinium alkali metals ion exchange on muscovite mica. *Journal of Colloid and Interface Science,* **1999**, 214, 400–406

14   Osman, M. A.; Suter, U. W. Determination of the cation-exchange capacity of muscovite mica. *Journal of Colloid and Interface Science,* **2000**, 224, 112–115

15   Osman, M. A.; Seyfang, G.; Suter, U. W. Two-dimensional melting of alkane monolayers ionically bonded to mica. *Journal of Physical Chemistry B,* **2000**, 104, 4433–4439

16   Heinz, H.; Koerner, H.; Vaia, R. A.; Anderson, K. L.; Farmer, B. L. Force field for mica-type silicates and dynamics of octadecylammonium chains grafted to montmorillonite. *Chemistry of Materials,* **2005**, 17, 5658

17   See http://www.laponite.com/pdfs/laponite.pdf. We suggest a modified formula $Na_{0.25}[Si_4O_8][Mg_{2.75} Li_{0.15} Na_{0.10}O_2(OH)_2]$ which offers better agreement between composition and CEC than the formula offered on the website $Na_{0.35}[Si_4O_8][Mg_{2.75} Li_{0.15}O_2(OH)_2]$. Nevertheless, the stoichiometry and distribution of the alkali ions between the interlayer and the octahedral sheet remain uncertain.

18   Heinz, H.; Castelijns, H. J.; Suter, U. W. Structure and phase transitions of alkyl chains on mica. *Journal of the American Chemical Society,* **2003**, 125, 9500–9510

19   Osman, M. A.; Ploetze, M.; Skrabal, P. Structure and properties of alkylammonium monolayers self-assembled on montmorillonite platelets. *Journal of Physical Chemistry B,* **2004**, 108, 2580–2588

20   Heinz, H.; Vaia, R. A.; Farmer, B. L. Relation between packing density and thermal transitions of alkyl chains on layered silicate and metal surfaces. *Langmuir* **2008**, 24, 3727–3733

21   Heinz, H.; Suter, U. W. Surface structure of organoclays. *Angewandte Chemie, International Edition,* **2004**, 43, 2239–2243

22   Heinz, H.; Vaia, R. A.; Krishnamoorti, R.; Farmer, B. L. Self-assembly of alkylammonium chains on montmorillonite: Effect of chain length, headgroup structure, and cation exchange capacity. *Chemistry of Materials,* **2007**, 19, 59–68

23   Teppen, B. J.; Rasmussen, K.; Bertsch, P. M.; Miller, D. M.; Schafer, L. Molecular dynamics modeling of clay minerals. 1. Gibbsite, kaolinite, pyrophyllite, and beidellite. *Journal of Physical Chemistry B,* **1997**, 101, 1579–1587

24   Hackett, E.; Manias, E.; Giannelis, E. P. Molecular dynamics simulations of organically modified layered silicates. *Journal of Chemical Physics,* **1998**, 108, 7410–7415

25   Zeng, Q. H.; Yu, A. B.; Lu, G. Q.; Standish, R. K. Molecular dynamics simulation of organic-inorganic nanocomposites: layering behaviour and interlayer structure of organoclays. *Chemistry of Materials,* **2003**, 15, 4732–4738

26   Heinz, H.; Vaia, R. A.; Farmer, B. L. Interaction energy and surface reconstruction between sheets of layered silicates. *Journal of Chemical Physics,* **2006**, 124, 224713

27   He, H. P.; Galy, J.; Gerard, J. F. Molecular simulation of the interlayer structure and the mobility of alkyl chains in HDTMA+/montmorillonite hybrids. *Journal of Physical Chemistry B,* **2005**, 109, 13301–13306

28   Drummy, L. F.; Koerner, H.; Farmer, K.; Tan, A.; Farmer, B. L.; Vaia, R. A. High-resolution electron microscopy of montmorillonite and montmorillonite/epoxy nanocomposites. *Journal of Physical Chemistry B,* **2005**, 109, 17868–17878

29   Osman, M. A.; Ernst, M.; Meier, B. H.; Suter, U. W. Structure and molecular dynamics of alkane monolayers self-assembled on mica platelets. *Journal of Physical Chemistry B,* **2002**, 106, 653–662

30   Zhu, J. X.; He, H. P.; Zhu, L. Z.; Wen, X. Y.; Deng, F. Characterization of organic phases in the interlayer of montmorillonite using FTIR and ^{13}C NMR. *Journal of Colloid and Interface Science,* **2005**, 286, 239–244

31   He, H. P.; Ding, Z.; Zhu, J. X.; Yuan, P.; Xi, Y. F.; Yang, D.; Frost, R. L. Thermal characterization of surfactant-modified montmorillonites. *Clays and Clay Minerals,* **2005**, 53, 287–293

32   Hayes, W. A.; Schwartz, D. K. Two-stage growth of octadecyltrimethyl-ammonium bromide monolayers at mica from aqueous solution below the Krafft point. *Langmuir* **1998**, 14, 5913–5917

33   Lagaly, G.; Weiss, A. Arrangement and orientation of cationic surfactants on silicate surfaces. IV. Arrangement of n-alkylammonium ions on weakly charged layer silicates. *Kolloid Zeitschrift und Zeitschrift für Polymere,* **1971**, 243, 48–55

34   Jacobs, J. D.; Koerner, H.; Heinz, H.; Farmer, B. L.; Mirau, P.; Garrett, P. H.; Vaia, R. A. Dynamics of alkyl ammonium intercalants within organically modified montmorillonite: dielectric relaxation and ionic conductivity. *Journal of Physical Chemistry B, 2006*, 110, 20143–20157

35   Mirau, P. A. *A Practical Guide to Understanding the NMR of Polymers*, Wiley VCH, Darmstadt, **2005**

36   Chen, B.; Evans, J. R. G. Elastic moduli of clay platelets. *Scripta Materialia,* **2006**, 54, 1581–1585

37   Nuzzo, R. G.; Dubois, L. H.; Allara, D. L. Fundamental studies of microscopic wetting on organic surfaces. 1. Formation and structural characterization of a self-consistent series of polyfunctional organic monolayers. *Journal of the American Chemical Society,* **1990**, 112, 558–569

38   Brovelli, D.; Caseri, W. R.; Hahner, G. Self-assembled monolayers of alkylammonium ions on mica: Direct determination of the orientation of the alkyl chains. *Journal of Colloid and Interface Science,* **1999**, 216, 418–423

39   Giese, R. F.; van Oss, C. J. *Colloid and Surface Properties of Clays and Related Minerals,* Dekker, New York, **2002**

40   *Organo-Clay Complexes and Interactions,* Yariv, S., Cross, H., Eds., Dekker, New York, **2002**

41   Lewin, M.; Mey-Marom, A.; Frank, R. Surface free energies of polymeric materials, additives and minerals. *Polymers for Advanced Technologies,* **2005**, 16, 429–441

42   *Polymer Handbook,* Brandrup, J., Immergut, E. H., Grulke, E. A., Editors, Wiley, New York, **1999**

Chapter 2
Polymer nanocomposites formulated with hectorite nanoclays

Günter Beyer

Kabelwerk Eupen AG, Eupen, Belgium

2.1   Introduction

Hectorite clay is a member of the smectite group of clay minerals, which are characterized by the fact that they swell easily in water [1]. In fact, smectite clay layers delaminate readily when the clay is dispersed in water and this phenomenon is closely related to the specific layer charge density of the clay layers. Like the more commonly used montmorillonite, hectorite clay is a so-called 2:1 hydrous phyllosilicate, where the clay layers comprise a central octahedral metal oxide sheet, which is condensed with two silicate sheets. Isomorphic substitutions in the clay crystal lattice create a net negative charge on the clay layer, which is neutralized with exchangeable cations that are positioned on the surfaces of the clay layers. The structural formula for hectorite clay as identified by Ross and Hendricks [2] is: $Na_{0.33}(Mg_{2.67}Li_{0.33})Si_4O_{10}(F,OH)_2$. As such, hectorite is further classified as a trioctahedral smectite clay while montmorillonite is a dioctahedral smectite clay; the difference is that all octahedral sites in the central octahedral sheet are occupied with metal ions for hectorite whereas only 2/3 of the sites are filled for montmorillonite [1]. Furthermore, hectorite clay is unique in that it contains unusually high levels of lithium for magnesium substitutions in the central octahedral sheet, yet it contains relatively few substitutions in the outer silicate sheets of the clay layer. In contrast, montmorillonite has substitutions in both the octahedral- and tetrahedral-sheets. This last difference has significant effects on preferred clay applications. For instance, montmorillonite is used as cracking catalyst because of these specific metal substitutions in the clay layer surface lattice. Hectorite clay is not used as a catalyst as this clay does not have the aforementioned surface lattice substitutions.

Hectorite is a relatively rare clay mineral. A large deposit of hectorite clay of hydrothermal origin is located in the central portion of the Mojave desert in San Bernardino county, near Hector in California. Other hectorite clay deposits have been described [3]. The hydrothermal hectorite clay has a slightly different clay layer shape when compared to montmorillonite clay. Hectorite clay layers have a rectangular, lath type shape of approximately 50 nm wide by 300 nm. Although montmorillonite clay varies greatly in layer shape, it is roughly ten times larger yet much squarer in shape. Given that hectorite clay has more clay layers than montmorillonite clay on an equal clay weight basis, hectorite clay has typically been used as a rheology modifier as it imparts thixotropic character to fluids. Such layer shape differences understandably also affect barrier properties when one is concerned with the barrier performance of nanoclay polymer composites. Using the tortuous path model, it may be surmised that montmorillonite would be the clay material of choice if one is concerned with gas barrier performance.

Hectorite clay has been synthesized using various approaches. Laponite is a synthetic, hydrothermal hectorite clay that is commercially available from the company Southern Clay Products. The clay layers are relatively small as they measure only 25 nm across, significantly smaller than the naturally occurring material. Fluorohectorite clay is another synthetic form of hectorite, yet this clay has very large clay layers [1]. The clay is readily produced in the lab via a solid state flux reaction [4]. Dow Corning discontinued production of such fluorohectorite clays but CO-OP Chemical Co. markets certain SOMASIF clays, which are synthetic hectorite type clays.

The exchangeable metal ions on the clay layers can be replaced with organophilic (quaternary) ammonium ions. The resulting organoclays find use in various industrial applications [5]. Various companies produce and sell commercial montmorillonite-based organoclays while hectorite-based organoclays are mainly sold by the company Elementis Specialties under the BENTONE trade name. A first application in plastics has been described by Nahin [6], but significant materials advantages were reported by Toyota researchers in the 1990s. They used a special organoclay, which was modified to be compatible with Nylon, and this work started research to specifically design organoclays for plastics applications, i.e. nanoclays.

2.2 Thermal stability of hectorite-based nanoclays and nanocomposites

When compounding nanoclays into thermoplastics, three items are important with respect to the thermal stability of nanoclay polymer composites:

- the thermal stability of the nanoclay
- the effect of the nanoclay on the thermal stability of the matrix polymer
- the thermal stability of the nanocomposite produced.

2.2.1 Nanoclay stability

The thermal stability of the organic component of the nanoclay is of interest, as it affects the nanoclay's ability to withstand polymer processing temperatures. A recent study compared the thermal degradation of untreated synthetic fluorohectorite, organic modifiers, and modified fluorohectorite [7]. The organic modifiers studied were methyl hydrogenated tallow dihydroxyalkylammonium (polar modifier) and dihydroxyethyl methyl hydrogenated tallow ammonium chloride (non-polar modifier). The untreated (pristine) fluorohectorite did not show any weight loss until a temperature of 800°C, and each organic modifier exhibited a single degradation peak.

A thermal stability analysis of the modified nanoclay showed a distinct change in the degradation process of the organic modifier. Several degradation peaks were observed, and the first degradation peak occurs at a lower temperature than for the pure organic material. This reduction in stability was greater for the nanoclay incorporating the polar modifier. The proposed mechanism for the catalytic effect of the clay on the degradation of the organic modifier is the presence of Lewis and/or Bronsted acid sites. Other degradation peaks occurred at a temperature higher than that observed for the pure organic material. This is proposed to be due to a barrier effect of the clay layers to the diffusion of the thermal degradation products.

The thermal stability of a synthetic fluorohectorite clay modified with dimethyl tallow quaternary ammonium ion was studied when excess surfactant was removed from the nanoclay complex [8]. The nanoclay was washed in a water/ethanol mixture to remove the surfactant that was physically adsorbed onto the external surfaces of the nanoclay. This washing increased the thermal stability of the nanoclay but decreased the surface energy.

The thermal stability of montmorillonite and hectorite nanoclays have recently been compared [9]. The clays were modified with an ion exchange capacity equivalent of dimethyl dihydrogenated tallow ammonium ion as organic modifier. Thermogravimetric analysis (TGA) as well as differential scanning calorimetry analysis (DSC) shown in Figures 2.1 and 2.2 illustrate that the (tetrahedral) surface layer charge on montmorillonite clay promotes organic modifier decomposition. The hectorite based nanoclay showed enhanced thermal stability.

2.2.2 Effect of the nanoclay on the degradation process of the matrix polymer

The effect of the nanoclay on the polymer stability must also be considered when evaluating the suitability of a particular nanocomposite system. For example, the incorporation of a standard dimethyl dihydrogenated tallow ammonium modified nanoclay (based either on hectorite or montmorillonite clay) into poly(vinyl chloride) (PVC) resin results in a rapid degradation of the PVC polymer through dehydrohalogenation reactions. These reactions are thought to be initiated by amine decomposition products, which are formed when the quaternary ammonium ion of the nanoclay degrades via a Hoffman elimination mechanism [10,11,12]. Nanoclays formulated with a triethanol tallow alkyl ammonium cation prove much more compatible with halogenated resins [13,14]. Amine degradation products that are formed when the triethanol tallow alkyl ammonium clay modifier degrades are less basic compared to amine degradation products derived from dimethyl dihydrogenated tallow ammonium ions. It is assumed that these less basic amines are less efficient in initiating PVC dehydrohalogenation reactions.

2.2.3   Thermal stability of the produced nanocomposites

PVC and composites will degrade after prolonged compounding and stabilizers are typically added to PVC formulations. Taking composite darkening as a measure for PVC compound degradation, the colour development during compounding with PVC resin is much less pronounced for triethanol tallow alkyl hectorite nanoclay based composites as compared to the montmorillonite nanoclay based composite with the same organic modifier. Again, these results reflect the difference in clay layer charge chemistry between the two clay types [9,10]. Similar effects can be observed for polpypropylene composites that contain either hectorite or montmorillonite nanoclay. When such composites are held at elevated temperatures (150 °C) for 24 hours or longer, one can observe significant degradation for certain montmorillonite based materials. As mentioned above, montmorillonites are common minerals and have a wide composition window and not all montmorillonite clays have similar layer surface substitutions (tetrahedral sites), so therefore this effect is limited to certain montmorillonites. The hectorite from the Hector region in California region however does not contain such surface substitutions and therefore polypropylene composites formulated with such nanoclays do not significantly degrade when exposed to elevated temperatures.

2.3   Flame retardant properties of hectorite-based nanocomposites

Intercalated plasticized PVC samples with montmorillonite and hectorite (both modified with the quaternary triethanol tallow ammonium) show lower peak heat release rates compared to the PVC compound with no filler measured by cone calorimeter [13]. The reduction is maximal for the montmorillonite based nanocomposite. Similarly, the smoke production rate has its lowest value for plasticized PVC with the organoclay montmorillonite. Times to ignition are very similar for both compounds. The reason for reduced peaks in heat release rate and also smoke production rate may be linked to the formation of more stable chars with less cracked surfaces using organoclays; see chapter by Beyer [this book]. The less cracked char is observed for PVC with montmorillonite. One possible explanation of the lower heat release rate and lower smoke production rate of plasticized PVC with montmorillonite in contrast to plasticized PVC with hectorite (both samples have the same quaternary alkyl ammonium cation) may be the quite different platelet dimensions of the clays. Montmorillonite is reported to have platelets of roughly 300 x 300 x 1 nm while hectorite has smaller rectangular layers of about 300 x 50 x 1 nm [9]. One can speculate that a better barrier effect is provided by the montmorillonite resulting in lower mass transport of degradation products and a better shielding effect against external heat [13].

2.4   Barrier properties of hectorite-based nanocomposites

One of the proposed uses for nanocomposites is in barrier applications, such as packaging. The inorganic silicate platelets are a barrier to the diffusion of gas or liquid through a nanocomposite article, so that the permeating molecules must undergo a "tortuous path".

Hectorite nanoclays have been evaluated in several polymer systems. The barrier performance of synthetic fluorohectorite, modified with octadecylamine, has been evaluated in a polypropylene (PP) matrix [15,16]. The compatibilizer used was maleated polypropylene, and both isotactic and syndiotactic PP resins were used. The permeants chosen were dichloromethane and n-pentane vapour, which differ in molecular dimensions, solubility parameters, and degree of interaction with the polymeric matrix. Several nanoclay loading levels were evaluated. Both the sorption of vapours by the nanocomposite and the diffusion through the sample film were measured.

The nanocomposites based on isotactic PP were similar in sorption characteristics to the neat polymer. The diffusion coefficient was significantly reduced as the nanoclay level increased, indicating better barrier performance. An increase in mechanical stiffness was seen as well. The nanocomposites based on syndiotactic PP also showed a decrease in the diffusion coefficient and an increase in stiffness, as compared to the neat polymer. The vapour sorption was unchanged for the dichloromethane, but was significantly reduced for the n-pentane.

A nanocomposite consisting of hydrogenated acrylonitrile butadiene rubber with a synthetic fluorohectorite has been studied for barrier performance [17]. Both unmodified fluorohectorite, and fluorohectorite modified with octadecylamine, were used. The permeating gas was oxygen, under dry and wet conditions (0% and 60% relative humidity, respectively). An improvement was obtained under both conditions, with better performance obtained with the organically modified fluorohectorite. The modified nanoclay is much better dispersed within the rubber matrix than the unmodified fluorohectorite, giving a superior "tortuous path" effect. Use of the organically modified nanoclay also resulted in greater improvements in material stiffness, as compared with the nanocomposite containing the unmodified nanoclay.

2.5 Nanocomposite foams formulated with hectorite nanoclay

One area of polymer based nanocomposites that has not been as extensively studied is polymeric foams. Polymeric foams are lightweight materials, with a strength/weight ratio that makes them attractive in various applications.

One group studying the performance of polymeric foams containing nanoclays is at Toyota Technological Institute, one of the original contributors to nanocomposite development. The polymer matrix used in their work is polycarbonate (PC), with a styrene-maleic anhydride (SMA) compatibilizer [18,19]. The nanoclay used is synthetic fluorohectorite modified with dimethyl dioctodecyl ammonium cation. The nanocomposite was foamed using supercritical carbon dioxide (CO_2). The loading level of the nanoclay was varied to study its influence on foam properties. Another parameter examined was the foaming pressure. Under most of the conditions evaluated the inclusion of nanoclays led to a higher cell density, indicating that the nanoclays might be acting as nucleating sites. The activation energy barrier decreases with increasing nanoclay content, and the foam modulus increases.

Other work on nanocomposite foams is based on a low density polyethylene (LDPE) matrix, with a maleated high density polyethylene compatibilizer [20,21]. The nanoclay used is hectorite modified with dimethyl dihydrogenated tallow ammonium chloride. The nanocomposite was foamed using azodicarbonamide (ADC). Two nanoclay loading levels were evaluated. In this work, the presence of hectorite was found not to affect the average cell size or cell wall thickness. The thermal stability and mechanical properties of the foam were improved by the addition of hectorite nanoclay. The coefficient of thermal expansion of the foamed material was not affected by the presence of hectorite nanoclay.

2.6 Nanoclay dispersion in thermoplastics

The dispersion of nanoclays in thermoplastics towards their nano state, i.e. individual layers, remains somewhat problematic until now. However, the desired materials benefits depend on the presence of delaminated (i.e. exfoliated) clay layers in polymer matrices, as they provide for a large two dimensional surface area that will affect the clay polymer interface. Ideally, all nanoclay layers are delaminated to their individual state, or when this is not feasible, at least a certain amount of layer dispersion is achieved, preferably in a repeatable fashion to achieve reproducible results.

Several approaches have been published that report on methods to assess and calculate degree of nanoclay delamination. Recently, a method was published to quickly measure nanoclay dispersion in plastics [22]. The approach is ideally suited for hectorite clay as it measures the silicate (Si-O bond) infrared (IR) absorption bandwidth, which depends on the degree of clay layer delamination. Since hectorite clay does not have isomorphous substitutions in the tetrahedral (surface) sheets of the clay layers, the silicate bonds absorb at distinct regions in the infrared depending whether the bond is within the clay layer plane or perpendicular to it. These absorption bands are more convoluted for montmorillonite clays, as isomorphous substitutions cause a disturbance in the absorption leading to several absorption bands overlapping each other. With the aid of polarized light, this analytical method can assess the orientation of delaminated hectorite clay layers in polymer films. Furthermore, the method is ideally suited for polyolefin plastics, or other polymers that do not significantly absorb infrared light in the 1000 – 1100 cm^{-1} silicate absorption region, although minor polymer absorption in this region can be subtracted from the pattern as was illustrated for Nylon composites [22]. Unpublished results for this paper showed that the Nylon 6/12 composite made with 12-aminododecanoic acid modified hectorite nanoclay has tethered platelets that were, and remained, randomly aligned despite various film processing approaches in an attempt to align the clay layers [23].

References

1   Güven, N., Smectites, in: Hydrous Phyllosilicates (exclusive of micas), Vol. 19 (edited by S. W. Bailey), *Reviews in mineralogy*, **1988**, 19, 497-559

2   Ross, C. S.; Hendricks, S. B. Minerals of the montmorillonite group, their origin and relations to soil and clays. *United States Geological Survey Professional paper*, **1945**, 205B, 23-79

3   Odom, I. E. Hectorite, Deposits, Properties and Uses, *Proceedings 10th Industrial Minerals International Conference*, (Industrial Minerals Division of Metal Bulletin) San Francisco, USA, May **1992**, 105-111,

4   Barrer, R. M.; Jones, D. L. Chemistry of soil minerals. Part VIII. Synthesis and properties of fluorhectorites. *Journal of the Chemical Society.,A,* **1970**, 1531 – 1537

5   Jones, T. R. The properties and uses of clays which swell in organic solvents. *Clay Minerals*, **1983**, 18, 399-410

6   Nahin, P. G.; Backlund, P. S. Organoclay-polyolefin compositions. *U.S. Patent*, **1963**, 3,084,117

7   Bellucci, F.; Camino, G.; Frache, A.; Sarra, A. Catalytic charring–volatilization competition in organoclay nanocomposites. *Polymer Degradation and Stability*, **2007**, 92, 425-436

8   He, H.; Duchet, J.; Galy, J.; Gérard, J.-F. Influence of cationic surfactant removal on the thermal stability of organoclays. *Journal of Colloid and Interface Science*, **2006**, 295, 202- 208

9   IJdo, W. L.; Benderly, D. *Nanocomposites 2007 conference* (Crain Communications), *Brussels*, 14-15 March **2007**

10   Benderly, D.; IJdo, W. L. Novel Organoclay for PVC Applications. *SPE Vinyltec conference, Iselin, New Jersey*, **2007**

11   Benderly, D.; Osorio, F. J.; IJdo, W. L. PVC Nanocomposites – Nanoclay chemistry and performance. *Journal of Vinyl and Additive Technology*, **2008**, 14, 155-162

12   Beyer, G. Flame retardancy of thermoplastic polyurethane and polyvinyl chloride by organoclays. *Journal of Fire Sciences*, **2007**, 25, 65-78

13   Beyer, G. Organoclays as flame retardants for PVC. *Polymers for Advanced Technologies*, **2008**, 19, 485-488

14   IJdo, W. L.; Mardis, W.; Benderly, D. Organoclay suitable for use in halogenated resin and composite systems thereof. *US Patent Application*, **2007**, 20070197711

15   Gorrasi, G.; Tortora, M.; Vittoria, V.; Kaempfer, D.; Mulhaupt, R. Transport properties of organic vapors in nanocomposites of organophilic layered silicate and syndiotactic polypropylene. *Polymer*, **2003**, 44, 3679–3685

16   Gorrasi, G.; Tammaro, L.; Tortora, M.; Vittoria, V.; Kaempfer, D.; Reichert. P.; Mulhaupt, R. Transport properties of organic vapors in nanocomposites of isotactic polypropylene. *Journal of Polymer Science: Part B: Polymer Physics*, **2003**, 41, 1798–1805

17   Gatos, K. G.; Karger-Kocsis, J. Effect of the aspect ratio of silicate platelets on the mechanical and barrier properties of hydrogenated acrylonitrile butadiene rubber (HNBR)/layered silicate nanocomposites. *European Polymer Journal*, **2007**, 43, 1097–1104

18   Ito, Y.; Yamashita, M.; Okamoto, M. Foam Processing and Cellular Structure of Polycarbonate-Based Nanocomposites. *Macromolecular Materials and Engineering*, **2006**, 291, 773 – 783

19   Mitsunaga, M.; Ito, Y.; Ray, S. S.; Okamoto, M.; Hironaka, K. Intercalated Polycarbonate/Clay Nanocomposites: Nanostructure Control and Foam Processing. *Macromolecular Materials and Engineering*, **2003**, 288, 543 – 548

20   Velasco, J. I.; Antunes, M.; Ayyad, O.; Saiz-Arroyo, C.; Rodrıguez-Perez, M. A.; Hidalgo, F.; de Saja, J. A. Foams based on low density polyethylene/hectorite nanocomposites: Thermal stability and thermo-mechanical properties. *Journal of Applied Polymer Science*, **2007**, 105, 1658–1667

21   Velasco, J. I.; Antunes, M.; Ayyad, O.; Lopez-Cuesta, J. M.; Gaudon, P.; Saiz-Arroyo, C.; Rodrıguez-Perez, M. A.; de Saja, J. A. Foaming behaviour and cellular structure of LDPE/hectorite nanocomposites. *Polymer*, **2007**, 48, 2098-2108

22   IJdo, W. L; Kemnetz, S.; Benderly, D. An infrared method to assess organoclay delamination and orientation in organoclay polymer nanocomposites. *Polymer Engineering and Science*, **2006**, 46, 1031 – 1039

[23]   IJdo, W.L. *private communication*, **2008**

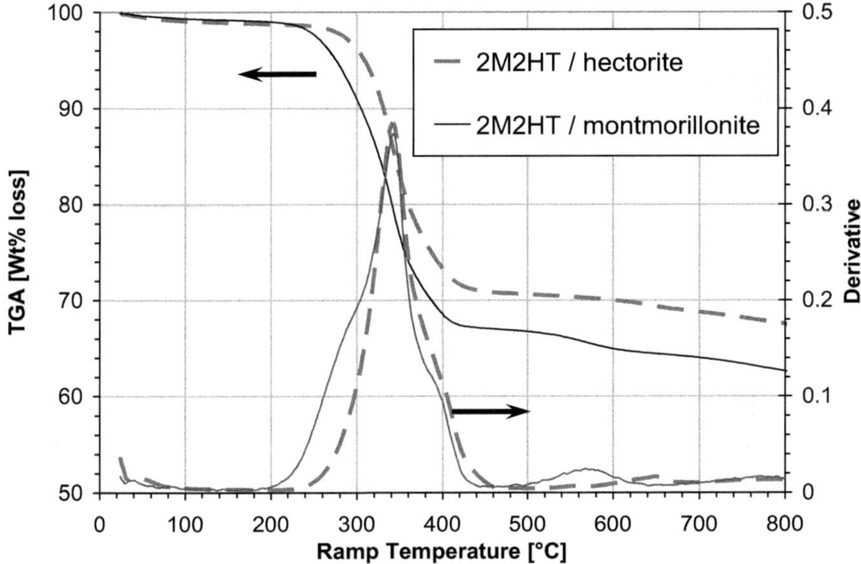

Figure 2.1

Thermogravimetric analysis (TGA) of hectorite and montmorillonite clay modified with
dimethyl dehydrogenated tallow ammonium (2M2HT) organic modifier. The
montmorillonite clay shows an earlier weight loss during the temperature ramp compared
to the hectorite-based nanoclay. Thermogravimetric analysis (TGA) was conducted
using a TA Instruments TGA 2950, run in nitrogen at a rate of 10 °C/minute
(Courtesy of Elementis Specialties)

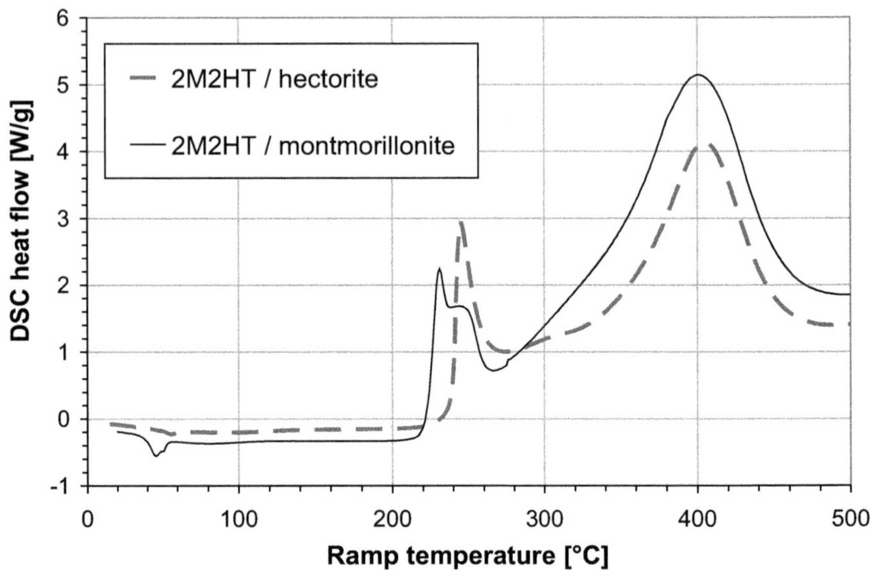

Figure 2.2
Differential scanning calorimetry (DSC) of hectorite and montmorillonite organoclays modified with dimethyl dehydrogenated tallow ammonium (2M2HT) organic modifier. The results show two distinct thermal events in the 200 – 300 °C region for the montmorillonite clay, while the hectorite clay only shows one, which is attributed to the different layer charge compositions of both clays. Differential scanning calorimetry (DSC) was conducted using a TA Instruments 2010, run in nitrogen at a rate of 10 °C/minute (Courtesy of Elementis Specialties)

Chapter 3
Nanocomposites based on carbon nanotubes

Olivier Decroly
Nanocyl S.A, Sambreville, Belgium

3.1   Introduction

Carbon nanotubes have been synthesised in bulk quantities for more than 20 years and used in commercial plastic applications since the early 1990s. The original company producing and selling them was Hyperion Catalysis International from USA. This company wrote the original patent describing their structure and synthesis using hydrocarbon feedstocks and catalysts [1]. At the same time as they were being launched commercially, their individual structure started to be investigated by Japanese researcher Sumio Iijima and his colleagues at Nippon Electric Company Corporation Fundamental Research Laboratories [2]. It is unlikely that there was any communication between the two laboratories at that time, resulting in a general confusion on the precise chronology of their commercialisation and discovery.

A carbon nanotube is a tube-shaped structure, made of carbon, that has a diameter measuring on the nanometre scale. A nanometre is one one-billionth of a metre, or about one ten-thousandth of the thickness of a human hair. Carbon nanotubes have many structures but only the multi-wall type has been successfully commercialised. . There is also a variety of carbon nanotubes called "single-wall" which attracts a lot of attention because of their much smaller diameter (1-2 nm) linked to specific properties such as electronic transport (metallic or semiconducting). Their commercial viability is still questionable but theycould possibly find applications in electronics or biomedical fields.

The global commercial volume for multi-wall carbon nanotubes (MWCNT) is approximately 50 tonnes but their production capacity is likely to be close to an order of magnitude higher. MWCNTs typically have a diameter ranging from 10 nm up to 50 nm, depending on manufacturer. The diameter is dependent on the number of walls and the hollow core thickness. Each wall is composed of a graphene layer. The length of carbon nanotubes is typically several microns but recent advances have made them much longer up to the centimetre range. Different powder textures are also available from various sources, based on more or less loose aggregates with an average size ranging from below 100 to several hundred of microns. Typical electron microscopy photographs are shown on Figure 3.1

Current industrial production processes for MWCNTs use chemical vapour deposition at high temperature (typically 700 °C) to decompose a source of hydrocarbon (e.g. methane, ethylene) in the presence of a catalyst system. The reactor technologies can vary (fluidised bed reactor or kiln). In most products, there is a residual catalyst mixture which can vary from less than 1% to up to 10%. If necessary, a further processing step can remove this residue.

Carbon nanotubes are known for exhibiting unique mechanical, electrical and thermal properties. A Young's modulus of 1000 GPa and tensile strength of 60 GPa have been measured on individual structures, which is several orders of magnitude higher than common engineering plastics. High electrical and thermal conductivities have also been determined experimentally, with values close to or better than metals. Tables 3.1 and 3.2 compare these properties to other structures with nano-dimensions used in polymer nanocomposites. Carbon nanotubes show a unique combination of stiffness, strength and tenacity compared with other high aspect ratio nanomaterials which typically lack one of these properties.

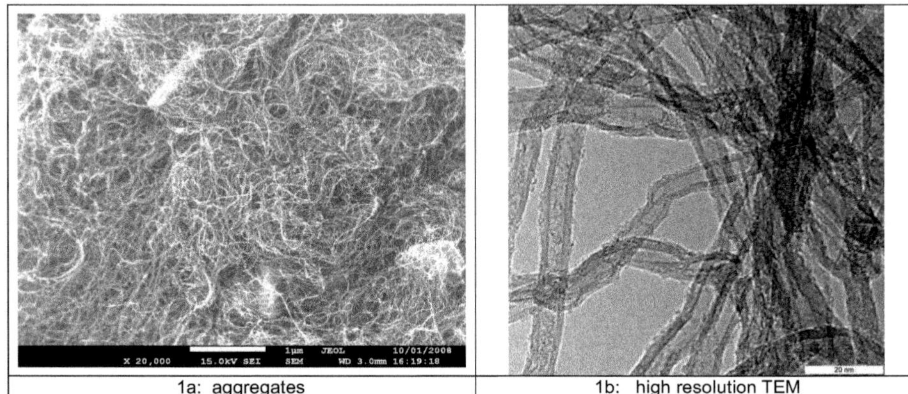

Figure 3.1
Electron microscopy pictures of commercial multi-wall carbon nanotubes
Figure 3.1a is a scanning electron microscope zoom on an entangled particle. 3.1b shows the hollow structure of multi-wall carbon nanotubes and their irregular shape using high resolution transmission electron microscopy

Table 3.1
Mechanical properties of high-aspect ratio nanomaterials

Material	Specific density	Young's modulus (E) (TPa)	Strength (GPa)	Strain at break (%)
Carbon nanotubes	1.3 – 2	1 – 1.7	10 – 60	10 – 30
Other nano-sized materials				
Single clay silicate layer (montmorillonite)	2.8 – 3	0.17	1	< 10
vapour grown carbon fibre (VGCF)	2	0.4 – 6	3 – 7	< 10
Graphite nanoplatelet	2	1	10 – 20	n.a

Table 3.2
Transport properties of conductive materials

Material	Thermal conductivity (W/m k)	Electrical conductivity (S/m)
Carbon nanotubes	> 3000	$10^5 - 10^7$
VGCF	1950	2×10^6

3.2   Carbon nanotube nanocomposites

The current commercial applications for multi-wall carbon nanotubes exploit mainly their high intrinsic electrical conductivity and high aspect ratio for the design of conductive materials (mainly thermoplastics). The technical benefits are of interest for end-uses in plastics for automotive and electronic applications. The recent launch of sporting goods based on carbon-fibre reinforced thermoset composites exploit the high strength of this material, but the volume of nanotubes concerned is relatively small. Dispersing a microstructure with high-surface area such as carbon nanotubes (>> 100 m²/g) is quite challenging but satisfactory results can be achieved at economical cost using current state of the art technologies for finely divided materials. There are more complex techniques described in the scientific literature for achieving high levels of dispersion such as in-situ polymerisation [3].

These composites are manufactured using conventional mechanical mixing devices such as twin-screw extruders for thermoplastic systems, high-shearing mixers, dispersers and ultrasonication for low viscosity media. For low viscosity systems it is usually necessary to use surfactant technology for improving and stabilising the dispersion during its manufacturing, and processing in the final stage of the composite. A masterbatch step might be required depending on the level of dispersion targeted. Control of the process is also required to address the different behaviour of the nanotubes such as alignment or reduction of aspect ratio (length/diameter), which can occur in high-shearing processes such as extrusion or thermoplastic injection moulding.

3.2.1   Conductive carbon nanotube composites

The use of carbon nanotubes for antistatic and conductive applications in polymers is already a commercial reality and is growing in industrial sectors such as electronics and the automotive industry. These applications currently consume the bulk of multi-wall carbon nanotubes produced. Figure 3.2 shows a typical conductivity plot for an engineering thermoplastic (polycarbonate). The loading for achieving electrical percolation with multi-wall carbon nanotubes can be 5-10 times lower compared to conductive carbon black grades. Similar comparisons are made in thermoset resins like epoxies but with much lower loading. This is explained by the theory of percolation: a path for electron flow is created when the particles are very close to each other or have reached a percolation threshold. Fibrous structures with high aspect ratio increase the number of electrical contacts and ensure a smoother pathway. The geometric aspect ratio of carbon nanotubes is typically greater than 100, compared with short carbon fibre (< 30) and carbon black (> 1) in the final product (e.g. an injection moulded part). This explains the lower content needed for a given resistivity. The percolation behaviour may vary depending on the type of resin, viscosity and processing of the polymers. There are many reports describing the percolation behaviour of different thermoplastic composites and their alloys prepared by melt-blending [4,5,6,7].

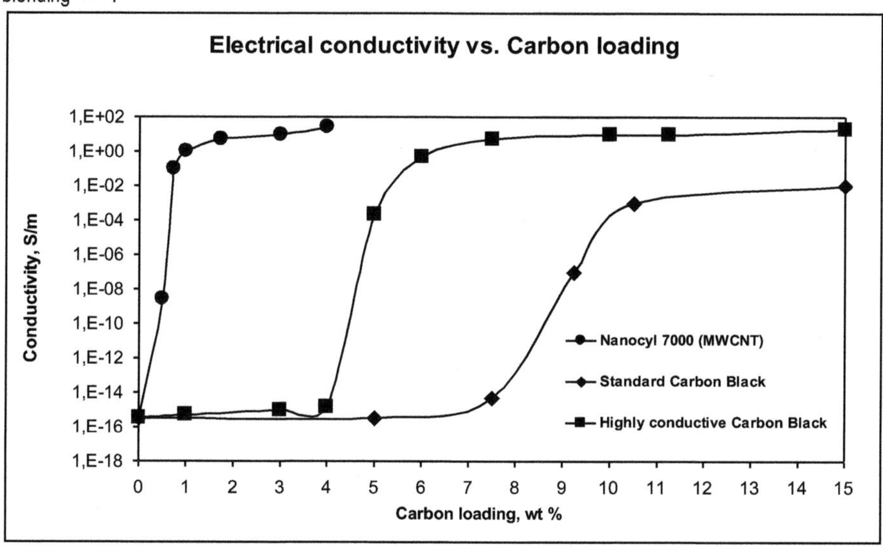

Figure 3.2

Conductivity plot of polycarbonate filled with carbon nanotubes (Nanocyl® 7000)

The lower loading of additives can offer several advantages including processability, surface finish, reduced sloughing and improved retention of the mechanical properties of the virgin polymer. These benefits have led to the use of multi-wall carbon nanotubes in the conductive polymer applications shown in Table 3.3. For these applications, they can compete with additives like highly conductive carbon black or carbon fibres on a price/performance basis. They can also offer superior performances providing previously unavailable product specifications:

Table 3.3		
Commercial conductive polymer applications for multi-wall carbon nanotubes		
Market	Applications	Features of formulations based on carbon nanotubes
Automotive	Fuel systems components & fuel lines (connectors, pump parts, o-rings, tubing), exterior body parts for electrostatic painting (fenders, mirror housing, fuel filler flap)	Better balance of properties vs. carbon black, processability for large parts, dimensional stability
Electronics	Manufacturing tools & equipment, wafer carriers, conveyor belts, trays, clean room equipments	Cleaner formulations vs. carbon fibres, better control of surface resistivity, processability for moulding thin parts, dimensional stability, balance of properties

3.2.2   Structural composite applications

The exceptional strength of carbon nanotubes benefits several sporting goods applications based on carbon fibre composites (baseball bats, bicycle frame and parts, hockey sticks, golf club shafts). The surface of carbon nanotubes (mainly multi-wall types in commercial products) are usually chemically modified for easier integration and better bonding to the polymer phase (e.g. epoxy or polyurethane) [8]. Typical improvements measured for the fibre-reinforced composite are in the range of a 10 to 50% increase in strength and impact resistance. The content of carbon nanotubes in this type of composites is limited (typically lower than 1 weight-% of the polymer) due to the resulting increase of the viscosity of the resin which can impede the processing of the composite by pre-impregnation of the carbon fibres or resin transfer moulding.

Additionally, carbon nanotubes show significant improvement in the mechanical properties of elastomers [9]. Carbon black and other particulate fillers are widely used for reinforcing elastomers in tyres and other industrial elastomer products. Formulations can contain high loading of traditional fillers (more than 50 weight-%.) for increased stiffness and strength. Substitution of these fillers by 5-10 % of multi-wall carbon nanotubes can offer similar level of stiffness and strength in elastomers with additional benefits in specific applications (o-rings, gaskets).

3.2.3   Coatings applications

A network of very thin conductive structures such as carbon nanotubes allows new applications in thin film technology, such as antistatic transparent and conductive coatings with permanent conductivity, better mechanical properties and chemical resistance. Highly conductive transparent films technologies are under development and could compete in the near future with the metal oxide-based systems such as indium tin oxide (ITO) sputtering used today for transparent electrodes in flat panel displays, and the more constraining design such as flexible displays. Very thin carbon nanotubes are preferred for this type of application for a better balance of transparency versus conductivity.

A cellulose paper composite using multi-wall carbon nanotubes has been developed and is used for designing more flexible heating resistance for car mirror defrosting, floor heating and other heating devices. EMI shielding applications are also targeted because of the paper high conductivity.

Other property improvements based on the use of carbon nanotubes, like flame retardancy [10,11] and antifouling are also under investigation. These could lead to new products more suited to new environmental regulations, with higher performance than existing products and reduced material cost (antifouling marine paint, thermal barrier coating for aerospace, fire proof coating for transportation).

In conclusion, the use of carbon nanotubes in commercial applications is already a reality, and gaining more and more attention. They are accepted by the industry as an added-value class of additives, competitive with other traditional options, and able to be handled by industry best practices. New useful and unexpected properties provided by carbon nanotubes are being investigated which will increase their penetration in the polymer industry.

REFERENCES

1. Tennent, H. G. Carbon fibrils, method for producing same and compositions containing same. U.S. Patent 4,663,230, 1987
2. Iijima, S. Helical microtubules of graphitic carbon. *Nature*, **1991**, 354, 56 – 58
3. Bonduel, D., Mainil, M., Alexandre, M., Monteverde, F., Dubois, Ph. Supported coordination polymerization: a unique way to potent polyolefin carbon nanotube nanocomposites. *Chemical Communications*, **2005**, 781-783
4. Pötschke, P., Dudkin, S. M., Alig, I. Dielectric spectroscopy on melt processed polycarbonate-multiwalled carbon nanotube composites. *Polymer*, **2003**, 44, 5023-5030
5. Meincke, O., Kaempfer, D., Weickmann, H., Friedrich, C., Vathauer, M., Warth, H. Mechanical properties and electrical conductivity of carbon-nanotube filled polyamide-6 and its blend with acrylonitrile/butadiene/styrene. *Polymer*, **2004**, 45, 739-748
6. Pötschke, P., Bhattacharyya, A. R., Janke, A. Carbon nanotube filled polycarbonate composites produced by melt mixing and their use in blends with Polyethylene. *Carbon*, **2004**, 42, 965-969
7. Wu, M., Shaw, L. On the improved properties of injection-molded carbon nanotubes-filled PET/PVDF blends. *Journal of Power Sources,* **2004**, 136, 37-44
8. Gojny, F. H., Wichmann, M., Fiedler, B., Schulte, K. Influence of different carbon nanotubes on the mechanical properties of epoxy composites - A comparative study. *Composites Science and Technology,* **2005**, 65, 2300-2313
9. Bokobza, L. Multiwall carbon nanotubes elastomeric composites: A review. *Polymer*, **2007**, 17, 4907-4920
10. Beyer, G. Improvements of the Fire Performance of Nanocomposites, presented at the 13th Annual BCC Conference on Flame Retardancy, Stamford CT, June 2002
11. Kashiwagi, T., Grulke, E., Hilding, J., Harris, R., Awad, W., Douglas, J. Thermal degradation and flammability properties of poly(propylene)/carbon nanotubes composites. *Macromolecular Rapid Communications*, **2002**, 23, 761-765

Chapter 4
Nanocomposite processing

Daniel Schmidt

Department of Plastics Engineering, University of Massachusetts Lowell, USA

In this chapter we focus on the processing techniques used to produce polymer nanocomposites (i.e. materials consisting of a polymeric matrix with discrete nanoparticles embedded in it), with an emphasis on the clay-based materials (nanoclays) described in the first chapter. Before addressing specifics, however, it is instructive to consider more general concerns associated with the production of polymer nanocomposites. We will approach these concerns by asking a series of (hopefully) relevant questions.

4.1   What is processing and why is it necessary?

While such a simple question is often dismissed as being self-evident, it is only by considering the most basic premises from which we start that we can develop new ideas about how to accomplish our goals. The general issue faced when preparing polymer nanocomposites is that the final product consists of a mixture of at least two distinct solid phases of matter, the polymer and the nanofiller, as well as any other additives that may be present in the formulation. To say that solids do not spontaneously mix under normal conditions is not entirely true; they can, over long periods of time. This raises a critical point: Processing is not *only* performed to combine multiple components, but to combine them *in a reasonable amount of time*. In addition to time, input of energy, output of waste and cost are all critical concerns as well, and will be considered throughout this chapter.

4.2   What is needed to process a polymer nanocomposite?

There are many ways to mix solid polymer and solid nanofiller, all of which take advantage of a limited number of physical phenomena, namely diffusion and flow. While these terms are often used to describe the behaviour of the polymer phase, these phenomena are equally applicable to the nanofiller as well. In a liquid medium, for instance, nanoparticles are certainly capable of diffusing from an area of high concentration to an area of low concentration, and nanofillers may be poured, or made to flow, just like any other powder. With that said, the timescales over which nanoparticles tend to diffuse and move are often longer than those over which polymer chains move, in spite of the fact that nanoparticles are often said to be similar in size to the polymer chains that surround them in a polymer nanocomposite. This can be explained by the ability of polymer to undergo reptation much more readily than nanoparticles are able to move through the polymer, allowing the polymer chains to snake their way through the material with minimal changes in the local structure. Given that this is the case, then most polymer nanocomposites are produced by enhancing the mobility in the polymer phase, allowing for diffusion, and (optionally) flow to occur.

4.2.1   Enhancing polymer mobility

Enhancing polymer mobility means more readily overcoming the interactions between neighbouring polymer chains. This may be accomplished through the input of energy to the system (i.e. thermally, mechanically), or by reducing the effectiveness of the aforementioned interactions (i.e. through the introduction of an additional compatible and more mobile species such as a solvent or a plasticizer).

One fundamental difference between these two approaches to enhanced polymer mobility is homogeneity. It is very difficult to suppress the formation of thermal gradients in any macroscopic piece of material, especially in a practical manufacturing process, and it is practically impossible to impart exactly the same amount of mechanical energy to every part of a sample when inducing flow (which, in turn, is an important orientation mechanism for high aspect ratio fillers). In contrast, the introduction of a compatible third phase with high mobility gives increased polymer mobility that can be homogeneous down to the molecular level.

4.2.2   The consequences of processing

At this point the reader may be led to believe that inducing more mobility in the polymer phase must be that much better for nanocomposite formation. This is only true to a point, however, for reasons that will become readily apparent. The input of too much thermal and/or mechanical energy (and subsequent shear heating, especially thanks to the higher viscosities typically associated with nanocomposites) can have detrimental effects on the organic components of the system, i.e. the nanoclay modifiers as well as the polymer, giving rise to coloration and eventually degrading materials properties. Alkylammonium-modified nanoclay nanofillers (i.e. those most commonly found in commercial production) experience the onset of modifier degradation around 180°C [1]. Alkylphosphonium nanoclays have been explored as more stable options, due to the greater stability of alkylphosphonium salts versus alkylammonium salts, but in practice their combination with the nanoclay results in much less pronounced improvements [2]. Imidazolium salts show much more promise in this regard, with onset temperatures of as high as 300-400°C measured in both nitrogen and air [3,4]. Onset temperatures in the same range (under nitrogen) have also been reported for pyridinium and quinolinium based modifiers [5], and one report also indicates exceptional thermal stability (5% mass loss at 600°C under nitrogen), using a clay modifier based on a highly stable carbocation [6]. Unfortunately, no commercially available nanoclays make use of these more stable modifiers as of this writing.

Modifiers aside, the polymer may also be thermally degraded via the introduction of the nanoclay. While many reports claim enhanced thermal stability in the presence of the nanoclay layers, this conclusion strongly depends on how thermal stability is measured. While some systems show across-the-board improvements, for instance, what is sometimes observed during the thermogravimetric analysis of polymer nanocomposites is that the temperature at which the maximum rate of degradation occurs is higher, but that the temperature at which degradation begins is actually *lower*. This may imply that the degradation kinetics are being slowed by the ability of the nanoclay to limit escape of degradation byproducts from the sample due to an improvement in barrier properties, but that the nanoclay itself, or some impurity associated with it, is at the same time catalyzing the degradation of the polymer. The latter should come as no surprise, given the known utility of nanoclays as catalysts in organic chemistry [7] or the catalytic tendencies of high surface area metal oxides in general.

Nanoclay-induced degradation of the matrix polymer during processing involving the input of thermal and/or mechanical energy can manifest itself in many ways, as demonstrated by a few relevant examples. In polyamide-6 nanocomposites, for instance, it has been observed that the presence of the nanoclay seems to enhance the rate of polymer degradation during injection moulding [8]. Regardless of mechanism, the results are the same – chain scission occurs, molecular weight is lost, and the properties of the matrix suffer. Interestingly, a separate study of polyamide-6/nanoclay nanocomposites indicates that this tendency towards process-induced degradation can actually *enhance* nanoclay dispersion levels [9], showing that a carefully controlled amount of degradation may nevertheless be beneficial under special circumstances. Another example of degradation during processing induced by the addition of nanoclay involves poly(vinyl chloride), where degradation of the alkylammonium modifiers may play a role [10]. In sum, then, there is evidence that both the nanoclay layers themselves (in the presence or absence of water, which they may anyway bring into the system) and nanoclay modifiers (either ionically bound to the clay layers or present as free salts with differing stability/chemical characteristics due to counterion effects) are both quite capable of degrading the matrix polymer with the input of sufficient thermal and/or mechanical energy.

Beyond degradation of the organic component, too much mechanical energy may even result in damage to the primary particles themselves and an associated loss in effective aspect ratio (and thus mechanical and barrier properties). While in the nanoclays this is generally not observed due to the very small absolute size of the primary particles, such an effect should not be neglected, especially when using nanoparticles with higher aspect ratios or larger absolute dimensions in combination with exceptionally harsh processing.

Systems where mobility has been enhanced via the addition of a highly mobile and compatible third phase may experience all of the aforementioned problems if heat and/or mechanical energy are applied to them. There are a few additional issues specific to these systems, however, regardless of input of heat and/or mechanical energy. One of the more general issues is the possibility of introducing impurities harmful to other components present in the system (polymer, nanoclay, or anything else). More to the point is the fact that solvent-based processing, often performed under fairly dilute conditions, will tend to concentrate any impurities, additives, or stabilizers present in the solvent in the final product following removal of the solvent. This is especially an issue if any reactive post-treatment is planned, since many solvents contain stabilizers intentionally added to act as radical scavengers and/or suppress reactions in the solvent, where such reactions are considered harmful. The concentration of these additives in a polymer that is to be subsequently crosslinked via electron beam irradiation or peroxide treatment may be substantially affected as a result.

Likewise, complete solvent removal, often assumed to be trivial, is actually extremely difficult to achieve, with some solvent-processed polymer films retaining more than 1 weight-% of what would normally be considered a highly volatile solvent after years (!) of vacuum drying [11]. This effect will only be exacerbated if the barrier properties of the nanocomposite are indeed substantially improved over the base polymer. On the other hand, the tendency of crystalline domains to exclude impurities during their formation will result in the concentration of impurities in the amorphous phase of a semicrystalline polymer nanocomposite. While this can have serious consequences if it is solvent additives, stabilizers, or other impurities that are being concentrated in the amorphous polymer domains, such a situation also concentrates the solvent itself, thus enhancing solvent removal [11].

4.2.3 A balanced approach

In sum, what is needed to process a polymer nanocomposite is to enhance the mobility of the polymer just enough to realize the desired structure and properties while minimizing the amount of harm done to the components of the formulation. Likewise, the impact of the presence of degradation byproducts, voids, and/or residual solvent on the properties of the final material should be considered pragmatically; as with any polymer processing operation, at the end of the day the goal is to produce consistently acceptable parts, not perfect ones.

4.3 Does the polymer have to be a solid at room temperature?

Indeed, we have made an assumption at the beginning of this section by stating that we begin with a polymer that is solid at room temperature. What if it is not? A distinctly different situation involves the combination of a liquid polymer with a solid nanofiller, something that may or may not be followed by some form of crosslinking or post-treatment depending on the application. An excellent example of this is high-vacuum grease, which is in effect a liquid nanocomposite composed of high molecular weight poly(dimethylsiloxane) in combination with a substantial amount of amorphous fumed silica nanoparticles. Such a material could be thought of as a nanocomposite melt, no different than any other polymer melt with nanoparticles in it.

4.4   Do we need to start with a polymer at all?

There is a more general assumption implied by all discussion to this point, namely that we begin our processing with a polymer. In fact, even this is not necessary for polymer nanocomposite formation. Instead, it is possible to directly combine monomeric species and nanofiller, then polymerize the former in the presence of the latter. For example, the work of the Toyota Corporation in the early 1990s responsible for inspiring a number of other groups to pursue research in this area involved the production of polyamide-6/layered silicate nanocomposites prepared in this fashion [12]. As an interesting historical note, this idea is not new; in fact, what may be the earliest literature report of an in-situ polymerized nanocomposite, synthesized through the polymerization of intercalated acrylic monomers between nanoclay layers, was published in 1961(!) [13].

4.5   Can we do away with the pre-formed nanofiller as well?

Indeed, while we have also assumed that the nanofillers in question were formed prior to nanocomposite processing, even this is not a requirement. The most common means of preparing nanofillers within a polymer involve the use of liquid metal oxide precursors like tetramethoxysilane (TMOS), tetraethoxysilane (TEOS), aluminum *sec*-butoxide, titanium *iso*-propoxide and the like. The polymer is swollen with these compounds, which are then made to undergo hydrolysis and condensation to form inorganic domains *in-situ*. Controlled mineralization, a technique inspired by the way in which nature produces nanocomposites like nacre, involves the combination of hydrophilic (generally charged) polymers with various water-soluble mineral species to form nanostructured calcium carbonate, hydroxyapatite and the like. The simultaneous formation of both the polymer and the nanofiller is also possible, and allows for the formation of single-phase organic/inorganic nanocomposites, something not readily achieved via the techniques described here. As we are more heavily focused on systems based on pre-formed nanoparticles, these materials will not be discussed in great detail; the reader should not be mislead, however, into thinking that they are scientifically uninteresting or industrially irrelevant as a result.

4.6   What are our options as far as pre-formed nanofillers?

While the focus of this chapter is primarily on the aforementioned nanoclays, it is worth noting that the processing techniques described here are relevant to a wide range of nanocomposites, including those based on other nanofillers. A partial list of nanofillers at or nearing large-scale commercial production at the time of writing is given in Table 4.1 with type (0-D = particle, 1-D = rod/fibre, 2-D = plate/sheet), typical aspect ratio (orders of magnitude given), and typical modifiers used identified.

Table 4.1

A partial list of nanofillers at or nearing large-scale commercial production, classified by type (0-D = particle, 1-D = rod/whisker, 2-D = plate/sheet) and typical aspect ratio of the primary particle (effective aspect ratios will depend on actual dispersion state), typical modifiers used to enhance dispersion, and selected sources for these materials

Nanofiller	Type	Aspect Ratios	Typical Modifiers	Selected Sources
Carbon black	0-D	~1	None; oxidation-induced functional groups (OH, COOH, epoxide, etc.)	Cabot, Columbian, Continental Carbon, Degussa, Mitsubishi, Sid Richardson, Tokai Carbon
Fumed silica/metal oxides	0-D	~1	None; hexamethyldisilazane	Cabot, Degussa, Wacker Chemie
Colloidal silica/metal oxide sols (aqueous and organic)	0-D	~1	Various acid, base, and salt treatments	Cabot, Degussa, Eka Chemicals, H. C. Starck, Nalco, Nissan, Nyacol, W. R. Grace
Multi-wall carbon nanotubes/carbon nanofibres	1-D	~10-10,000	None; oxidation-induced functional groups (OH, COOH, epoxide, etc.); other specific organic functionalities	ASI, Arkema, Bayer MaterialScience, Electrovac, Hyperion Catalysis, M.E.R., Nanocyl, Nanoledge, Unidym, Zyvex Performance Materials
2:1 Phyllosilicates ("nanoclays")	2-D	~10-10,000	Alkylammonium cations; other surfactant cations; alkali metal cations	CBC Japan, Elementis Specialties, Kunimine Industries, Laviosa Chimica Mineraria, Nanocor, Southern Clay Products
Hydrotalcite	2-D	~10-10,000	Alkylcarboxylate anions; other surfactant anions; nitrate anions; carbonate anions	Akzo-Nobel, Kisuma Chemicals/Kyowa Chemical, Süd-Chemie
Expanded/exfoliated graphite/graphene	2-D	~10-10,000	None; oxidation-induced functional groups (OH, COOH, epoxide, etc.); other specific organic functionalities	Vorbeck Materials, XG Sciences
NOTE: Fullerenes and polyhedral oligomeric silsesquioxanes (POSS) are well-defined molecular species; as it is not clear whether they fit the usual definition of fillers as opposed to chemical or (macro)molecular additives, they have been omitted from this table as a result.				

4.7   What makes a nanofiller disperse in a particular polymer during processing?

Whole books could be written attempting to answer this question in detail. The simple answer is that one must have, first and foremost, a system in which there is some level of compatibility between the polymer and the nanofiller. If the polymer will not "wet out" the nanofiller, dispersion will be difficult to obtain, and even if good dispersion is achieved, it may be difficult to retain this structure upon further processing or to realize the desired property enhancements. This is because in most polymer nanocomposites the vast majority of the polymer is at or within a few nanometres of the surface of a nanofiller particle; if favourable interactions are not present, therefore, the vast majority of the polymer is affected. An important example of this comes from the work of Alexandre et al, who managed to achieve exceptionally high levels of nanoscale dispersion in an ultra-high molecular weight polyethylene/nanoclay system through in-situ polymerization, only to find that the mechanical properties were actually *degraded*, tensile modulus included, and that further processing resulted in re-aggregation of the clay layers [14]. *An absolutely essential point when thinking about the processing of polymer nanocomposites, therefore, is that if the nanofiller and the polymer matrix are poorly matched, no amount of processing will change this.*

To expand on this idea of compatibility and the process by which dispersion is obtained, it is important to distinguish thermodynamic factors from kinetic ones. In thinking about the thermodynamics of a particular combination of polymer and nanofiller, both entropic and enthalpic effects are relevant, and are developed below for the case of a single-phase amorphous thermoplastic.

4.7.1   The thermodynamics of dispersion: entropy

While it is difficult to be completely general across all polymer nanocomposites, entropy may work in two ways in these systems. On the molecular level, polymer chains immediately adjacent to a nanofiller surface or trapped between two closely spaced nanofiller particles are often said to be confined or otherwise altered in conformation versus normal ("bulk") polymer. In general, any restrictions in accessible chain conformations should translate into a loss of conformational entropy, which is energetically unfavourable and becomes more so with increasing temperature – an explanation very much in line with the theory of rubber elasticity, which predicts the fact that a rubber band will resist stretching more effectively at higher temperatures based on a purely entropic argument. This would seem to bode ill for nanofiller dispersion; however, there is an opposing entropic factor that must be taken into account, namely the configurational entropy or entropy of mixing, which tends to push the system towards a dispersed state simply because there are many more equivalent configurations with many dispersed particles than with a few poorly dispersed aggregates. This effect also becomes more significant with temperature.

With respect to the nanoclays in particular, a third entropic factor comes into play, namely the increase in conformational entropy of the surfactant tails ionically bound to the nanoclay layer surface. In the pure organoclay, these tails lie flat (or nearly so) and are heavily constrained by the adjacent clay layers. Once polymer intercalation occurs, however, they are able to move much more freely and have access to more energetically equivalent configurations, resulting in an entropy increase that favours intercalation. Similar effects would be expected in other modified nanofillers, although *only* if the modifier experiences an increase in accessible conformations as a result of dispersion. This typically means that the modifier must contain chains of some length and flexibility, and that the modifier grafting density on the surface of the nanoparticles must not be so high as to induce close-packing. In this situation the modifier should have both the ability and the room to move once nanofiller dispersion is achieved.

4.7.2 *The thermodynamics of dispersion: enthalpy*

From an enthalpic standpoint, the situation is somewhat simpler, at least qualitatively. Previously we spoke about the attractive interactions between polymer chains; needless to say, nanoparticles also experience attractive interactions with one another. Both statements imply that it is enthalpically favourable for polymer chains to be with other polymer chains and nanoparticles to be with other nanoparticles. In order to achieve dispersion, then, from an enthalpic standpoint at least we must apply a very similar approach as applied to that of the polymers. Specifically, we must either add energy to the system to overcome the attractive interactions between the nanoparticles, or we must add something to the system to reduce these attractive interactions. We can accomplish the former through the input of thermal or mechanical energy to a system. Back to the idea that no amount of processing will make an incompatible system compatible, however, this alone is not enough; the surface-to-volume-ratios of nanofillers are sufficiently high that there are many, many more interparticle interactions per unit mass of nanofiller than would be found in the same mass of a chemically equivalent filler consisting of much larger particles. While these interparticle interactions are individually very weak, they are so numerous that physically pulling or shearing apart agglomerated nanoparticles is quite difficult, and the amount of thermal energy needed to effectively disrupt them is generally more than sufficient to cause unwanted chemical reactions, thermal degradation of various components of the system, etc.

The key to overcoming this issue is to select a system based on a polymer capable of favourable interactions with the nanofiller of interest, thus screening interactions between nanoparticles and limiting their effectiveness. Some level of favourable interactions are therefore necessary (but not sufficient) for inducing nanoparticle dispersion. It is the combination of these favourable interactions with the input of thermal and/or mechanical energy that is both necessary *and* sufficient for nanoparticle dispersion in a polymer nanocomposite.

Again with reference to nanoclays in particular an additional enthalpic effect becomes relevant. As implied previously, the mechanism of polymer intercalation in the nanoclays involves the swelling of the interlayer gallery as the polymer chains diffuse in, coupled with the extension of the modifier tails away from the layer surface as a result of their newfound freedom. In effect, the modifiers are chosen based on the incompatibility of their alkyl tails with the more polar clay surface on which they are forced to rest. The result is that they are quite happy to be displaced in the presence of molecules more compatible with the clay surface than they are. It is worth noting that it is exactly for this reason that nanoclay dispersion in polyolefins is so difficult to achieve – there is no driving force to displace one alkyl chain by another. By this logic, it would seem that intercalation, at least, should be enthalpically favourable with a wide range of polymers, regardless of whether the modifiers contained one, two, or even three alkyl tails, so long as they retained the ability to "stand up" in the presence of the intercalating polymer. In fact, however, another enthalpic factor that must not be neglected is the interaction of the polymer chains with the modifier tails; if this interaction is sufficiently unfavourable, intercalation will not occur. Therefore, the answer to the enthalpic question in nanoclay based nanocomposite systems is that the polymer-clay and polymer-modifier interactions should be favourable, while the modifier-clay interactions should be unfavourable. Such a situation will maximize the efficiency of any subsequent processing.

4.7.3 *Complications: crystallinity*

As stated earlier in this section, the previous discussion is most accurately applied to single-phase amorphous thermoplastics. In the case of semi-crystalline polymers, one potential difference relates to the issue of reduced conformational entropy in polymer chains associated with the surface of the nanofiller particles. Similar restrictions are placed on polymer chains that become part of crystalline polymer domains in such a system. In spite of this, crystallization is still favourable, implying that the enthalpy of crystallization is sufficient to overcome this entropic effect, at least up to the melting point of the crystalline polymer. For the same reasons, crystals tend to exclude impurities from their interior as they grow. The consequences are that, depending on where nucleation occurs, any nanofiller particles not acting as nuclei may be literally pushed around by crystal growth fronts moving through the system. In practice, the ability of nanofillers to act as nucleating agents greatly depends on the nature of the system.

In polyamide-6, for instance, it is well-known that the presence of dispersed nanoclay enhances the formation of the otherwise metastable γ phase [15], an apparent result of nanoclay-mediated changes in the hydrogen bonding pattern of the polymer [16]. In this case, then, the nanoclays not only act as effective nucleants, they alter the favoured crystalline phase as well. Similar results have been reported in poly(vinylidene fluoride), with both the preferred crystalline phase and the shape and size of the crystallites affected [17]. In these cases it would appear thermodynamically favourable not to exclude the nanofillers from the crystalline polymer phase. Instead, upon cooling, each nanofiller particle (primary or agglomerated) should be rapidly enveloped by a layer of crystalline polymer that grows outward from the particle as it is suspended in the melt, thus preventing the development of additional contacts between nanofiller particles during cooling.

This assumes, of course, that the surface characteristics of the primary and the agglomerated nanofiller particles are similar enough that they give rise to heterogeneous nucleation equally effectively (per unit surface area of course). If this is not the case, one may then find that some particles serve as nuclei while others tend to be excluded from the crystalline zones and are pushed to the amorphous phase as a result, with the dispersion, distribution, and orientation of such particles affected accordingly. Likewise, nanoclay modification matters; in both of the aforementioned systems, for instance, it has also been shown that the observed behaviour clearly depends on the modifiers decorating the nanoclay surface [17,18].

This idea can be taken one step further. In nanoclay-based polypropylene nanocomposites, nanoclay dispersion will not occur unless an additional third component is added (polypropylene-*graft*-maleic anhydride (PP-*g*-MAH) is most often used for this purpose). This implies that the PP-*g*-MAH, which tends to show reduced crystallinity versus the base polypropylene, is the primary polymeric component responsible for intercalating into and surrounding the nanoclay stacks. In addition to the lack of strong interactions between polypropylene and the nanoclay surface, then, the base resin will be physically separated from the nanoclay surface as well. The result is that the crystallinity of the base resin is much less strongly affected by the presence of the nanoclay. Whether the hypothesized PP-*g*-MAH-rich domains surrounding the nanoclay particles will crystallize more readily than in the absence of the nanoclay has not been heavily studied and would certainly depend on copolymer composition. Regardless, such domains will inevitably be pushed around and/or sandwiched in between growing crystalline domains of the polypropylene base resin during cooling of the nanocomposite. This situation has distinctly different consequences with respect to dispersion, distribution, and orientation of the nanofiller particles, in addition to the microstructure of the polymer itself.

4.7.4   Complications: multi-phase systems

In the case of truly multi-phase systems (versus the aforementioned PP/PP-*g*-MAH systems, for instance, which can be selected for compatibility), things become even more complicated, especially if one or more phases are capable of crystallizing. With that said, some general statements may nevertheless be made relevant to the thermodynamics of dispersion.

In polymer blends, the increase in entropy as a result of mixing is much less than for small molecules due to the restraints already placed on the conformations available to an individual polymer chain. Therefore, unless the enthalpic interactions between the two polymers are very favourable, they tend to avoid one another as much as possible. This makes phase-separation the rule rather than the exception, and ensures that the interface between most polymers costs significant energy to create. Work must be done to reduce the size of the domains present in such a system, and the end result is always metastable, something that can have severe consequences when it comes to fire properties, for instance. With that said, if a nanofiller is added whose surface is compatible with more than one of the polymer phases present, it can effectively "erase" the interface in a thermodynamic sense by interposing itself between two incompatible polymers, allowing the overall system to lower its energy as a result. This compatibilizing or emulsifying effect is very much akin to what is observed in so-called Pickering emulsions [19,20], and a number of examples in literature demonstrate its utility in the realm of polymer nanocomposites as well [21,22,23,24,25,26,27,28]. As the energy of the system is lowered by placing the nanoparticles at the interface, this driving force may be enough to enhance the level of dispersion in some cases, and this effect will certainly alter nanoparticle distribution and orientation in these systems.

In copolymers, on the other hand, nanoparticles cannot interpose themselves between segments with a tendency to phase-separate from one another due to the fact that such segments are part of the same molecule, and the aforementioned emulsifying effects would not be expected as a result. With that said, one issue relevant to nanoclay dispersion that has been neglected in our thermodynamic discussion up to now is why exfoliation would be favourable if a polymer has intercalated between two nanoclay sheets. After all, one might ask, is it not happy to enjoy positive interactions on both sides?

The thermodynamics of exfoliation are not well-understood, but it is apparent that some systems respond to the input of mechanical force and undergo an intercalated to exfoliated transition, while others do not. In systems where crystallization is highly favoured, this may act as a driving force for exfoliation, since the polymer would very much like to become part of a crystal, and crystal growth from one of the nanoclay sheets should push the other away. Still, poly(ethylene oxide)/nanoclay systems have not been observed to exfoliate, in spite of the readily observed crystallinity in these systems, so clearly this is not a sufficient explanation.

A second possibility is that the outcome of such a situation depends on the strength of the interactions between the polymer and the nanoclay surface. Such interactions are certainly important, but this cannot be the only explanation, since there are a number of reports indicating that variations in modifier structure alone can cause intercalation but varying degrees of exfoliation in the same polymer matrix [29,30,31,32,33]. Therefore, the polymer/modifier interactions must come into play as well. While the concentration of modifier chains near the nanoclay surface is always higher than elsewhere in the system, for instance, more favourable polymer/modifier interactions might result in a stronger driving force for the system to attempt to "dilute" these chains as much as possible by supplying those regions of the sample with more polymer.

With that said, polymer/nanoclay interactions remain important, and properly tuning them in multi-phase copolymeric systems may offer an additional means of affecting dispersion. In particular, theoretical studies indicate that a block copolymer with one segment that interacts favourably with the surface and one that does not might be able to more effectively induce exfoliation than either homopolymer [34]. The basic idea in this case is that the nanoclay-compatible segments are strongly driven to intercalate, and drag the incompatible segments in as well. At this point, the system has lowered its energy by allowing for intercalation, but, in contrast with the case where the entire polymer experiences favourable interactions with the nanoclay, the incompatible segments can only lower their energy if the nanoclay layers separate, giving them a means of getting away from one of the two surfaces. Here the implications for dispersion are clear, although less so for distribution and orientation.

4.7.5 *Achieving thermodynamic compatibility – practical considerations*

Because the entropic effects discussed up to now do not change substantially when one nanofiller modifier is chosen from a family of similar modifiers, thermodynamic compatibility is often achieved by optimizing the interactions of the modifier with the polymer. A very simplistic example is that of fumed silica. Bare fumed silica has a polar surface, making it more likely to display thermodynamic compatibility with more polar polymers. When fumed silica is treated using hexamethyldisilazane, however, giving rise to a heavily methylated surface, the material becomes much more hydrophobic, making it more suitable for dispersion in more hydrophobic polymers. This situation applies to a number of different classes of nanofillers – but *not* the nanoclays.

As noted previously when discussing issues of enthalpy, the intercalation mechanism in nanoclay systems involves the polymer displacing the long, hydrophobic tail of the typical alkylammonium modifiers used and interacting directly with the more polar surface of the nanoclay layers. Under normal circumstances the surfaces of the nanoclay layers are devoid of hydroxyls [35], and might therefore be said to resemble the surface of crystalline silica (although the underlying material is quite different and does indeed affect surface interactions). What this means, paradoxically, is that in the nanoclay case we actually have very little control over the chemistry of the surface the polymer eventually ends up interacting with! The edges of the clay layers possess relatively small numbers of reactive hydroxyls and may be chemically modified as a result [35], but thermodynamically favoured intercalation and/or exfoliation will only be achieved in a nanoclay-based nanocomposite if the polymer in question is able to interact favourably with the silica-like surfaces of the nanoclay layers themselves.

In systems where some level of favourable interaction occurs between the polymer and the surface of the nanoclay layers, modifier structure plays a role, with more polar modifiers (i.e. those with only one long alkyl substituent and/or containing polar substituents such as hydroxyethyl groups) tending to give greater compatibility with more polar polymers and less polar modifiers (i.e. those with two long alkyl substituents and no polar substitutents). With that said, more important with respect to compatibility than the number of long alkyl substituents the modifier possesses is the effective grafting density of substituents on the surface of a single nanoclay layer, which is a product not only of modifier structure but of cation exchange capacity (CEC). Work with polydimethylsiloxane/nanoclay nanocomposites has shown, for instance, that the range of nanoclay compatibility can be described in terms of this grafting density concept [36], although it is fair to say that the distribution of the long alkyl substituents will not be the same in the two "equivalent" cases of a twin-tail modifier on nanoclay versus a single-tail modifier on another nanoclay with twice the CEC.

That complication aside, an excessively low grafting density will tend to reduce compatibility in the case where the polymer and the modifier chains interact at least as favourably as the polymer and the nanoclay surface. On the other hand, an excessively high grafting density will result in a situation where the modifier chains cannot effectively "stand up" and be displaced from the surface by the polymer due to steric hindrance, resulting once more in a reduction in thermodynamic compatibility, at least through the described nanoclay intercalation mechanism. It should be noted here that a sufficiently low grafting densities of long chains on the surface of other platy nanofillers (hydrotalcite, graphene) should also allow for intercalation and polymer/nanofiller interactions along the same lines as have been described for the nanoclays. Once too much material has been grafted to the nanoclay surface to allow for intercalation according to the mechanism previously described, however, the nanoclay layers will act like any other nanofiller particles that have been heavily surface treated, which in practice makes dispersion somewhat difficult when the usual alkylammonium modifiers are involved.

With that said, promising results may nevertheless be achieved in nanoclay systems where the surface coverage is very high through the use of polymeric modifiers of higher molecular weights and with structures sufficiently similar to the matrix polymer to allow both for favourable enthalpic interactions and local mixing of the polymer and modifier chains, something impossible in the case of the paraffinic structure that a dense coverage of alkyl chains tends to give. The success of this technique has been demonstrated in epoxies [37,38] and more recently in polypropylene [33], although one disadvantage of this approach is that a much larger mass of modified nanoclay must be added to the polymer to achieve dispersion of the same number of clay layers as in the case of the traditional low molecular weight alkylammonium modifiers. In addition, the use of polymeric modifers with multiple cationic groups may limit dispersion through grafting of single modifier chains to multiple nanoclay layers.

Another approach to this issue is to graft the polymer itself to the clay surface using reactive modifiers. This can be accomplished through the modification of the nanoclay surface with cationic free-radical initiators (first reported in 1967! [39]) or comonomers, although it should be noted that the use of such small molecules as the sole modifying species may mean that the resultant nanoclays are quite polar in nature and do not disperse well in the desired medium (be it monomer, solution, or polymer). The use of surfmers (surfactant-monomers, i.e. polymerizable surfactants [40,41]) combines in the same molecule the reactivity of the aforementioned comonomers with the original function of the traditional alkylammonium modifiers to address this issue, however, and may be readily applied to a wide range of polymerization chemistries so long as the appropriate functional cationic surfactants are available or can be synthesized for use as reactive clay modifiers. An alternative to this would be to modify the nanoclays with a mixture of different reactive and compatibilizing species, although this approach is complicated by the fact that mixtures of interlayer cations in such systems have shown the ability to phase-separate from one another [42], which may alter their accessibility and reactivity as a result. Another commonly used grafting technique takes advantage of the ready availability of hydroxyethyl-substituted alkylammonium nanoclay modifiers to achieve chemical interactions/covalent bond formation with a wide range of polar, hydroxyl-reactive resins. Indeed, the other end of the modifier tail may be functionalized, for that matter, as in the case of the 12-aminododecanoic acid modifier used by the Toyota group in their original work on polyamide-6 nanocomposites [12] and demonstrated with diamine modifiers in epoxy systems as well [38].

A third approach is the addition of a polymeric compatibilizer – a polymer or copolymer that is reasonably compatible with the matrix polymer but that is able to more favourably interact with the nanofiller. This approach, in combination with the use of alkylammonium modified nanoclays, finds almost universal acceptance by groups working with polyolefin nanocomposites, and can give rise to significant property enhancements when properly applied [29]. This situation is actually quite similar to the aforementioned situation of covalent grafting, except that the polymeric modifiers (or co-modifiers, in the case of clays that are already alkylammonium-modified) are bound to the nanoclay layers by intermolecular forces over a large area rather than by a small number of covalent and ionic bonds. In practice it is also more convenient in that it allows for the use of commercial organoclays and readily available compatibilizing agents, although as noted previously in the sections on crystallinity and multi-phase systems, this is likely to have additional consequences.

While much of the discussion above has been qualitative in nature, it should be noted that one means of quantifying the level of compatibility between a polymer and a given nanofiller involves a comparison of solubility parameters. The solubility parameters of many polymers are available in the literature, from sources like the Polymer Handbook [43], for instance. Such data is much less commonly available for nanofillers, although it may be readily estimated by dispersing the nanofiller of interest in a range of common solvents with known solubility parameters and examining the severity of settling with time.

It should be noted here that the solvent viscosities are assumed to be similar enough not to significantly affect the results and specific interactions between the solvent and the nanofiller surface are neglected. Likewise, all dispersions must be mixed for sufficient time to ensure maximum dispersion in a particular solvent before settling is allowed to occur, and solvent evaporation and agitation of the sample during the experiment must be prevented. With that said, the solvent that gives the slowest settling should have the closest solubility parameter to the nanofiller in question (in tandem, the solvent that gives the highest level of transparency in the dispersion prior to settling should be the closest match with respect to refractive index). Solvent blends may then be used more closely bracketing the exact solubility parameter (and refractive index) of the nanofiller. In the case of the nanoclays this result represents an *effective* solubility parameter, since every nanoclay platelet possesses chemically distinct regions that will interact differently with the solvent (layer edges, layer surface, cationic modifier head-group, non-polar modifier tail). Even so, this technique may serve as a useful guide for selecting systems to explore.

4.7.6 The kinetics of physical dispersion

Having offered a general overview of the thermodynamics of dispersion during nanocomposite processing, we now come to the issue of kinetics. Dispersion may be thermodynamically favourable, but, going back to the need for processing, mixing will not occur spontaneously. The input of thermal and/or mechanical energy remains a must, therefore, once thermodynamic compatibility is established, in order to overcome interactions between nanoparticles and achieve dispersion. As covered in previous discussions, more heat and mechanical energy over longer periods of time result in more diffusion, flow, mixing, and, in compatible systems, dispersion, but too much energy over too much time will make degradation likely to occur.

In the nanoclays in particular the process of dispersion down to the single sheet level (exfoliation) is thought to occur or be accelerated by a shearing or peeling mechanism induced by flow within the system, as shown in Figure 4.1. Analogous processes may be imagined for platy fillers in general. All things being equal, it would seem that separating rod-like or spherical particles should be easier, given the reduction in contact area due to their curvature; given the extreme difficulties commonly faced by those attempting to disperse carbon nanotubes, however, it is clear that all things are not equal, and that the thermodynamics and strength of interactions in these systems does indeed come into play.

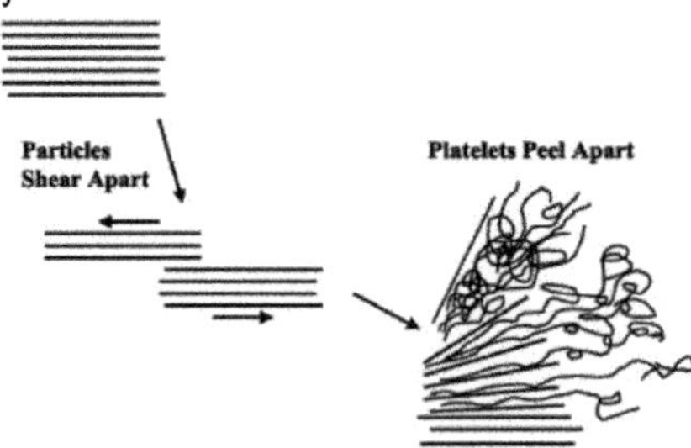

Figure 4.1

A schematic illustration of the process by which intercalated nanoclay platelets are thought to separate through the action of mechanical shear

A critical point here is that, regardless of dispersion mechanism, once thermodynamic compatibility is realized between the polymer and the nanofiller, the input of mechanical energy in some form or another is the sole means available to us to achieve dispersive and/or distributive mixing. We enhance the mobility of the polymer (either thermally or through the introduction of a third component) as a means to an end, and as a way to maximize the effectiveness of the mechanical energy being put into the system. When heat is used as a means of increasing the polymer mobility, therefore, an optimal process will minimize both the amount of heat and the time over which it is applied such that there is just enough time and mobility to make the most efficient use of the mechanical energy being put into the system and to achieve the desired levels of dispersive and distributive mixing.

With that said, it is instructive to consider just how readily some systems will show exfoliation when the thermodynamics are favourable. In particular, there is some evidence (x-ray diffraction and transmission electron microscopy) that not only intercalation (which is effectively instantaneous) but exfoliation as well may be realized in a highly compatible but otherwise "static" polyamide-6/5 weight-% nanoclay (Cloisite 93A) system if the system is given mobility for a sufficient amount of time – 2 hours at 240°C, with complete intercalation seen in the first x-ray diffraction pattern, taken after 30 seconds of heating [44]. Although this would seem to imply that exfoliation in highly compatible systems requires no shear whatsoever, it is important to note that this was performed in a hot-press, which may result in some small amount of shear/melt flow due to lateral expansion of the sample, and to recall that convective flow is a common feature of heated fluids, molten polymer included. While more work is therefore required to prove definitively that zero-shear exfoliation is possible, this shows that exfoliation can be achieved in compatible systems with far lower levels of shear than are commonly thought of when nanocomposite processing is discussed. At this point one might be tempted to ask why we spend so much energy on mechanical mixing if exfoliation is possible without it. This brings us back to a key point made at the beginning of this chapter: We must not only to mix components, but do so *in a reasonable amount of time*. If, as implied here, the relevant quantity for dispersion in a compatible system is the product of shear rate and time, and there is no minimum shear rate below which exfoliation cannot occur, we will nevertheless tend to favour short times and high shear rates in most production processes in order to produce materials quickly. Nevertheless, systems that are highly compatible and routinely left in the liquid state for long periods of time, such as epoxy and urethane resins, for instance, may give industrial relevance to "static" exfoliation.

4.7.7 *Dispersion kinetics in the presence of chemical reactions*

Up to now we have focused on the nature of process-induced nanofiller dispersion in polymer nanocomposites based on pre-formed polymer and nanofiller. As noted previously, however, we are not required to begin with a solid or even a liquid high molecular weight polymer. If, instead, one phase is formed in the presence of the other, while the thermodynamic principles described above still apply, the kinetics of dispersion are definitely affected. In the situation where the polymer is being formed in the presence of nanofiller particles, the end result all depends on local variations in polymerization rate versus proximity to the filler surface. If there is no variation, dispersion will depend on interactions with the polymer phase being formed, as previously described, and the results will be not unlike those obtained through the introduction of a third component designed to enhance polymer mobility, except that in this case the third component will consist of monomers and/or oligomers, not solvent or plasticizer.

In the event that the nanofiller surface affects the rate of polymerization, on the other hand, the nature and magnitude of this effect can significantly alter the kinetics of nanofiller dispersion. Simply put, if the nanofiller surface locally enhances the rate of polymerization, some enhancement in dispersion may be observed (so long as the monomers wet the particle surface) due to the fact that material will form more rapidly between the particles than elsewhere, thus pushing them apart. On the other hand, if the nanofiller surface locally depresses the rate of polymerization, an increase in agglomeration may be observed as material forming elsewhere pushes the particles together. While this effect is not as severe as the exclusion of impurities from a crystalline domain as it grows, there is nevertheless a clear analogy between this situation and the various situations described in semi-crystalline systems. Changes in dispersion due to these nanofiller-induced variations in local polymerization rate effects rely on the fact that, under the right conditions, gel-like high polymer domains may be formed in a lower viscosity medium. If the nanofiller particles are trapped at the center of such domains, as the domains grow the nanofiller particles will tend to separate. If the nanofiller particles are initially excluded from such domains, on the other hand, their concentration in the remaining low viscosity material will increase and agglomeration will be more likely.

The issue of viscosity represents a key point in this discussion. Specifically, since the nanofiller particles mentioned here require some time to move, controlling the evolution of the viscosity in these systems (which, in turn, means controlling the polymerization process) will substantially impact whether these effects are seen or not. For instance, if polymerization is enhanced at the nanofiller surface, a situation that has been described for epoxy/nanoclay nanocomposites [45], but the overall polymerization rate is exceptionally rapid, the system will solidify before any of the nanofiller particles have time to move, and no enhancements in dispersion will be observed. In fact, similar arguments could be applied to the effects of crystallization from the melt on nanofiller dispersion, although the much higher viscosities one expects to find in the melt versus in a reactive system in the early stages of polymerization are likely to make this effect proportionally less significant in that case.

As a brief aside, nanofiller dispersion in those systems where the polymer is pre-formed but the nanofiller is being created *in-situ* is primarily controlled by local concentration variations of the nanofiller precursor, coupled with the tendency of the nanofiller particles to phase-separate upon formation and act as nuclei for further addition of material. In semi-crystalline systems, one would expect such precursors to be localized primarily in the amorphous phase. In copolymers and blends such precursors should segregate depending on their level of compatibility with the various polymer phases present, allowing for the formation of very interesting nanostructures if the polymer is blocky in nature [46].

Finally, in systems where both nanofiller and polymer are being formed simultaneously, the relative rates of reaction coupled with the tendency of the reaction products to phase-separate will determine the final structure. If the tendency to phase-separate is strong and the nanofiller forms first, the situation becomes very similar to that of polymerization in the presence of pre-formed nanofillers. If the tendency to phase-separate is strong but the polymer forms first, the situation is then very similar to that of nanofiller formation within a polymer matrix. If little or no phase-separation occurs, however, single-phase nanocomposites consisting of hybrid organic/inorganic interpentrating, semi-interpenetrating, or co-crosslinked networks may be formed. The class of solids referred to as organically modified silicas (ORMOSILs) and organically modified ceramics (ORMOCERs) more generally are examples of such materials [47,48,49].

4.8 What should a "well-processed" polymer nanocomposite look like?

This is an extremely important question whose answer greatly depends on the nature of the properties desired. While complete nanoscale dispersion is often regarded as the "best" situation, and maximizing the developed surface area and effective aspect ratio of the nanoparticles should in theory maximize transparency and stress transfer from the matrix to the particles and give the greatest enhancement in barrier properties (in the case of nanoclays and other platy nanofillers), in reality things are a bit more complicated.

4.8.1 The realities of nanocomposite processing

First, even when the nanofiller and the polymer are well-matched, perfect nanoscale dispersion is never achieved in real nanocomposite systems; in fact, such a structure may not be necessary or even desirable. Lower levels of interaction between polymer and nanofiller and/or the presence of weakly bound clusters may allow for debonding as a toughening mechanism during irreversible deformation of the nanocomposite, for instance [50, 51, 52, 53]. Likewise, polymer/nanoclay nanocomposites with intercalated structures show enhancements in fire and thermal properties that can approach or even exceed those of exfoliated systems [54]. Additionally, perfect nanoscale dispersion resulting in the highest achievable aspect ratio in a given system will also result in the lowest achievable percolation threshold [55], meaning that the viscosity of the system is likely to increase more substantially as a result, with all of the associated complications coming into play. One mitigating factor with respect to viscosity in these systems in general is the tendency of the nanoclay platelets to shear-align [56,57,58]. The resultant thixotropic properties of these materials mean that the viscosity increase all but disappears at higher shear rates [56,58], although as noted before, the input of mechanical energy must not be too great or degradation may occur.

Second, a completely random orientational distribution of anisotropic nanofillers is (almost) never achieved in real nanocomposite systems. One possible exception would be systems where the material has solidified during a mixing process so that random orientation is effectively "frozen-in" as a result, although this requires that the flow field present in the sample be essentially random up to the point of solidification, something very hard to achieve in practice. A second exception would involve the sintering of a polymer/nanofiller powder compact containing nanofiller particles having completely random orientations – experimentally this is less difficult to achieve than the previous example, but very low levels of nanoscale dispersion would be expected, making such materials less attractive from the standpoint of a number of materials' properties. The end result is that almost all polymer nanocomposites show some degree of anisotropy. With that said the alignment of the nanoclay layers in a particular direction may be of particular value if the goal is to specifically enhance a given property along a single axis or in a single plane. For example, in-plane alignment of nanoclay layers in a thin nanocomposite sheet would be expected to give noticeably higher in-plane mechanical reinforcement, directly akin to what is observed in the fibre composite world, coupled with enhanced transparency and barrier properties perpendicular to the plane of the sheet.

Third, larger structures cannot be ignored. The typical commercial nanoclay is delivered as a powder with an average particle size of microns to tens of microns. Due to the rather weak bonding present within these large agglomerates, it takes very few of them to substantially compromise the ultimate mechanical properties of a polymer nanocomposite (stress and strain at break, impact strength, etc.), *even with high levels of nanoscale dispersion*. On the other hand, controlled aggregation can be extremely beneficial under the right circumstances. For instance, a process that gives rise to a continuous network of nanofiller particles in contact with one another, rather than disrupting all such contacts in the name of maximizing nanoscale dispersion, will produce a material with substantially higher thermal and (in the case of electrically conductive nanofillers) electrical conductivity at the same nanofiller loading level.

Nanoscale dispersion, then, is not the end-all of the nanocomposite world, nor should it be the only concern during nanocomposite processing; a broader view is required to successfully process and make full use of this unique class of nanomaterials.

4.9 What are our options for nanocomposite processing?

Having addressed some of the basic questions that come to mind when faced with the task of processing a polymer nanocomposite formulation, we will now describe in detail a range of specific techniques used to prepare these materials. This list of processing techniques is not meant to be all-inclusive, or to imply that no other means for making these materials exists, but is intended instead to provide both breadth and depth in the area of nanocomposite processing, and to show how those in the field have taken the step from general principles to engineering practice. Likewise, the fact that these processes are listed separately does not mean that they must be applied separately; indeed, in some cases the combination of multiple types of processing will provide better results than any one alone.

4.9.1 The importance of pre-processing

Here, *pre-processing* refers to controlling the form of the materials to be combined, and, in some cases, performing some pre-mixing, so as to maximize the effectiveness of the subsequent processing techniques. Pre-processing can make a substantial difference in the outcome of a given approach to nanocomposite formation, and may even be performed by the supplier before materials are even received – it could be something as simple as controlling the particle size of the agglomerated nanofiller, for instance, or choosing suspension-grade polymers (coarse powder) instead of the usual pellets. With that said, if the starting materials are not available in the most optimal form, it may be sensible to address this issue first, rather than dealing with it only after it becomes apparent that things are not working as well as they might.

At this point, the reader may question the necessity of pre-processing, when a number of well-known polymer processing techniques are highly effective for the mixing of polymers and fillers. In order to answer this complaint, we will consider a few basic features of nanocomposites in particular that separate them from conventionally filled polymers. Firstly, in polymer nanocomposites, the distances between adjacent nanofiller particles tend to be much, much smaller than those observed in typical filled systems, even when nanofiller dispersion is imperfect. Secondly, the very high surface to volume ratios of the primary nanofiller particles coupled with the increases in surface energy observed with decreasing particle size tend to greatly enhance the tendency of these particles to agglomerate – particularly significant, since the particles are already very close together. Thirdly, in the case of high aspect ratio nanofillers in particular – nanoclays, for instance – the percolation threshold is quite low, meaning that dispersed high aspect ratio nanoparticles will begin to "feel" one another's presence at exceptionally low concentrations in the polymer matrix. While this last effect is not unique to nanocomposites, it should be noted that nanofillers, due to their small size, are capable of undergoing significant processing without the loss of aspect ratio that larger high aspect ratio fillers would experience under the same conditions(i.e. due to physical damage).

Having said all that, we now imagine the processing of two identical nanocomposite systems. In one case, nanofiller particles are distributed on the surfaces of the usual polymer pellets. In the other case, the same polymer, in powder form, has been mixed with the nanofiller, also in powder form. If we were to melt blend these two materials, what would happen? Given the low surface to volume ratio of the polymer pellets vs. the polymer powder and the fact that the nanofiller can only sit on the surface of the polymer prior to melt blending, it is clear that much larger domains of concentrated nanoclay will be present in the case of the pellets. In contrast, in the case of the powder pre-mix, the vast majority of the distributive mixing has already been done.

Why does this matter? If we imagine the case of an agglomerate of nanoparticles that would like to disperse, it will have little chance to do so if trapped within a nanoparticle-rich domain. Even if our unhappy agglomerate were able to begin undergoing dispersion, re-agglomeration is almost certain, regardless of how much mechanical energy is put into the system, so long as the local nanoparticle concentration remains high, due to small interparticle distances and local percolation coupled with significant interactions between nanoparticles and nanoparticle agglomerates. This effect will be even more severe when the aspect ratio of the nanoparticles is high. While nanoparticles stuck in regions of high nanoparticle concentration suffer low dispersion levels as a result, those in area of low nanoparticle concentration are not numerous enough (although they will be dispersed more effectively) to make up for this fact. This is especially true if the overall nanoparticle concentration is low enough that dispersion of the majority of the nanoparticles is theoretically possible (i.e. the system is below the percolation threshold). The optimal situation, then, is to achieve maximum distributive mixing *before* significant dispersive mixing occurs, such that domains of high nanoparticle concentration are small (ideally no larger than the average size of a single nanofiller agglomerate), thus allowing the nanoparticles to separate as efficiently as possible without encountering their peers and tending to re-agglomerate as a result. This should be kept in mind during all subsequent discussions of nanocomposite processing, as many otherwise useful mixing techniques may not meet this criterion without some careful optimization or pre-processing.

Finally, a statement is in order regarding the use of trade-names. In the plastics processing community, many pieces of equipment are commonly referred to using the names of the manufacturers who make them. This trend is followed throughout the discussions below so as to make them as intelligible as possible to those in the field. This should not be misconstrued as an endorsement of any particular product on the part of the author, simply an attempt to give some examples of relevant process equipment available at the time of writing. For the same reasons it should be noted that the pictures presented are representative images only, and are not intended to imply that all instruments of a particular type will share the same appearance, or that the sizes shown represent the minimum or maximum sizes available from any one manufacturer.

4.10 What processing techniques involve just polymer and nanofiller?

4.10.1 Physical mixing/dry blending

The simplest form of nanocomposite processing involves the physical combination of nanofiller and polymer in the form of a pre-mix. Dry blending is a generic term that implies mixing under conditions below the softening or melting point of the polymer in question. A large number of processes could be described by these terms. Finer powders, when effectively mixed, will give finer, more homogeneous nanocomposites, and nanofillers are typically available as such, with commercial nanoclays generally available as free-flowing powders with average particle sizes in the order of 5-25 μm. This size range is probably optimal from a handling standpoint, as the powders are fine, but not so fine that are very readily blow about and end up everywhere (as in the case of finer carbon blacks and fumed metal oxides, for instance) or are strongly affected by static electricity (as in the case of fumed metal oxides in particular). With the exception of suspension-grade polymers or spray-dried polymer emulsions, however, most polymers are provided in the form of pellets, making them generally inoptimal if fine structure is desired. Milling (or cryomilling, if necessary) represents an effective means to address this problem, and is therefore recommended as a *pre-processing* technique when finer structures are required, both here and in general (i.e. where complete dispersion and homogeneous distribution must be obtained as rapidly as possible, for instance). This goes not only for the polymer but for the nanofillers as well; in the case of nanoclays, for instance, custom-modified nanoclays prepared in-house must be ground to powders at least as fine as the commercial nanoclays if equivalent or better levels of dispersion are hoped for, and in a consistent fashion as well, to ensure the repeatability of the nanocomposites being produced.

Once the nanofiller and the polymer are available in the desired form, they must be mixed together. While one common lab-scale technique involve placing both in a plastic bag and shaking the bag, this method is unlikely to give reproducible results without the addition of a small amount of a third phase (solvent, lubricant, plasticizer) that softens the surface of the polymer particles slightly so as to encourage the nanofiller particles to adhere to their surface. Such additives must be used very sparingly, however, otherwise wholesale caking or excessive softening of the polymer may occur. It must also be remembered that the introduction of any additives may cause problems downstream as well. A well-known improvement on this process in terms of both reproducibility and scalability can be achieved through the use of a twin shell ("V") blender. Along the same lines, the use of high-shear/high-intensity mixers (Henschel or Welex mixers, for instance) for the preparation of dry-blends has long been carried out in the vinyl industry, not only as a means of combining filler and polymer powders but also as a method of introducing plasticizer, through the use of shear and temperature. An example of such a mixer is shown in Figure 4.2.

Figure 4.2
A small-capacity Henschel high-intensity mixer for preparing dry-blends, shown open
Photo taken in the Department of Plastics Engineering at the University of Massachusetts Lowell

Alternatively, the use of orbital mixers (sometimes referred to as Hauschild, dental, or SpeedMixers, not to be confused with orbital shakers used in the biological sciences) represents a highly effective technique for forming homogeneous powder pre-mixes that is especially amenable to R&D efforts and small-scale product development (see Figure 4.3). The use of disposable mixing vessels represents an added cost, but the benefit is that no cleaning is necessary between mixing cycles runs and the chances of contamination are minimized. Acoustic or sonic mixing represents another alternative for powder blending; such mixers are rapid, effective, scalable to larger capacities than orbital mixers, and relatively easy to clean, although some cleaning is nevertheless required between mixing runs. Examples of such instruments from MacroSonix and Resodyn are shown in Figure 4.4 and Figure 4.5, respectively.

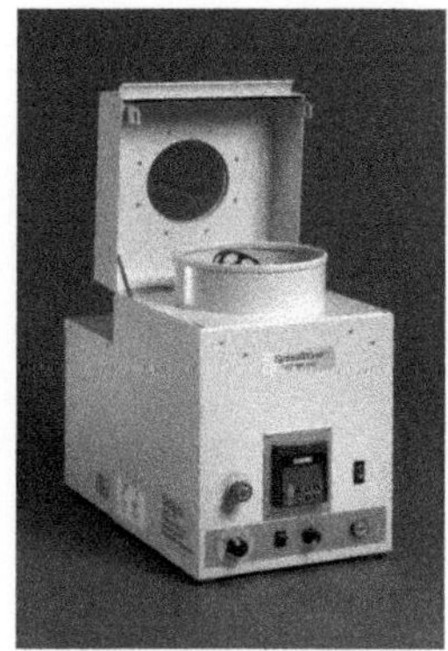

Figure 4.3
A SpeedMixer DAC-150 orbital mixer (110-140 g capacity), shown with the lid open
Image courtesy of FlackTek Inc., received 3/2008

Figure 4.4
A MacroSonix low-frequency acoustic mixer (20 L capacity) manufactured by Design Integrated Technology
Image courtesy of Design Integrated Technology, received 4/2008

Figure 4.5
A Resodyn LabRAM low-frequency acoustic mixer (500 g capacity)
Image courtesy of Resodyn Acoustic Mixers Inc., received 3/2008

4.10.2 Compaction

In this processing technique, a powder/pellet pre-mix is introduced into a mould or between the platens of a press and compressed mechanically. At its most basic, this process is almost entirely mechanical in nature (neglecting frictional heating) and involves minimal shear (neglecting edge effects), although heat can be applied to the sample at the same time to accelerate densification of the sample. If sufficient heat is applied to soften or melt the polymer, flow will be enhance and the process becomes the widely used hot-pressing/compression moulding technique. In the absence of such heating, however, this process is primarily diffusional in nature, and while the particles will reorganize during initial densification, little flow or particle breakup is expected to occur. This technique does not, therefore, produce high levels of dispersion, although it can give uniform distribution and is very useful for intentionally forming a percolated network structure, since the smaller nanofiller particles will tend to surround each of the larger polymer grains.

4.10.3 Solid state shear processing

Solid state shear processing is a technique somewhat akin to dry blending in a high shear mixer but taken to the next level. While this is certainly a form of physical mixing or dry blending, distributive mixing tends to be the primary purpose of these techniques, whereas the focus with solid state shear compounding tends to be more towards dispersive mixing, with sufficient input of mechanical energy in the latter case to literally break apart the pellets or particles being processed, or even to induce chemical reactions in some cases. So far, solid state shear processing has been carried out by feeding polymer and nanofiller (separately or as a particle/powder pre-mix) into a twin-screw extruder (pulverizer) with active cooling at ambient or sub-ambient temperatures [59,60], and through the use of a specially designed pan-mill [61]. More generally, it would appear that such processing could be applied with almost any piece of equipment suited to the application of shear to a powder or molten phase, if suitably modified, although the technique is new enough that as yet it is difficult to say what would be optimal from the standpoint of cost and performance. In spite of the lower degree of mobility one would expect if the polymer phase were truly in the solid state, intercalation and partial exfoliation have been observed in nanoclay-based polymer nanocomposites processed in this fashion [62], and the realization of polymer microstructures not readily attainable via other means may be possible with this technique, given the large amounts of shear the materials experience and the potential for shear-induced reactions ("mechanochemistry"). For the same reasons, however, a reduction in the aspect ratio of high aspect ratio nanofiller particles may also be observed, depending on the process conditions and the materials in question. Finally, subsequent processing required to form the desired part may serve to at least partially erase some of the interesting and useful microstructural features developed, although with careful engineering of the production process as a whole it should certainly be possible to minimize such effects if desired.

The use of roll mills (two roll, three roll, etc.; a two-roll mill is shown in Figure 4.6) could also be described as a form of solid-state shear processing when applied to thermoplastic systems, since it effectively combines the compressive forces described in the discussion of compaction-based nanocomposite processing with a substantial amount of shear. Roll mills have found utility for many years, and have long been used for the preparation of what may have been the world's first polymer nanocomposites, namely carbon black filled rubber. Such mills may also be used for the processing of polymer/nanofiller powder pre-mix, with their shearing action inducing significant break-up of the nanofiller aggregates. Shear heating may be supplemented by heated rollers, and the end result is consolidated strips or sheets of polymer nanocomposite material. As implied by the form of the product, some flow does occur during this process but the mixing action is nevertheless likely to be more dispersive than distributive, and the results of such milling tend to be more operator-dependent and difficult to reproduce than other mixing processes.

Figure 4.6
A Farrel Corporation Polymill lab-scale two-roll mill
Image courtesy of Farrel Corporation, received 5/2008

4.10.4 Melt blending

In contrast with the processing techniques described previously, melt blending is the first in which the *entire* polymer phase is intentionally taken to temperatures sufficient to induce softening/melting and flow – as compared to processes where such effects are more localized or flow does not occur at all. As nanofillers have a tendency to increase the viscosity of the polymer melt, substantially in some cases, systems with high volume fractions of nanofiller relative to the theoretical percolation threshold [55] (taking into account the *effective* aspect ratio of the nanoparticles achieved during mixing) will be likely to require increased process temperatures to allow for successful processing. Formulations based on lower nanofiller concentrations, however, can generally be processed at temperatures similar to those used for the unfilled polymer, and are recommended for most applications in any event, since dispersion will be easier to achieve and less process-induced degradation will occur. Such formulations generally represent the optimum with respect to both processing and properties as a result.

For the batch processing of small amounts of material via melt blending, internal mixers are a convenient solution, with selection of screw type and mixing chamber (often referred to as a bowl) allowing for a fair amount of control over the type of mixing and the amount of shear. In this process the mixing bowl is heated to a temperature appropriate for the melt processing of the polymer of interest, after which the polymer and nanofiller are introduced. While order of addition is of clear importance in internal mixing, and taking the initial step of melting the polymer, then gradually adding the nanofiller would be likely to produce better results than adding everything at once, this requires that the polymer be exposed to additional mechanical and thermal energy as a result of the longer mixing times such processing requires. It is generally preferable, therefore, to introduce a pre-mixed powder or dry blend of polymer and nanofiller if at all possible, as this should favour the realization of good dispersion (assuming thermodynamic compatibility) and uniform distribution in a minimum amount of time. The importance of optimizing the fill factor in such mixers when processing polymer nanocomposites has also been reported [63]. Common internal (batch) mixers are often referred to simply by the name of their manufacturer, with Banbury, Brabender, and Haake mixers shown in Figure 4.7, Figure 4.8, and Figure 4.9, respectively. As implied previously, one serious disadvantage when using internal mixers is that residence times are typically on the order of 5-10 minutes, longer in some cases, and the input of mechanical energy in the form of shear tends to be quite substantial as well (although this does depend on screw and mixing bowl selection). Nanocomposite preparation using internal mixers is often subject to some thermal degradation of the polymer and/or the nanofiller modifier as a result. In addition, experimental comparisons with twin-screw extruders indicate that the latter give higher levels of dispersion in both polymer/nanoparticle [64] and polymer/nanoclay nanocomposites [65,66]. Internal mixers are also less suited to large-scale production, although they are nevertheless very useful for development work, especially when working with thermally stable nanocomposite formulations or, alternatively, for testing the thermal stability of such materials under process conditions.

Figure 4.7
A Farrel Corporation F270 Banbury® internal mixer
Image courtesy of Farrel Corporation, received 3/2008

Figure 4.8
A Brabender Plastograph EC internal mixer
Image courtesy of Brabender GmbH & Co. KG, received 4/2008

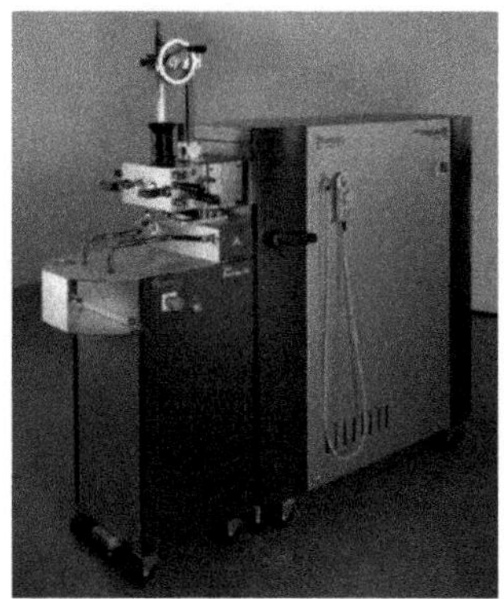

Figure 4.9
A Thermo Scientific HAAKE PolyLab OS internal mixer
From http://www.thermo.com/eThermo/CMA/PDFs/Various/File_831.pdf.,
Viewed 1/2008, reproduced with permission

For continuous melt blending of both small and large quantities of polymer nanocomposites, extrusion remains the technique of choice, with twin-screw extrusion strongly preferred over single-screw extrusion due to its reputation for enhanced mixing efficiency (examples of single- and twin-screw extruders shown in Figure 4.10 and Figure 4.11). This reputation has been confirmed in nanoclay-based polymer nanocomposites in particular, with twin-screw extruders giving better dispersion than single-screw extruders in systems with both moderate [67] and high levels of compatibility [65,67,68]. With that said, the use of specially designed dispersive mixing elements [69,70] (see Figure 4.12) and/or reciprocating extruders/kneaders (see Figure 4.13) [71] may allow for single-screw extrusion processes to approach the mixing efficiency of twin-screw mixing in some instances.

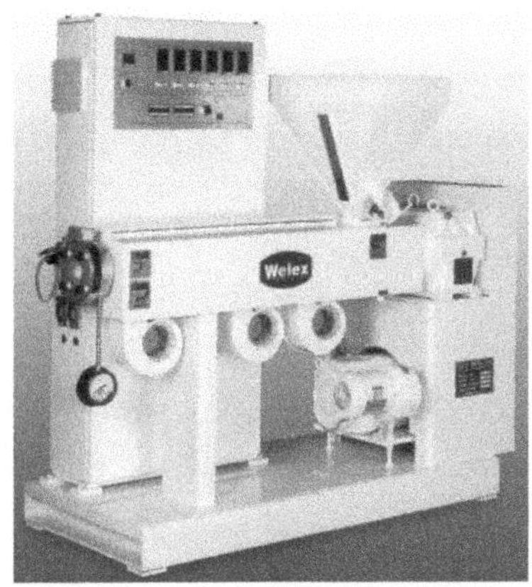

Figure 4.10
A Welex Mark I general purpose single-screw extruder
From http://www.welex.com/pdf/Mark%20I.pdf, viewed 1/2008, reproduced with permission

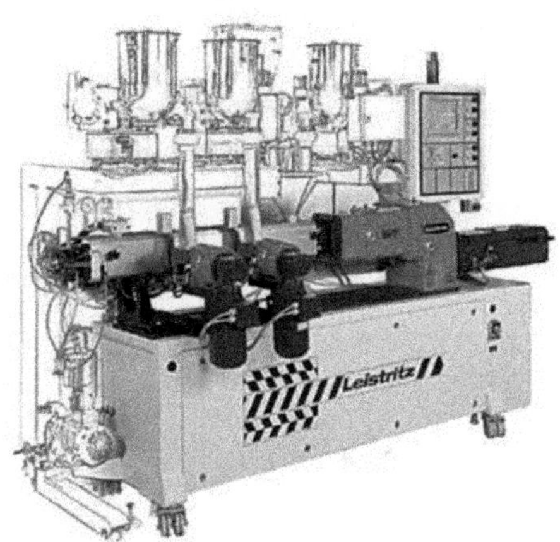

Figure 4.11
A Leistritz ZSE-27 HP twin-screw extruder
From http://www.leistritz.com/alec/en/04_products/twin%20screw%20extruder/00_img/zse_27.jpg,
viewed 1/2008, reproduced with permission

Figure 4.12
Xaloy Nano™ Mixer single-screw mixing element (U.S. Patent no. 6,497,508) designed for
nanocomposite processing
From http://www.xaloy.com/pdf/nanomix-infosheet.pdf, viewed 1/2008, reproduced with permission

Figure 4.13
A BUSS Quantec 110-10C single-screw kneader
From http://www.busscorp.com/documents/database/seiten/52640/produkte_def.pdf, viewed
2/2008, reproduced with permission

Differences in dispersion state are also observed when changing between non-intermeshing,
tangential and intermeshing twin-screw systems or between counter- and co-rotating screws, and
screw design is clearly an important parameter, but these factors appear to be less important in
systems with high levels of thermodynamic compatibility [67]. In such cases, a general purpose
screw profile and the same melt temperatures as used for the unfilled polymer are often sufficient
to realize good results in systems with relatively low volume fractions of nanofiller relative to the
percolation threshold, without unmanageable increases in torque and with residence times much
shorter than those common to internal/batch mixing processes (thus minimizing degradation as
well). With that said, longer residence times generally result in better mixing [67]. Residence time, in
turn, may be maximized by feeding both the polymer and the nanofiller at the primary (upstream)
feed port [72] – an approach that ensures that all of the materials see all sections of the screw profile.
With that said, as the following paragraphs will show, this is not meant to imply that this is the best
way to proceed – as is often the case, complications arise.

Along these lines, feeding is a critical issue when performing twin-screw extrusion with polymer nanocomposite formulations. The simultaneous feeding of polymer pellets and nanofiller powders may be difficult to accomplish in a uniform and reproducible manner, and, going back to the need for distribution prior to dispersion, any variations in composition in such an already inoptimal mix from the standpoint of nanofiller distribution will further reduce the chances of producing high dispersion levels in these systems. As such, the need for pre-processing here is substantial. One solution is the feeding of a powder pre-mix, since there is less of a tendency for the components to separate and much of the distributive mixing has already been achieved prior to introduction into the extruder. Another possibility that avoids the handling of powders altogether is the use of pelletized nanocomposite masterbatches with high nanofiller content. These pellets are homogeneously mixed with pellets of the base resin of interest (readily achieved through the use of a twin shell blender, again emphasizing the need for at least some pre-processing). This mixture of pellets is then fed into the extruder, where the masterbatch is effectively diluted, or "let down", into the base resin. The major disadvantage of this technique is that not all compositions of interest will be available in masterbatch form. Furthermore, at a more fundamental level this approach does not solve the problem of powder handling or feeding, or any of the other issues associated with compounding the masterbatch in the first place; rather, it makes them someone else's problem, with the masterbatch user necessarily reliant on the masterbatch producer to do things properly.

If pre-processing is not possible, on the other hand, and polymer pellets and nanoclay powder must be combined, in spite of what has been said before regarding residence times and such, co-feeding through the primary port may actually risk the formation of dense agglomerates, as the clay powder is put under substantial compression prior to the point where the polymer melts [44]. The use of a side-stuffer to feed the nanoclay at a secondary feed port risks similar problems, and in fact, it has been observed that the best results, even in the absence of dispersive mixing elements downstream of the nanoclay feed, come from direct feeding of the nanoclay into a vent port [73], given that the extruder output rate is sufficient to allow the feeder to provide nanoclay in a smooth and continuous fashion, which is then rapidly dispersed into and carried along by the melt [44]. To be clear, this approach will not induce intercalation or exfoliation if they are not thermodynamically favourable. Regardless of the level of polymer/nanoclay compatibility, however, proper feeding will help to improve macroscale dispersion levels and prevent the formation of large defects that would otherwise compromise properties by minimizing re-agglomeration during feeding [44]. Taking a step back, it is reasonable to expect that the issues addressed here, while focused on nanoclays in particular, are probably more generally applicable to a range of nanofillers.

Returning to nanoclay-based systems, even when feeding is properly addressed, a number of other parameters affect the levels of intercalation and exfoliation achieved during polymer nanocomposite extrusion. In particular, screw profile, feed rate, and screw speed would all be expected to play a role in the final outcome of a nanocomposite processing operation. For a polypropylene/organoclay nanocomposite, for instance, Lertwimolnun et al find that intercalation and some level of exfoliation occur within a fraction of a second following melting and that intercalation is otherwise independent of process conditions [74]. Improvements in exfoliation were found to be tied to increasing the ratio of the screw speed to the feed rate and the total amount of mechanical energy introduced into the system (i.e. due to the combination of shear rate and residence time), consistent with previous discussions presented here, although quite interestingly, the most aggressive screw profile was actually found to *reduce* the level of exfoliation when used at high feed rates [74]. As alluded to in previous comparisons of single and twin-screw extrusion techniques, such effects may be less significant in systems with higher levels of thermodynamic compatibility than this one, again in line with observations that dispersive mixing elements and high shear may not be necessary and exfoliation may be very rapid if the nanoclay is fed properly into a highly compatible melt [44,73]. With that said, fast intercalation (when thermodynamically favoured) appears to be a more general phenomenon; as of the time of writing, no studies have managed to observe the intermediate states one might expect between unintercalated and intercalated, and with good reason – diffusion between the interlayer galleries in organically modified layered silicates appears to be quite rapid [75,76]. One conclusion of all of this is that, depending on the level of polymer/nanofiller compatibility and the level of dispersion desired (which in turn depends on the properties being targeted), it may not be necessary to resort to lengthy, highly intensive mixing processes to achieve success.

Up to this point, our discussions of extrusion as a means of melt compounding have focused primarily on industrial-scale single- and twin-screw extruders (shown previously in Figure 4.10 and Figure 4.11). Such equipment is well-suited to production, but less so for small-scale development work. For this purpose a number of companies have developed bench-top recirculating twin-screw extruders, or microcompounders, with fixed (solid) screw profiles capable of processing as little as 2 cm^3 of material. An example of one such instrument is shown in Figure 4.14. One very recent study reports on the effects of a range of processing parameters on the preparation of polyamide-6/acrylonitrile-butadiene-styrene (ABS)/nanoclay nanocomposites with a specific focus on one such microcompounder [77]. Material is fed at the top of the screw profile, mixed and expelled at the bottom, where a valve controls whether it is then directed back to the top of the screw profile once more or out of the extrusion nozzle. As the screw profiles used are often low to moderate shear, the effective L/D ratio for these instruments is not large, recirculation is necessary in order to achieve reasonable levels of nanofiller dispersion, and residence times are generally higher than in larger programmable twin-screw extruders as a result. While nanocomposites are processed effectively by these instruments [73], it is nevertheless important to keep these differences in mind, since they will tend to result in a different level of mixing and, in some cases, greater process-induced degradation. Another issue with these instruments is that the specific throughput (Q/N, where Q = feed rate and N = screw speed) tends to be far lower than is practically achievable in large-scale twin-screw extruders, making scale-up a challenge. Still, without such instruments, small-scale development work would be very difficult and would likely rely primarily on internal/batch mixers. These small-scale lab extruders therefore serve as an important tool and are at least one step closer to their larger brethren, even if they do not always produce identical results.

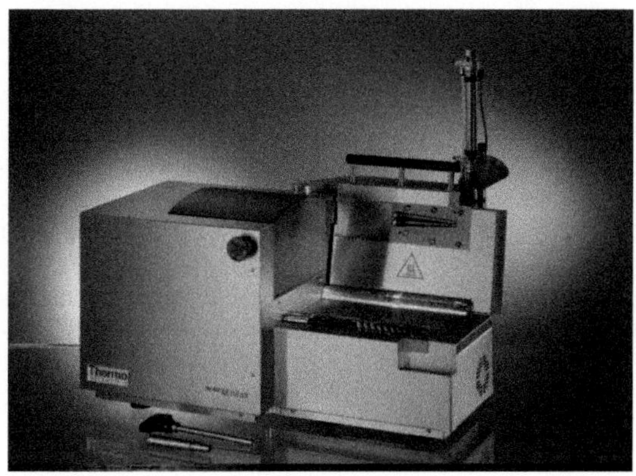

Figure 4.14
A Thermo Scientific HAAKE MiniLab II benchtop recirculating twin-screw extruder (5 cm³ capacity)
Image courtesy of Thermo Fisher Scientific, received 3/2008

In addition to the various techniques described above, the introduction of additional mechanical energy, in the form of ultrasonic vibrations, has also been utilized as a means of processing polymer nanocomposites. In particular, ultrasonication has been applied to polymer/nanoclay nanocomposites in the melt state [78], and successfully applied during both single-screw [79,80,81,82,83,84,85] and twin-screw melt-blending operations [86,87], with improved dispersion observed in all cases; mechanical properties were not measured in the twin-screw studies, but improved across the board in the single-screw studies. Finally, while the introduction of too much ultrasonic energy can result in significant polymer degradation, and some degree of chain scission, radical formation and reaction within the system are the norm during ultrasonic melt processing in any event, and this may be taken advantage of to induce polymer/nanofiller grafting reactions.

4.11   What additional options do we have with solutions and dispersions?

4.11.1   Physical mixing/"wet blending"

This subsection has been titled in an analogous manner to the previous subsection describing the preparation of physical blends of solid polymer and nanofiller because of the importance here, as there, of physical mixing as a form of *pre-processing*. In this case, however, the goal is to produce a compatible mixture of polymer, nanofiller, and solvent – in the case of the nanoclays, an approach to nanocomposite formation successfully demonstrated as far back as 1961(!) [88]. Clearly, the first criterion is the identification of a suitable solvent. Solvents for polymers are generally known, and may be found in references like the Polymer Handbook [43]. Solvents appropriate for dispersion of a particular nanofiller, on the other hand, are less readily available and must generally be identified based on the research literature, or, failing that, by trial and error. The dispersion and settling method mentioned previously in the discussion of how to achieve thermodynamic compatibility in practice represents a simple means of assessing solvents for this purpose.

Once an appropriate solvent has been identified, mixing must be performed. Ultrasonication has been shown to be exceptionally effective for forming aqueous dispersions of fine mineral particles in general [89] and nanoclays in particular [90], although these reports also indicate that dispersion is dependent on the dimensions of the mixing vessel [90], the frequency and duration of treatment and that improper ultrasonication can actually induce aggregation [89]. Nevertheless, because of the potential for excellent dispersion, ultrasonication is often used to disperse various nanofillers in other solvents as a result. While an ultrasonic bath can be used for this purpose, an ultrasonic probe (Figure 4.15) is generally considered to be superior thanks to its higher power output and more focused delivery of mechanical (ultrasonic) energy. As implied by previous discussions covering ultrasonication in the melt, however, one significant disadvantage to this technique is the potential to damage the organic components present in the system. Radical generation has been shown to occur during the ultrasonication of water, for instance [91], and has also been shown to degrade both surfactants[92] and polymers in solution [93]. Furthermore, media with viscosities above a few thousand centipoise (mPa·s) are not effectively mixed in this fashion due to their inability to flow back fast enough to re-establish contact with typical ultrasonic probe tips between oscillations. In contrast, the acoustic mixing technology described previously (see Figure 4.4 and Figure 4.5) appears to be able to circumvent this viscosity limitation, and would seem less likely to induce degradation versus ultrasonication thanks to the much lower operating frequency, but this point has not been confirmed experimentally.

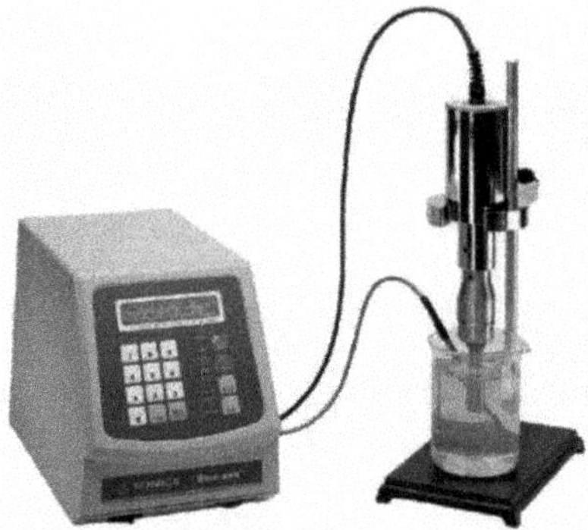

Figure 4.15
A Sonics and Materials Vibracell VCX Series ultrasonic processor
From http://www.sonics.biz/liquid-new-sheet/VCX500-750.pdf, viewed 1/2008, reproduced with perrmission

Alternative techniques for the preparation of such dispersions include the use of so-called homogenizers or high-shear mixers, common examples of which are shown in Figure 4.16 and Figure 4.17. Such mixers are highly effective, but tend to introduce substantial amounts of air into the liquid, making degassing necessary in some cases, and must be cleaned during runs (not easy, given the enclosed mixing head) in order to avoid contamination. This problem can be particularly severe if the viscosity starts high or increases substantially as a result of nanofiller dispersion during mixing, however, and in both situations substantial heat may be generated as well, resulting in unwanted solvent loss or side reactions (in the case of reactive systems). The use of an orbital mixer (Figure 4.3), represents a more gentle option versus the high-shear mixers that also avoids the cleaning and contamination issues, although for heavily agglomerated particles the short mixing cycles typically employed with such units may not be sufficient. In fact, however, if the solvent has been properly chosen and is compatible with the nanofiller, simple overnight stirring using a magnetic stirrer and a Teflon-coated stir-bar is probably the easiest means of achieving high levels of dispersion without too much trouble; an overhead stirrer is effective as well, but more difficult to clean.

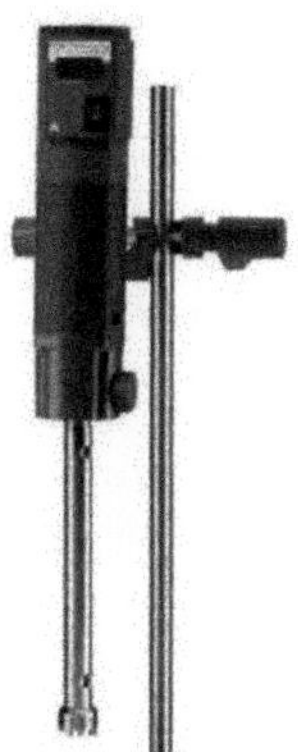

Figure 4.16
An IKA T 25 digital Ultra Turrax high-performance disperser/homogenizer
From
http://www.ika.net/ika/datasheet/datasheet.php?iProduct=3565000&iLang=1&filename=DataSheet
_T_25_digital_ULTRA-TURRAX_.pdf, viewed 3/2008, reproduced with permission

Figure 4.17
A Silverson L4RT high-shear laboratory mixer
From http://www.silverson.com/USA/Assets/Images/labmix_small.jpg, viewed 1/2008, reproduced
with permission

Once an appropriate mixture of polymer, nanofiller and solvent has been prepared, a variety of processing techniques are available. Film casting is probably the most commonly applied, and involves pouring the mixture into a casting dish and allowing it to evaporate, thus forming a nanocomposite film. High density polyethylene (HDPE) or perfluoroalkoxy (PFA) Teflon casting dishes are preferred for such work, being non-porous, non-adherent and chemical-resistant. In the best of cases, smooth, robust, transparent, free-standing films containing well dispersed nanofiller particles may be readily produce via this technique, although as noted previously, the removal of the majority of the solvent can require weeks even in a vacuum oven [11], a problem made worse by any enhanced barrier properties realized in the nanocomposite. Common problems associated with the technique include film break-up and the formation of voids or bubbles, although this can be controlled to some extent by controlling the rate of evaporation from the sample. Relative humidity may also play a role in film quality, depending on the nature of the polymer, nanofiller, and solvent involved, while local changes in airflow can have an effect as well. For these reasons, reproducibility can sometimes be difficult to achieve. Finally, in the case of high aspect ratio nanofillers, some degree of orientation may be observed in the plane of the film, depending on the casting conditions, the evolution of viscosity with solvent loss and the like. Such film casting techniques can be time-consuming but can be adapted for commercial utility when applied in dip- or continuous coating lines.

In adding to simple evaporation, spray and jet techniques may also be utilized. Spray-coating (i.e. using an airbrush or equivalent) represents an alternate means of producing nanocomposite thin films or coatings, the major difference being that evaporation is much more rapid, giving high aspect ratio nanofillers much less time to orient with respect to the substrate. Spray drying, on the other hand, involves a drying column or tower and a nozzle that produces a fine aerosol of the polymer/nanofiller/solvent mixture and allows for the production of a powdered nanocomposite product, making it potentially useful as a *pre-processing* technique. Finally, electrospinning represents a potentially useful jet technique for the production of non-woven nanocomposite fibre mats. It is also worth noting that, the expense of the solvent aside, all of the techniques mentioned in this paragraph are scalable, rapid, and find industrial application.

Moving away from a simple mixture of polymer, nanofiller and solvent, two other possibilities stand out as potentially interesting means of nanocomposite processing. The first is the combination of an aqueous nanofiller dispersion and an aqueous polymer dispersion (latex) to form a nanocomposite latex dispersion. This may be applied from the liquid state as a cast or sprayed nanocomposite film or coating, or spray-dried to form nanocomposite powders, so long as the nanofiller and polymer latex retain some degree of colloidal stability. Alternatively, if the co-dispersion is intentionally destabilized, it is possible to coagulate the system, filter and dry it to produce a nanocomposite pre-mix appropriate for further processing – an approach first noted in the patent literature in 1947(!) [94]. Both approaches are therefore of interest as *pre-processing* techniques as well, and are advantageous in that no organic solvents are needed in order to carry them out, and a wide range of polymer latex dispersions are available, making many polymer nanocomposite systems amenable to processing via these methods.

The final nanocomposite processing method presented in this section is the layer-by-layer (LbL) deposition technique. As above, this technique involves the use of an aqueous nanofiller dispersion; it may involve the use of an aqueous polymer latex as well, although aqueous polymer solutions are more common. Most work with this technique involves species with opposite charges – for instance, anionic nanoclays and cationic polyelectrolytes. In this scheme, a substrate might be coating with the polymer via simple dip-coating in the polymer solution, then rinsed to remove the excess. The same substrate would then be dipped into the dispersion of the oppositely charged nanoclay, some of which would adhere due to electrostatic interactions with the polyelectrolyte on the substrate. The excess nanoclay would be rinsed off, and the process then repeated many times to build up a solid film that will no longer dissolve in water. Because each coating cycle results in an increase in coating thickness on the order of 1 nm, this is not a rapid technique; nevertheless, it can be automated [95] or adapted to a continuous coating line [96]. The advantage of this technique is not that it gives dispersion or distribution – far from it, LbL produces an alternating laminate structure of layers consisting solely of polymer and nanoclay. Such films show a host of interesting properties due to the high nanofiller loading levels and dictated by the technique, and have since been applied to uncharged polymers as well, relying on hydrogen bonding with the nanoclays rather than electrostatic interactions [97]. While the free-standing films and thick parts are difficult to produce and some degradation in properties is observed in highly humid environments, this processing technique shows significant promise for the formation of nanocomposite coatings.

4.12   What about reactive processing?

The final class of nanocomposite processing techniques to be listed here are those that involve some form of materials synthesis, as opposed to the mixing of pre-formed polymer and nanofiller. While excellent dispersion and interesting properties have been repeatedly demonstrated via in-situ polymerization and all of these methods find utility as *pre-processing* techniques, they have received a relatively attention industrially because of the changes they would require in established polymerization processes.

The most well-known of these techniques is in-situ (bulk) polymerization, mentioned previously as the means by which the Toyota group first prepared the polyamide-6 nanocomposites, in the early 1990s[12], that spurred many others to look at polymer nanocomposites based on nanoclays. This process is quite simple, and involves dispersion of the nanofiller of choice directly in liquid monomer(s), followed by polymerization of the monomer(s) to form a polymer nanocomposite. The nanofiller may be modified solely for enhanced dispersion or by various reactive modifiers as described in section 4.7.5. The end result of such modifications greatly depends on the relative rates of polymerization at and away from the filler surface, as described in the section on dispersion kinetics in the presence of chemical reactions. This technique may be applied not just to thermoplastics like polyamide-6 but to thermosets as well. Epoxy/nanoclay nanocomposites have received the most attention in this regard [98], although a number of other systems have also been studied.

In terms of multiphase systems, nanocomposites may be produced via suspension polymerization in much the same way as bulk polymerization. Here, the nanofiller particles may be dispersed either in the aqueous phase or in the monomer droplets, which are then polymerized to form a coarse nanocomposite powder pre-mix. Emulsion polymerization can also be used as a means of nanocomposite formation, the end result being the formation of a co-dispersion of nanofiller and polymer particles. Here, too, techniques described in the previous section are relevant; in particular those related to dispersion based processing.

As might be expected, given the high specific surface area of the nanofillers involved in these processes, their presence during polymerization is not without its effects. In particular, one can expect changes in average molecular weights and polydispersity in the presence of the nanofiller. As a result, it is generally advisable to compare the molecular weight distributions of any polymers produced in the presence of nanofiller to the same polymer produced in the absence of any nanofiller, so as to avoid unfair comparisons between the two with respect to properties.

4.13   Are there any additional considerations?

Finally, it should be noted that processing does not end with nanocomposite formation. Rather, the products of almost all of the processes described here may be further formed through a wide range of conventional plastics processing techniques. While a detailed discussion of these issues is beyond the scope of this chapter, it is emphasized that subsequent forming operations of polymer nanocomposites based on high aspect ratio nanofillers (injection moulding, blown or cast film production, foaming, etc.) will induce important structural changes with respect to the properties of the parts being produced. In particular, it must be remembered that anisotropic properties in polymer nanocomposites based on anisotropic nanofillers are the rule not the exception, and that melt flow patterns orient high aspect ratio fillers like nanoclays.

To give an example, an injection-moulded tensile bar will show some level of nanoclay orientation from the gate through to the end of the gauge, but towards the end of the mould, where the flow-front can no longer move parallel to the dogbone, substantial disorientation of the nanoclay will occur. Likewise, a much greater degree of nanoclay orientation will be observed in the plane of thinner blown and cast film products than in the aforementioned injection-moulded tensile dogbone, meaning that the mechanical properties of the dogbone will reflect the true mechanical properties of the film products even less accurately than in the case of an unfilled or spherical particle filled polymer. Nanofillers often act as nucleating agents during foam processing, due to the large amount of interfacial area in such systems, although compatibility at the polymer/nanofiller surface makes a substantial difference in the magnitude of this effect [99]. Nanoclays have also been shown to orient within the cell walls of nanoclay-based nanocomposite foams [100], and are likely to reduce the rate of gas diffusion out of the cells and increase the melt strength of the polymer, further changing the foaming process as a result.

While flow-induced orientation can be a nuisance if not properly controlled, it can also show substantial utility. In the aforementioned examples involving film products, for instance, in-plane orientation is likely to improve the barrier properties of these films, an excellent outcome given the need for high barrier packaging materials. In some cases, however, flow-based orientation may not be achievable in the desired fashion. One potential means of overcoming this issue involves the use of applied electrical [101,102] or magnetic [103] fields, which may be used to control the orientation of the nanoclay layers or to enhance dispersion. In spite of the fact that the field strengths required to achieve significant orientation in a large part may be difficult to achieve in practice, depending on the size and shape of the part, the fact that different nanofillers orient in different ways may allow for the design of interesting and useful nanostructures leading to substantial control over the local properties of a given part.

This is just one example in a growing trend. As time goes by, we will become more and more sophisticated in our understanding of polymer nanocomposites as an interesting and useful class of materials, and we will make more and more of our ability to produce and use them. We will also become more sophisticated in our attempts to realize desired properties via the combination of multiple nanofillers, fillers, and other additives and their consideration as a package, rather than discrete, non-interacting components, and to look to nanofillers not as magical materials that no one understands but as a class of interesting and useful polymer additives that offer unique sets of properties. Throughout these evolutions, the issue of nanocomposite processing, and our ability to understand the underlying physics and chemistry at work, will only grow in importance.

REFERENCES

1   Xie, W.; Gao, Z.; Pan, W.-P.; Hunter, D.; Singh, A.; Vaia, R. Thermal Degradation Chemistry of Alkyl Quaternary Ammonium Montmorillonite. *Chemistry of Materials*, **2001**, 13, 2979-2990.

2   Xie, W.; Xie, R.; Pan, W.-P.; Hunter, D.; Koene, B.; Tan, L.-S.; Vaia, R. Thermal Stability of Quaternary Phosphonium Modified Montmorillonites. *Chemistry of Materials*, **2002**, *14*, 4837-4845.

3   Gilman, J. W.; Awad, W. H.; Davis, R. D.; Shields, J.; Harris Jr., R. H.; Davis, C.; Morgan, A. B.; Sutto, T. E.; Callahan, J.; Trulove, P. C.; DeLong, H. C. Polymer/Layered Silicate Nanocomposites from Thermally Stable Trialkylimidazolium-Treated Montmorillonite. *Chemistry of Materials*, **2002**, 14, 3776-3785.

4   Awad, W. H.; Gilman, J. W.; Nyden, M.; Harris Jr., R. H.; Sutto, T. E.; Callahan, J.; Trulove, P. C.; Delong, H. C.; Fox, D. M. Thermal degradation studies of alkyl-imidazolium salts and their application in nanocomposites. *Thermochimica Acta*, **2004**, 409, 3-11.

5   Chigwada, G.; Wang, D.; Wilkie, C. A. Polystyrene nanocomposites based on quinolinium and pyridinium surfactants. *Polymer Degradation and Stability*, **2006**, 91, 848-855.

6   Zhang, J.; Wilkie, C. A. A carbocation substituted clay and its styrene nanocomposite. *Polymer Degradation and Stability*, **2004**, 83, 301-307.

7   Ortgeo, J. D.; Kowalska, M.; Cocke, D. L. Interactions of montmorillonite with organic compounds-adsorptive and catalytic properties. *Chemosphere*, **1991**, 22, 769-798.

8   Davis, R. D.; Gilman, J. W.; VanderHart, D. L. Processing degradation of polyamide 6/montmorillonite clay nanocomposites and clay organic modifier. *Polymer Degradation and Stability*, **2003**, 79, 111-121.

9   VanderHart, D. L.; Asano, A.; Gilman, J. W. Solid-State NMR Investigation of Paramagnetic Nylon-6 Clay Nanocomposites. 2. Measurement of Clay Dispersion, Crystal Stratification, and Stability of Organic Modifiers. *Chemistry of Materials*, **2001**, 13, 3796-3809.

10   Wan, C.; Zhang, Y.; Zhang, Y. Effect of alkyl quaternary ammonium on processing discoloration of melt-intercalated PVC-montmorillonite composites. *Polymer Testing*, **2004**, 23, 299-306.

11   Freier, T.; Kunze, C.; Schmitz, K.-P. Solvent removal from solution-cast films of biodegradable polymers. *Journal of Materials Science Letters*, **2001**, 20, 1929-1931.

12   Usuki, A.; Kojima, Y.; Kawasumi, M.; Okada, A.; Fukushima, Y.; Kurauchi, T.; Kamigaito, O. Synthesis of nylon 6-clay hybrid. *Journal of Materials Research*, **1993**, 8, 1179-1184.

13   Blumstein, A. Étude des polymérisations en couche adsorbée I. *Bulletin de la Société Chimique de France*, **1961**, 899-905.

14   Alexandre, M.; Dubois, P.; Sun, T.; Garces, J. M.; Jérôme, R. Polyethylene-layered silicate nanocomposites prepared by the polymerization-filling technique: synthesis and mechanical properties. *Polymer*, **2002**, 43, 2123-2132.

15   Gupta, B.; Lacrampe, M.-F.; Krawczak, P. Polyamide-6/clay nanocomposites: a critical review. *Polymers & Polymer Composites*, **2006**, 14, 13-38.

16   Wu, Q.; Liu, X.; Berglund, L. A. FT-IR spectroscopic study of hydrogen bonding in PA6/clay nanocomposites. *Polymer*, **2002**, 43, 2445-2449.

17   Shah, D.; Maiti, P.; Gunn, E.; Schmidt, D. F.; Jiang, D. D.; Batt, C. A.; Giannelis, E. P. Dramatic enhancements in toughness of polyvinylidene fluoride nanocomposites via nanoclay-directed crystal structure and morphology. *Advanced Materials*, **2004**, 16, 1173-1177.

18   Liu, A.; Xie, T.; Yang, G. Comparison of Polyamide-6 Nanocomposites Based on Pristine and Organic Montmorillonite Obtained via Anionic Ring-Opening Polymerization. *Macromolecular Chemistry and Physics*, **2006**, 207, 1174-1181.

19   Corcorran, S.; Lochhead, R. Y.; McKay, T. Particle-stabilized emulsions: A brief overview. *Cosmetics & Toiletries*, **2004**, 119, 47-50 & 52.

20   Miller, R.; Fainerman, V. B.; Kovalchuk, V. I.; Grigoriev, D. O.; Leser, M. E.; Michel, M. Composite interfacial layers containing micro-size and nano-size particles. *Advances in Colloid and Interface Science*, **2006**, 128-130, 17-26.

21   Wang, H.; Zeng, C.; Elkovitch, M.; Lee, L. J.; Koelling, K. W. Processing and Properties of Polymeric Nano-Composites. *Polymer Engineering and Science*, **2001**, 41, 2036-2046.

22   Voulgaris, D.; Petridis, D. Emulsifying effect of dimethyldioctadecylammonium-hectorite in polystyrene/poly(ethyl methacrylate) blends. *Polymer*, **2002**, 43, 2213-2218.

23   Gelfer, M. Y.; Song, H. H.; Liu, L.; Hsiao, B. S.; Chu, B.; Rafailovich, M.; Si, M.; Zaitsev, V. Effects of Organoclays on Morphology and Thermal and Rheological Properties of Polystyrene and Poly(methyl methacrylate) Blends. *Journal of Polymer Science: Part B: Polymer Physics*, **2003**, 41, 44-54.

24   Ray, S. S.; Pouliot, S.; Bousmina, M.; Utracki, L. A. Role of organically modified layered silicate as an active interfacial modifier in immiscible polystyrene/polypropylene blends. *Polymer*, **2004**, 45, 8403-8413.

25   Ray, S. S.; Bousmina, M. Effect of Organic Modification on the Compatibilization Efficiency of Clay in an Immiscible Polymer Blend. *Macromolecular Rapid Communications*, **2005**, 26, 1639-1646.

26   Fang, Z.; Harrats, C.; Moussaif, N.; Groenickx, G. Location of a Nanoclay at the Interface in an Immiscible Poly(e-caprolactone)/Poly(ethylene oxide) Blend and Its Effect on the Compatibility of the Components. *Journal of Applied Polymer Science*, **2007**, 106, 3125-3135.

27   Kelarakis, A.; Giannelis, E. P.; Yoon, K. Structure–properties relationships in clay nanocomposites based on PVDF/(ethylene–vinyl acetate) copolymer blends. *Polymer*, **2007**, 48, 7567-7572.

28   Vo, L. T.; Giannelis, E. P. Compatibilizing Poly(vinylidene fluoride)/Nylon-6 Blends with Nanoclay. *Macromolecules*, **2007**, 40, 8271-8276.

29   Reichert, P.; Nitz, H.; Klinke, S.; Brandsch, R.; Thomann, R.; Mülhaupt, R. Poly(propylene)/organoclay nanocomposite formation: Influence of compatibilizer functionality and organoclay modification. *Macromolecular Materials and Engineering*, **2000**, 275, 8-17.

30   Kim, Y.; White, J. L. Formation of Polymer Nanocomposites with Various Organoclays. *Journal of Applied Polymer Science*, **2005**, 96, 1888-1896.

31   Borse, N. K.; Kamal, M. R. Melt Processing Effects on the Structure and Mechanical Properties of PA-6/Clay Nanocomposites. *Polymer Engineering and Science*, **2006**, 46, 1094-1103.

32   Kráčalík, M.; Mikešová, J.; Puffr, R.; Baldrian, J.; Thomann, R.; Freidrich, C. Effect of 3D structures on recycled PET/organoclay nanocomposites. *Polymer Bulletin*, **2007**, 58, 313-319.

33   Pospiech, D.; Kretzschmar, B.; Willeke, M.; Leutritz, A.; Jehnichen, D.; Janke, A.; Omastová, M. Exfoliation Behaviour of Montmorillonite Modified by Poly(oxyalkylene)s in Polypropylene and the Properties of the Resulting Nanocomposites. *Polymer Engineering and Science*, **2007**, 47, 1262-1271.

34   Balazs, A. C.; Singh, C.; Zhulina, E. Modeling the Interactions between Polymers and Clay Surfaces through Self-Consistent Field Theory. *Macromolecules*, **1998**, 31, 8370-8381.

35   Rausell-Colom, J. A.; Serratosa, J. M. *Chemistry of Clays and Clay Minerals* Newman, A. C. D. (editor), ISBN 0471-01141-X, Wiley-Interscience, New York, **1987**, page p. 371.

36   Schmidt, D. F.; Clément, F.; Giannelis, E. P. On the origins of silicate dispersion in polysiloxane/layered-silicate nanocomposites. *Advanced Functional Materials*, **2006**, 16, 417-425.

37   Zilg, C.; Mülhaupt, R.; Finter, J. Morphology and toughness/stiffness balance of nanocomposites based upon anhydride-cured epoxy resins and layered silicates. *Macromolecular Chemistry and Physics*, **1999**, 200, 661-670.

38   Triantafillidis, C. S.; LeBaron, P. C.; Pinnavaia, T. J. Thermoset Epoxy-Clay Nanocomposites: The Dual Role of α,ω-Diamines as Clay Surface Modifiers and Polymer Curing Agents. *Journal of Solid State Chemistry*, **2002**, 167, 354-362.

39   Dekking, H. G. G. Propagation of vinyl polymers on clay surfaces. II. Polymerization of monomers initiated by free radicals attached to clay. *Journal of Applied Polymer Science*, **1967**, 11, 23-36.

40   Summers, M.; Eastoe, J. Applications of Polymerizable Surfactants. *Advances in Colloid and Interface Science*, **2003**, 100-102, 137-152.

41   Bognolo, G. Polymerizable Surfactants. *Chemistry and Technology of Surfactants*, Farn, R. J. (editor), ISBN 1-4051-2696-0, Wiley-Blackwell, Oxford, **2006**, pages 204-226.

42   Ijdo, W. L.; Pinnavaia, T. J. Solid Solution Formation in Amphiphilic Organic-Inorganic Clay Heterostructures. *Chemistry of Materials*, **1999**, 11, 3227-3231.

43   *Polymer Handbook (4th Edition)*, Brandrup, J.; Immergut, E. H.; Grulke, E. A.; Abe, A.; Bloch, D. R. (editors), ISBN 0-471-47936-9, John Wiley & Sons, New York, **2003**. Online version available at http://www.knovel.com/knovel2/Toc.jsp?BookID=1163

44   Hunter, D. L., *private communications*, April-May **2008**.

45   Lan, T.; Kaviratna, P. D.; Pinnavaia, T. J. Epoxy self-polymerization in smectite clays. *Journal of Physics and Chemistry of Solids*, **1996**, 57, 1005-1010.

46   Verploegen, E.; Dworken, B. T.; Faught, M.; Kamperman, M.; Zhang, Y.; Wiesner, U. Tuning Mechanical Properties of Block Copolymer/Aluminosilicate Hybrid Materials. *Macromolecular Rapid Communications*, **2007**, 28, 572-578.

47   Mackenzie, J. D.; Bescher, E. P. Structures, Properties and Potential Applications of Ormosils. *Journal of Sol-Gel Science and Technology*, **1998**, 13, 371-377.

48   Haas, K.-H.; Wolter, H. Synthesis, properties and applications of inorganic–organic copolymers (ORMOCER®s). *Current Opinion in Solid State and Materials Science*, **1999**, 4, 571-580.

49   Pagliaro, M.; Ciriminna, R.; Man, M. W. C.; Campestrini, S. Better Chemistry through Ceramics: The Physical Bases of the Outstanding Chemistry of ORMOSIL. *Journal of Physical Chemistry B*, **2006**, 110, 1976-1988.

50   Zerda, A. S.; Lesser, A. J. Intercalated Clay Nanocomposites: Morphology, Mechanics, and Fracture Behaviour. *Journal of Polymer Science: Part B: Polymer Physics*, **2001**, 39, 1137-1146.

51   Li, Y.; Wei, G.-X.; Sue, H.-J. Morphology and toughening mechanisms in clay-modified styrene-butadiene-styrene rubber-toughened polypropylene. *Journal of Materials Science*, **2002**, 37, 2447-2459.

52   Chen, L.; Wong, S.-C.; Pisharath, S. Fracture Properties of Nanoclay-Filled Polypropylene. *Journal of Applied Polymer Science*, **2003**, 88, 3298-3305.

53   Lu, T.; Tjiu, W. C.; Tong, Y.; He, C.; Goh, S. S.; Chung, T.-S. Morphology and Fracture Behaviour of Intercalated Epoxy/Clay Nanocomposites. *Journal of Applied Polymer Science*, **2004**, 94, 1236-1244.

54   Zhu, J.; Uhl, F. M.; Morgan, A. B.; Wilkie, C. A. Studies on the Mechanism by Which the Formation of Nanocomposites Enhances Thermal Stability. *Chemistry of Materials*, **2001**, 13, 4649-4654.

55   Garboczi, E. J.; Snyder, K. A.; Douglas, J. F.; Thorpe, M. F. Geometrical percolation threshold of overlapping ellipsoids. *Physical Review E*, **1995**, 52, 819-828.

56   Krishnamoorti, R.; Ren, J.; Silva, A. S. Shear response of layered silicate nanocomposites. *Journal of Chemical Physics*, **2001**, 114, 4968-4973.

57   Varlot, K.; Reynaud, E.; Kloppfer, M. H.; Vigier, G.; Varlet, J. Clay-Reinforced Polyamide: Preferential Orientation of the Montmorillonite Sheets and the Polyamide Crystalline Lamellae. *Journal of Polymer Science: Part B: Polymer Physics*, **2001**, 39, 1360-1370.

58   Lele, A.; Mackley, M.; Galgali, G.; Ramesh, C. In situ rheo-x-ray investigation of flow-induced orientation in layered silicate–syndiotactic polypropylene nanocomposite melt. *Journal of Rheology*, **2002**, 46, 1091-1110.

59   Khait, K. Advanced reclamation technology for mixed polymeric waste. *Polymer Recycling*, **1996**, 2, 299-301.

60   Khait, K.; Torkelson, J. M. Solid-state shear pulverization of plastics: a green recycling process. *Polymer-Plastics Technology and Engineering*, **1999**, 38, 445-457.

61   Wang, Q.; Cao, J.; Huang, J.; Xu, X. A Study on the Pan-Milling Process and the Pulverizing Efficiency of Pan-Mill Type Equipment. *Polymer Engineering and Science*, **1997**, 37, 1091-1101.

62   Shao, W.; Wang, Q.; Ma, H. Study of polypropylene/montmorillonite nanocomposites prepared by solid-state shear compounding (S3C) using pan-mill equipment: the morphology of montmorillonite and thermal properties of the nanocomposites. *Polymer International*, **2005**, 54, 336-341.

63   Kim, D.; Lee, J. S.; Barry, C. M. F.; Mead, J. L. Effect of Fill Factor and Validation of Characterizing the Degree of Mixing in Polymer Nanocomposites. *Polymer Engineering and Science*, **2007**, 47, 2049-2056.

64   Kim, D.; Lee, J. D.; Barry, C. M. F.; Mead, J. L. Microscopic Measurement of the Degree of Mixing for Nanoparticles in Polymer Nanocomposites by TEM Images. *Microscopy Research and Technique*, **2007**, 70, 539-546.

65   Gianelli, W.; Camino, G.; Dintcheva, N. T.; Lo Verso, S.; La Mantia, F. P. EVA-Montmorillonite Nanocomposites: Effect of Processing Conditions. *Macromolecular Materials and Engineering*, **2004**, 289, 238-244.

66   Ryu, S. H.; Chang, Y.-W. Factors affecting the dispersion of montmorillonite in LLDPE nanocomposite. *Polymer Bulletin*, **2005**, 55, 385-392.

67   Dennis, H. R.; Hunter, D. L.; Chang, D.; Kim, S.; White, J. L., Cho, J. W.; Paul, D. R. Effect of melt processing conditions on the extent of exfoliation in organoclay-based nanocomposites. *Polymer*, **2001**, 42, 9513-9522.

68   Cho, J. W.; Paul, D. R. Nylon 6 nanocomposites by melt compounding. *Polymer*, **2001**, 42, 1083-1094.

69   Cho, J. W.; Logsdon, S.; Omachinski, S.; Qian, G.; Lan, T.; Womer, T. W.; Smith, W. S. Nanocomposites: A Single Screw Mixing Study of Nanoclay-filled Polypropylene. *ANTEC 2002 Proceedings*, Vol. 1, pages 214-218, 60th SPE Annual Technical Conference, San Francisco, CA, U. S. A., 6-9 May **2002**.

70   Markarian, J. New mixing developments move single screw compounding forward. *Plastics Additives & Compounding*, **2004**, September/October issue, 6, 50-53.

71   Anderson, P. G. Processing nanocomposites on a kneader reciprocating single screw compounding system. *ANTEC 2003 Proceedings*, Vol. 1, pages 283-287, 61st SPE Annual Technical Conference, Nashville, TN, U. S. A., 4-8 May **2003**.

72   Anderson, P. G. Twin Screw Extrusion Guidelines for Compounding Nanocomposites. *ANTEC 2002 Proceedings*, Vol. 1, pages 1-5, 60th SPE Annual Technical Conference, San Francisco, CA, U. S. A., 6-9 May **2002**.

73   Chavarria, F.; Shah, R. K.; Hunter, D. L.; Paul, D. R. Effect of Melt Processing Conditions on the Morphology and Properties of Nylon 6 Nanocomposites. *Polymer Engineering and Science*, **2007**, 47, 1847-1864.

74   Lertwimolnun, W.; Vergnes, B. Influence of Screw Profile and Extrusion Conditions on the Microstructure of Polypropylene/Organoclay Nanocomposites. *Polymer Engineering and Science*, **2007**, 47, 2100-2109.

75   Vaia, R. A.; Jandt, K. D.; Kramer, E. J.; Giannelis, E. P. Kinetics of Polymer Melt Intercalation. *Macromolecules*, **1995**, 28, 8080-8085.

76   Manias, E.; Chen, H.; Krishnamoorti, R.; Genzer, J.; Kramer, E. J.; Giannelis, E. P. Intercalation Kinetics of Long Polymers in 2 nm Confinements. *Macromolecules*, **2000**, 33, 7955-7966.

77   Özkoç, G.; Bayram, G.; Quaedflieg, M. Effects of Microcompounding Process Parameters on the Properties of ABS/Polyamide-6 Blends Based Nanocomposites. *Journal of Applied Polymer Science*. **2008**, 107, 3058-3070.

78   Lee, E. C.; Mielewski, D. F.; Baird, R. J. Exfoliation and Dispersion Enhancement in Polypropylene Nanocomposites by In-Situ Melt Phase Ultrasonication. *Polymer Engineering and Science*, **2004**, 44, 1773-1792.

79   Li, J.; Zhao, L.; Guo, S. Ultrasonic Preparation of Polymer/Layered Silicate Nanocomposites during Extrusion. *Polymer Bulletin*, **2005**, 55, 217-223.

80   Zhao, L.; Li, J.; Guo, S.; Du, Q. Ultrasonic oscillations induced morphology and property development of polypropylene/montmorillonite nanocomposites. *Polymer*, **2006**, 47, 2460-2469.

81   Lapshin, S.; Isayev, A. I. Continuous Process for Melt Intercalation of PP–Clay Nanocomposites With Aid of Power Ultrasound. *Journal of Vinyl & Additive Technology*, **2006**, 12, 78-82.

82   Swain, S. K.; Isayev, A. I.; Effect of ultrasound on HDPE/clay nanocomposites: Rheology, structure and properties. *Polymer*, **2007**, 48, 281-289.

83   Li, J.; Zhao, L.; Guo, S. Ultrasound Assisted Development of Structure and Properties of Polyamide 6/Montmorillonite Nanocomposites. *Journal of Macromolecular Science, Part B: Physics*, **2007**, 46, 423-439.

84   Lapshin, S.; Isayev, A. I. Ultrasound-Aided Extrusion Process for Preparation of Polypropylene–Clay Nanocomposites. *Journal of Vinyl & Additive Technology*, **2007**, 13, 40-45.

85   Li, J.; Jiang, G. J.; Guo, S. Y.; Zhao, L. J. Effects of ultrasonic oscillations on structure and properties of HDPE/montmorillonite nanocomposites. *Plastics, Rubber and Composites*, **2007**, 36, 308-313.

86   Kim, K. Y.; Ju, D. U.; Nam, G. J.; Lee, J. W. Ultrasonic Effects on PP/PS/Clay Nanocomposites during Continuous Melt Compounding Process. *Macromolecular Symposia*, **2007**, 249-250, 283-288.

87   Kim, K. Y.; Nam, G. J.; Lee, J. W. Continuous extrusion of long-chain-branched polypropylene/clay nanocomposites with high-intensity ultrasonic waves. *Composite Interfaces*, **2007**, 14, 533-544.

88   Uskov, I. A.; Tarasenko, Yu. G.; Kusnitsyna, T. A. Compounding of Dissolved Polymethylmethacrylates with Fillers. *International Chemical Engineering*, **1961**, 1, 132-134.

89   Lu, S. C.; Song, S. X.; Dai, Z. F. Dispersion of fine mineral particles in water. *Advanced Powder Technology*, **1992**, 3, 89-96.

90   Lapides, I.; Heller-Kallai, L. Novel features of smectite settling. *Colloid and Polymer Science*, **2002**, 280, 554-561.

91   Didenko, Y. T.; Suslick, K. S. The energy efficiency of formation of photons, radicals and ions during single-bubble cavitation. *Nature*, **2002**, 418, 394-397.

92   Weavers, L. K.; Pee, G. Y.; Frim, J. A.; Yang, L.; Rathman, J. F. Ultrasonic Destruction of Surfactants: Application to Industrial Wastewaters. *Water Environment Research*, **2005**, 77, 259-265.

93   Price, G. J. Applications of high intensity ultrasound in polymer chemistry. *Chemistry under Extreme or Non-classical Conditions*, Van Eldick, R.; Hubbard, C. D. (editors), ISBN 0-471-16561-3, Wiley-Interscience, New York, **1997**, pages 381-427.

94   Carter, L. W.; Hendricks, J. G.; Bolley, D. S. Elastomer Reinforced with a Modified Clay. *U. S. Patent*, **1950**, 2,531,396.

95   Jang, W.-S.; Grunlan, J. C. Robotic dipping system for layer-by-layer assembly of multifunctional thin films. *Review of Scientific Instruments*, **2005**, 76, 103904.

96   Mehrabi, A.; Akhave, J. R.; Licon, M.; Koch, C. A. Continuous process for manufacturing electrostatically self-assembled coatings. *International Patent*, **2004**, WO 2004/071677.

97   Podsiadlo, P.; Kaushik, A. K.; Arruda, E. M.; Waas, A. M.; Shim, B. S.; Xu, J.; Nandivada, H.; Pumplin, B. G.; Lahann, J.; Ramamoorthy, A.; Kotov, N. A. Ultrastrong and Stiff Layered Polymer Nanocomposites. *Science*, **2007**, 318, 80-83.

98   Becker, O.; Simon, G. P., Dusek, K., Epoxy Layered Silicate Nanocomposites. *Advances in Polymer Science*, **2005**, 179, 29-82.

99   Zeng, C.; Han, X.; Lee, L. J.; Koelling, K. W.; Tomasko, D. L. Polymer-Clay Nanocomposite Foams Prepared Using Carbon Dioxide. *Advanced Materials*, **2003**, 15, 1743-1747.

100   Okamoto, M.; Nam, P. H.; Maiti, P.; Kotaka, T.; Nakayama, T.; Takada, M.; Ohshima, M.; Usuki, A.; Hasegawa, N.; Okamoto, H. Biaxial Flow-Induced Alignment of Silicate Layers in Polypropylene/Clay Nanocomposite Foam. *Nano Letters*, **2001**, 1, 503-505.

101 Park, J. U.; Choi, Y. S.; Cho, K. S.; Kim, D. H.; Ahn, K. H.; Lee, S. J. Time–electric field superposition in electrically activated polypropylene/layered silicate nanocomposites. *Polymer*, **2006**, 47, 5145-5153.

102 Kim, D. H.; Choo, K. S.; Mitsumata, T.; Ahn. K. H.; Lee, S. J. Microstructural evolution of electrically activated polypropylene/layered silicate nanocomposites investigated by in situ synchrotron wide-angle X-ray scattering and dielectric relaxation analysis. *Polymer*, **2006**, 47, 5938-5945.

103 Koerner, H.; Hampton, E.; Dean, D.; Turgut, Z.; Drummy, L.; Mirau, P.; Vaia, R. Generating Triaxial Reinforced Epoxy/Montmorillonite Nanocomposites with Uniaxial Magnetic Fields. *Chemistry of Materials*, **2005**, 17, 1990-1996.

Chapter 5
Stabilisation of polymer nanocomposites
Rudolf Pfaendner

Ciba Lampertheim GmbH, Lampertheim, Germany

5.1   Introduction

Polymer nanocomposites are an area with substantial scientific interest and emerging industrial practice. However, despite the published and proven benefits of nanocomposites such as mechanical properties, barrier properties and contribution to fire retardancy, like all kind of organic materials polymer nanocomposites are sensitive to oxidation. Oxidation will take place at any time of the lifetime of the polymer (and of the nanocomposite) during thermal transformation, storage or application. Furthermore, the combined action of light and oxygen results in photooxidation. Degradation of the polymer and irreversible changes in the polymer structure are the consequence of photooxidation resulting finally in insufficient mechanical properties, cracking, failure of the part, and changes in visual appearance (e.g. discoloration) and other properties. Fortunately suitable stabilisers inhibit the oxidation process and can protect polymers against both oxidative and photo-oxidative damage and maintain their properties during processing as well as during the product life cycle in the intended application [1]. Stabilisation of polymer nanocomposites depends on the stabilisation of the particular polymer material under consideration, and also consideration of the influence of a filler with a huge surface area. Therefore, the interactions between stabilisers and "traditional" fillers and the current state-of-the-art in the stabilisation of filled systems are summarised first in this chapter, before turning more specifically to nano-sized materials.

5.2   Challenges of stabilisation of filled polymers

Fillers fulfil a variety of functions in plastic applications such as enhancing stiffness, improving dimensional stability, reducing shrinkage and also cost.   Materials used are very often calcium carbonate, talc and kaolin, with the main application in polyvinyl chloride (PVC), elastomers and polyolefins [2]. It can be shown experimentally, irrespective of the particular structure, that inorganic fillers can have a negative effect on the oxidative stability of the polymer, to a variable extent. There is no doubt that it is mainly interactions between the stabiliser and the filler and adsorption/desorption mechanisms which are responsible for this influence. The surface area of the filler and pore volumes, surface functionality, hydrophilicity, thermal and photosensitisation properties of the filler, transition metal ion content (manganese, iron, titanium) have all been identified as potential elements of the interaction [3]. Not negligible is the mechanical impact of fillers during processing in the equipment (mechanodegradation) leading to the formation of radical species and resulting in additional consumption of stabilisers. Furthermore, efficient stabilisation of polymers will be even more complicated when other materials such as carbon black or pigments such as TiO_2 are incorporated in the formulation in addition to natural inorganic fillers.

Standard test methods to analyze stability of polymers cover multiple extrusion, accelerated heat ageing and UV light exposure. Multiple extrusion evaluates the melt processing stability of a polymer or polymer formulation as high temperature, shear forces and oxygen in the polymer transformation process cause degradation. The melt properties are usually characterised by melt flow rate (MFR) or melt volume rate (MVR). Accelerated heat ageing is simulating long-term thermal stability. Samples are oven-aged at defined temperatures and conditions and the change of the properties, e.g. visual appearance, colour, and mechanical values are recorded over ageing time. A simple test of long-term stability for quality control purposes is to determine the oxygen induction time (OIT). The induction time for the onset of the oxidation is measured e.g. with differential scanning calorimetry (DSC) or differential thermoanalysis (DTA) Photooxidation is tested by UV exposure, artificial or natural weathering. Changes in properties e.g. visual appearance such as chalking, crazes, gloss and/or mechanical properties such as tensile strength, elongation, impact strength can be measured. More detailed information on testing methods, mechanisms of degradation, activities of stabilisers and technical aspects of stabilisation for different polymers can be found elsewhere [1].

A few published examples illustrate that the influence of fillers on the stability of polymers is very significant:

Calcium carbonate and talc decrease the photostabilising efficiencies of all light stabilisers (hindered amines, benzophenones) and antioxidants dramatically. Polypropylene containing 0.05% of a typical phenolic antioxidant (AO-1, structures at the end of this chapter, "list of stabilisers") and heated to 120 °C, took 3508 hours to deteriorate to 50% retention of tensile strength, vs. only 470 hours in the presence of 10% $CaCO_3$ (talc results in 2180 hours) [4].

Polypropylene films, stabilised with 2,6-ditert.-butyl-4-methylphenol and a low molecular weight hindered amine light stabiliser (HALS, LS-1, see structures at the end of the chapter) undergoing 130 °C heat ageing showed a stability of 181 days until embrittlement i.e. until complete loss of mechanical properties. The same formulation containing 40% of talc (metal impurities: 60 ppm Mn, 5 ppm Cu) degraded after only 4 days [5].

The effect of processing is also measurable, and this is simulated by multiple extrusion experiments. Highly filled materials and high shear strength have been shown to cause a higher degradation of the filled material compared to neat polypropylene (PP) [6].

Photooxidation is even more severe: a typical stabilised polypropylene film (0.2% LS-1) achieved 2660 hours to 50% retention of tensile strength, compared with only 364 hours in the presence of 10% $CaCO_3$ [7].

The influence of silica (SiO_2) on the oxidative and photooxidative stability of polypropylene can be correlated with the metal ion content (Fe, Al, Ti) of impurities present in natural silicas. Most of the silicas act as photosensitisers and accelerate the degradation of the polymer even at low concentrations (0.1%). Thermal oxidation at 130 °C (shown as time to embrittlement) mainly resulted in reduced lifetime if SiO_2 is present. 0.1% AO-1 achieved 456 hours without and 336 hours with SiO_2. Some HALS e.g. LS-2 or LS-1 seem to be an exception as these can increase the thermal stability of SiO_2-containing PP, e.g. with 0.1% LS-2 values of 312 hours in the presence of SiO_2 and only 91 hours without SiO_2 were measured. With the exception of LS-1 as a light stabiliser, photooxidation resulted in a considerable reduction of lifetime of all individual stabilisers and combinations tested if SiO_2 was present, e.g. 0.1% of LS-2 achieved 1512 hours without and 540 hours in the presence of SiO_2. The results with different HALS stabilisers indicated that not just adsorption but also desorption mechanisms and release of stabilisers are important [3]. This might be an explanation why macrocyclic HALS compounds can achieve a very high protection level in filled polymers [8].

In this context, to achieve the desired processing and long-term properties of filled polymers, the adsorption effect of the filler and the effect of filler impurities have to be compensated. One simple solution is to increase the stabiliser level to the necessary extent. However this might be an insufficient or a costly approach and it is limited because of issues of compatibility between the additives and the polymer. As already mentioned in the examples above, particularly light stabilisers of the HALS type show considerable performance differences in the presence of fillers depending on the molecular weight and structure [3,5]. Therefore, the selection of an adjusted stabiliser or stabiliser combination for the selected filler polymer pair is decisive. Furthermore, a number of chemical compounds have been suggested which as a general rule show a high affinity to the filler surface and are preferentially adsorbed over the stabilisers.

Agents that modify the surface of fillers, sometimes referred to as filler deactivators or coupling agents, can be chosen from different chemical classes. Already coated fillers (stearic acid, stearates) show a reduced impact on thermal stability in combination with efficient antioxidants compared to untreated fillers [9].

The surface of fillers is more efficiently blocked when filler deactivators with a reactive chemical group are used e.g. based on oligomeric epoxides (0.5 - 2.5 weight-% of the entire formulation) [5,9,10]. Other additives mentioned as coupling agents are silanes [11], titanates [12] and polymeric additives e.g. based on polypropylene-graft-maleic anhydride [13-16] or polypropylene-graft-acrylic acid [17,18]. The latter polymeric additives have more influence on the mechanical properties of the filled polymer and less on the oxidative stability of the system. Excellent results with regard to the oxidative stability of filled polymers can be achieved by using amphiphilic modifiers such as acrylates with long carbon chain side groups [19]. For example the gloss of polypropylene after artificial weathering was maintained by using only 0.1% of the amphiphilic acrylate in the presence of 10% talc. Other amphiphilic molecules to be used for filler deactivation are ethylene bisstearylamide, dodecenylsuccinic anhydride or low molecular weight polyethyleneglycol. A very pronounced synergistic effect on the stability of filled polymers (talc/polypropylene) was found using a combination of 2 different modifiers at 0.2% concentration, in preference to either one [20].

Technically there are two ways to introduce the surface modifiers: Either the filler is surface treated at the manufacturing stage of the filler (purification, milling) or the modifier is applied directly at the compounding step of the polymer filler formulation.

The foregoing facts on the degradation and stability of filled polymers in general certainly indicate that the stabilisation of nanocomposites against oxidative and photooxidative degradation to maintain the performance for the required lifetime is a challenging target, because further aspects of the nanosized material and the entire nanocomposite composition have to be taken into consideration.

5.3   Processing and long-term thermal stabilisation of polymer nanocomposites

Stabilisation of a nanocomposite both during the processing step and for long-term thermal stability has to address the stability of the polymer substrate, the influence of the nano-structured filler (nanofiller) and in addition the influence of modifiers used to intercalate and/or exfoliate the nanofiller, as well as other ingredients/additives of the formulation e.g. compatibilisers. The stability of the individual polymer substrate, its degradation mechanisms and the action of stabilisers to protect the polymer are in general very well known [1]. The nanofillers of interest to the industry and dealt within this chapter are mainly layered silicates, most often from natural sources such as montmorillonite. Unlike synthetic minerals and like traditional fillers, the clays from natural origin will contain metal ions as contaminants. These will act in the same way as in other fillers but are likely to be more homogeneously distributed and therefore will be more crucial to the polymer stability than metal ions in micro sized fillers.

Layered silicates as such are only compatible with polar polymers such as ethylene-vinyl acetate (EVA), polyamide (PA), and thermoplastic polyurethane (TPU). To improve the compatibility of layered silicates with most polymers, in order to allow intercalation or exfoliation a modification step is necessary to make the filler organophilic. Modification of the clay is usually achieved by cation exchange with a long chain amine or a quaternary ammonium salt. The amount of these organic modifiers can be up to 40 % weight of the inorganic clay, and the thermal stability of the organic modifier is less than the stability of the clay itself. In particular the thermal stability of the ammonium salts is very limited at processing temperatures used for extrusion or injection moulding of most polymers. Most of the ammonium structures tend to undergo a Hofmann elimination resulting in volatile olefins and amines. The thermal degradation of ammonium salts starts already at 180 °C and is furthermore reduced by catalytically active sites on the silicate layer [21]. Improved thermal stabilities compared to ammonium ions are shown by phosphonium, imidazolium, pyridinium, stibonium and tropylium ions and others [22].

Some nanocomposites based on hydrophobic polymers such as polyolefins often require an additional compatibiliser on top of the organic modification of the filler e.g. polypropylene-g-maleic anhydride, which is used in concentrations of 5-25%. These lower molecular weight modified polypropylene materials usually possess an inferior stability compared to the parent polymer and are, therefore, a further reason for a potential insufficient long-term performance of polymer nanocomposites.

More recently it has been demonstrated that both the organic modification of the layered silicate and the compatibiliser can be avoided. In this example direct manufacturing of nanocomposites was shown to be feasible if selected block copolymer structures are used as an additive at the compounding stage [23].

There are many publications describing the stability of nanocomposites based on thermogravimetric analysis experiments (TGA). In these "short-term" TGA investigations the findings on thermal stability are not very conclusive. It seems that some nanocomposites e.g. from polymers like PP, polyethylene (PE) and polystyrene (PS) are more stable than the parent polymer. Others polymers like PVC [24] and PA are found to be less stable. The explanation of an increased stability is the increased barrier effects of nanocomposites to oxygen due to a more intensive char formation on the surface of the sample exposed to heat. Through this insulating layer the heat transport to the bulk is reduced and the mass loss during thermal decomposition is decreased [22,25]. Furthermore, the shielding effect of the clay layers could increase the activation energy of the thermal degradation of the polymer matrix [26]. Lower stabilities may be attributed to the instability of the modifier, or to hydrolytic degradation of condensation polymers due to trapped water within the nanoclay layers.

For PP with unmodified or modified clay with and without compatibiliser (polypropylene-graft-maleic anhydride (PP-g-MAH) or copolymer) it was demonstrated by TGA that under air the thermal stability of the composite with unmodified clay was a little less than the virgin PP but it is improved by using compatibilisers (Table 5.1). The thermal stability under nitrogen depends clearly on the compatibiliser used. It can be seen that PP-g-MAH makes no contribution to thermal stability, but other block copolymer structures can [23].

These short-term tests can only give an indication of the overall thermal stability and are not suitable for drawing conclusions on the processing behaviour and long-term thermal properties closer to the usage temperatures. Furthermore the potential criteria of improved thermal stability of nanocomposites (barrier formation, char formation) are not relevant to long-term properties, as the nanocomposite has to stay intact in order to be of use, and there is sufficient time for oxygen diffusion. Therefore, the typical evaluation experiments for stability during processing, i.e. multiple extrusion and for long-term heat ageing (forced air oven) are essential to the prediction of polymer nanocomposite' behaviour in real life applications.

Table 5.1
Thermal stability of PP nanocomposites
(Evaluation: TGA, heating rate at 10 °C/min)
PE: Polyethylene, PEO: polyethylene oxide, ODA: octadecylacrylate, NVP: N-Vinylpyrrolidinone, MAH: maleic anhydride

Composition	T [°C] of 10% weight loss (air)	T [°C]of 10% weight loss (N$_2$)
PP	308	428
PP + 5% Na-Montmorillonite	300	434
PP + 5% Na-Montmorillonite + 1% PE-block-PEO	319	435
PP + 5% Na-Montmorillonite + 1% Poly(ODA-grad-MAH)	328	447
PP + 5% Na-Montmorillonite + 1% Poly(ODA-grad-NVP)	330	444
PP + 7.5% PP-g-MAH	322	426
PP + 5% Na-Montmorillonite + 7.5% PP-g-MAH	337	436
PP + 5% organic modified montmorillonite + 7.5.% PP-g-MAH	338	434
PP + 5% organic modified montmorillonite + 7.5% PP-g-MAH + 1% poly(ODA-grad-MAH)	350	446

A multiple extrusion experiment was carried out on a common state-of-the-art polypropylene nanocomposite consisting of PP, maleated PP (15% Polybond® 3200, supplier: Chemtura) and nanoclay (5% Nanofil® 15, which is modified by distearyl dimethyl ammonium chloride from Südchemie, now Southern Clay). This shows a strong increase of MVR (melt volume rate), even when stabilised (0.2% blend (1:1) of AO-1 and PS-1) indicating degradation of the PP matrix because of thermal stress

(Figure 5.1). As expected the degradation can be reduced by adding a higher level of conventional stabilisers (blends of phenolic antioxidant and phosphite). However, the degradation can be almost completely prevented (shown by the lowest MVR value) even after 5 extrusion cycles using the newly developed proprietary stabiliser system AO-2, which contains only 0.15% of processing stabiliser. AO-2 addresses the specific stabilisation needs of polyolefin nanocomposites, maintains protection of the polymer under challenging processing conditions and extends the processing window. In addition, odour formation (e.g. amines from ammonium degradation) in the finished article is minimized and discoloration during processing and application is reduced. Finally, the mechanical properties of the nanocomposite are somewhat improved.

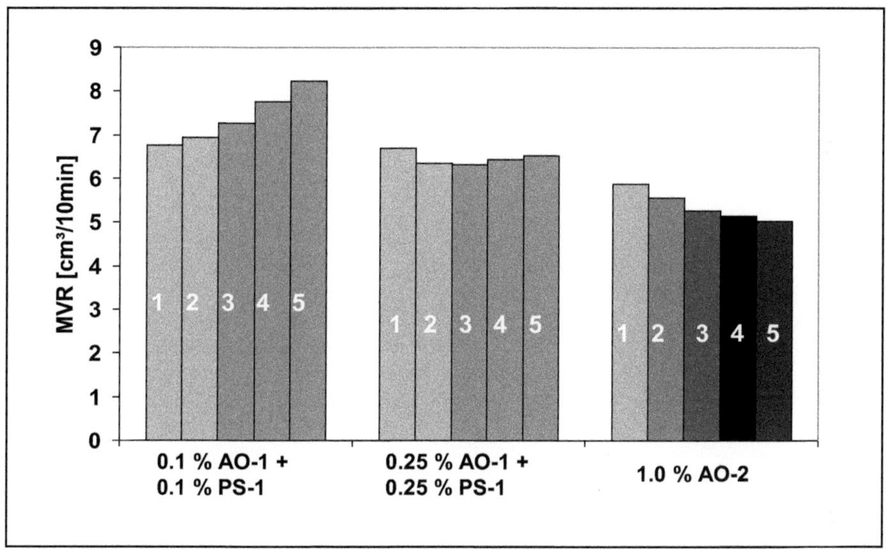

Figure 5.1

Multiple extrusion steps (from 1 to 5 steps) of stabilised PP nanocomposite (5% modified clay, 15% PP-g-MAH)

As expected from the poor processing stability of the state-of-the-art nanocomposites the impact of the nanofiller and of the modifier on the long-term thermal stability should be even more crucial.

An indication of the long-term thermal stability of stabilised nanocomposites based on layered silicates may be gained from OIT (oxygen induction time) measurements. Already it is quite obvious that conventionally stabilised nano materials are rather limited in thermal stability, e.g. PP stabilised with 0.05% phenolic antioxidant (AO-1) and 0.05% phosphite (PS-1) achieves only a lifetime of 1.8 minutes (fluorohectorite) or 2.3 minutes (montmorillonite) at 190 °C. With proper stabilisation including filler deactivators the lifetime can be prolonged to more than 90 minutes. The key factor for this improvement is the combination of phenolic antioxidants, phosphite processing stabilisers and reactive molecules of selected oxirane or dianhydride structures [27].

Although there have been a number of publications about the stability of nanocomposites in heating (as mentioned above) or combustion processes and a number of publications on photooxidation, there is practically no published knowledge on the long-term thermal properties of nanocomposites according to standard evaluation protocols such as accelerated heat ageing (oven ageing). However, these data are essential in order to assess the properties of the materials in service.

The above mentioned stabiliser system AO-2, which was shown to offer improvements in processing stability, has been used to study the performance of the system in long-term heat ageing. A conventionally stabilised PP (0.1% phenolic antioxidant + 0.1% phosphite) achieves a thermal stability until embrittlement of more than 40 days at 135 °C (Table 5.2). The thermal stability of the same stabilised polymer in the presence of 5% modified clay (distearyl dimethyl ammonium chloride) and PP-g-MAH (15%) is reduced to only 14 days, which is completely inadequate for example for automotive applications. With the addition of the additive package AO-2, and a combination of phenolic antioxidants, processing stabilisers and filler deactivators at an optimised loading, the long-term thermal stability of a nanocomposite can be decisively extended up to values equivalent to those seen in unfilled materials.

Table 5.2
Long-term thermal stability of PP nanocomposites
(5% modified clay, 15% PP-g-MAH)

Stabilisation	Days until embrittlement at 135 °C
Virgin PP (no filler and no compatibiliser) 0.1% AO-1 + 0.1% PS-1	> 40
PP nanocomposite with 0.1% AO-1 + 0.1% PS-1	14
PP nanocomposite with 0.50% AO-2	23
PP nanocomposite with 0.75% AO-2	33

5.4 Light stabilisation of polymer nanocomposites

Scientific literature shows that the degradability of nanocomposites by UV light is a serious problem [28]. Polypropylene/montmorillonite-based nanocomposites (stabilised with antioxidant) degrade much faster under photooxidative conditions than pure polypropylene [29,30,31]. This behaviour has been attributed to active species from the clay generated by photolysis or photooxidation e.g. from alkylammonium modifiers, and by interactions between antioxidant, montmorillonite and maleic anhydride modified PP.

The photooxidation of ethylene-propylene-diene rubber (EPDM) montmorillonite nanocomposites showed drastically reduced induction time for photooxidation in the presence of the nanoclay, in particular if an exfoliated structure was present [32]. Similarly EPDM nanocomposites synthesised by layered double hydroxides (LDH) showed a faster increase of carbonyl and hydroxyl group formation, indicating degradation, compared with pristine EPDM as measured by Fourier transform infrared spectroscopy (FTIR) [33].

The photooxidation of syndiotactic PP/synthetic clay nanocomposites (fluorohectorite modified by octadecylammonium) showed that the nanoparticles catalyzed the decomposition process [34]. The presence of PP-g-MAH as compatibiliser accelerated the degradation further. In natural clays iron impurities play an active role in the dramatic modification of the oxidation kinetics [35]. Similarly boehmite, modified by long-chain alkyl benzene sulfonic acid showed a photodegradation effect in isotactic PP and syndiotactic PP [36].

Photooxidation of low density polyethylene (LDPE) nanocomposites with modified clay resulted in a much shorter lifetime than the virgin polymer. Complete loss of mechanical properties occurred after just 40 h of irradiation at 8% nanoclay loading compared to 100 h for the pure polymer in the Q-UV test [37].

Photooxidation of polycarbonate and polycarbonate nanocomposites resulted in yellowing via the well-known Fries rearrangement. However, the rate was much faster in the presence of 2.5% clay. This was explained by a higher UV absorption of the dark natural clay and increased formation of radical species thereof [38].

Because of the inferior light stability of polymer nanocomposites it is obvious to evaluate the effect of light stabilisers on the stability, which is addressed in several patents and papers. As expected the influence of antioxidants on the photostability is not very pronounced. Nevertheless the data show a certain delay of photooxidation [39]. In linear low density polyethylene (LLDPE) based nanocomposites it has been clearly shown that the addition of UV absorbers of benzotriazole, benzophenone and hydroxyphenyltriazine structures decisively extended the lifetime of the nanocomposite film. However, a metal deactivator (MD-1) alone outperformed the UV absorber in these experiments, indicating that the influence of metal impurities is particularly crucial. A combination of UV absorber with metal deactivator showed only a minor additional improvement. When MD-1 is added to the LLDPE nanocomposite elongation at break and tensile strength of the tested film samples were still at 70% of their starting values after 300 hours of QU-V compared to a complete loss of mechanical properties of the unstabilised sample after 70 hours [37].

An improvement in the photostability of syndiotactic PP/synthetic organoclay nanocomposites could be shown by using a combination of phenolic antioxidant (AO-1), phosphite (PS-1) processing stabiliser and oligomeric HALS (LS-2). However, the stabilisation effect was reduced with increasing filler concentration [34]. At 5% organoclay the stability of the nanocomposite was lowered to one quarter in comparison with the unfilled syndiotactic PP at an LS-2 loading of 0.3%, and the compatibiliser (PP-g-MAH) supported the photooxidation further. The stabilization of boehmite-based PP nanocomposites with the same system (AO-1, PS-1, LS-2) did not show any improvement of photostability at higher filler concentrations; however a low molecular weight HALS was more efficient [36]. It was proven by adsorption experiments in solution that the oligomeric HALS interacts with the boehmite and, therefore, is no longer available for protecting the polymer.

Table 5.3

Light stability of polypropylene nanocomposites with different stabiliser systems containing 5% organic modified clay and 10% PP-g-MAH

Processing (extrusion/injection moulding temperatures in °C)	Stabilisation of PP nanocomposites	Tensile impact strength [kJ/m²] after 1000 hours of artificial weathering
200/230	0.25% AO-1 + 0.25% PS-1	0
200/230	0.9% AO-2	0
200/230	0.9% AO-2 +0.5% UV-1	149
200/230	0.9% AO-2 +0.5% UV-2	152
200/230	0.9% AO-2 + 0.5% LS-1/LS-2 (1:1)	128
200/230	0.9% AO-2 + 0.25% UV-2 + 0.25% LS-1/LS-2 (1:1)	136
230/250	0.25% AO-1 + 0.25% PS-1	0
230/250	0.9% AO-2	0
230/250	0.9% AO-2 +0.5% UV-1	144
230/250	0.9% AO-2 +0.5% UV-2	146
230/250	0.9% AO-2 + 0.5% LS-1/LS-2 (1:1)	125
230/250	0.9% AO-2 + 0.25% UV-2 + 0.25% LS-1/LS-2 (1:1)	134

The latter results are in line with the beneficial use of combinations of low and high molecular weight light stabilisers to balance availability and migration to the surface. In addition to polyolefins it was experimentally demonstrated that polyamide nanocomposites can be synergistically stabilised by low and high molecular weight light stabilisers (LS-1, LS-2) in combination with processing and long-term heat stabilisers. In this case the criterion was yellowing after UV exposure [40].

A much more powerful solution to achieving sufficient light stability of polymer nanocomposites is to combine the above mentioned long-term thermal stabiliser system (AO-2) containing processing, heat stabilisers and filler deactivators with light stabilisers of UV-absorber or hindered amine (HALS) types (Table 5.3). These experiments show that light stability of nanocomposites is enhanced with different classes of light stabilisers and is independent of the processing temperature. After 1000 hours of artificial weathering the starting value of the tensile impact strength is more or less unchanged for all samples containing the combination.

Extended artificial weathering testing of these stabiliser compositions reveals (Figure 5.2) that the filler deactivator containing AO-2 contributes by itself to the extended light stability of the PP nanocomposite (5% of distearyl dimethyl ammonium chloride modified clay and 10% PP-g-MAH). The combination of efficient UV absorbers (UV-2) and AO-2 decisively improves the light stability of PP nanocomposites. After 4500 hours of artificial weathering 50% of tensile impact strength is retained. A further extension of the stability, if needed, can be obtained by increasing and adjusting the loading of the individual stabiliser components.

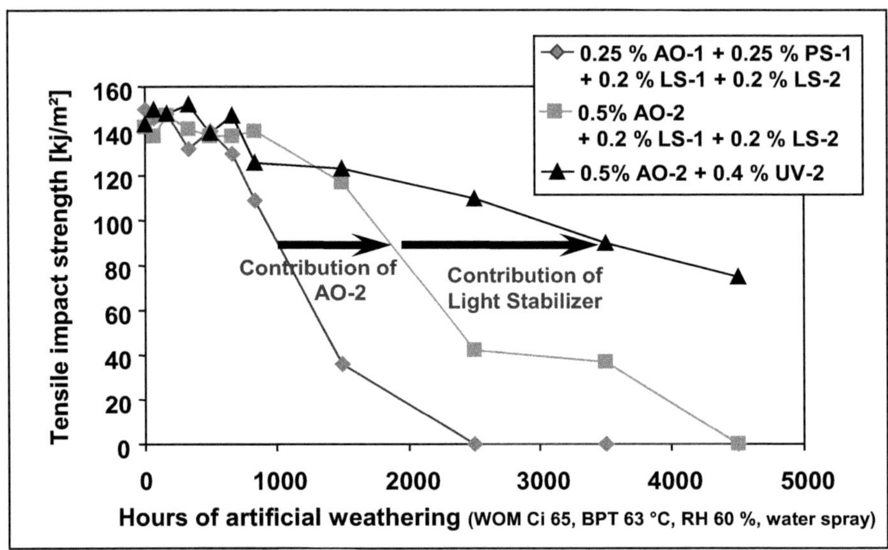

Figure 5.2

Artificial weathering of stabilised polypropylene nanocomposites containing 5% modified clay and 10% PP-g-MAH

(Evaluation: Weather-O-Meter Ci 65, black panel temperature (BPT) 63 °C, relative humidity (RH) 60 %, water spray)

An alternative stabiliser concept to using filler deactivation and adjusted stabiliser selection has been published more recently [41,42]. In this case structurally modified additives and stabilisers were proposed to improve the compatibility between stabiliser and filler. However, it has still to be shown in application tests whether this concept offers any advantage compared with present stabiliser solutions.

5.5   Summary and outlook

The stabilisation of polymer nanocomposites against oxidation and photooxidation for processing and long-term use time in selected applications is a complex target. Most published data relate to the thermal stabilities of polypropylene and other polyolefins as the polymer matrix, which represent polymers with high susceptibility to oxidation and photooxidation. The polymer matrix must be stabilised in the presence of a filler (often from natural origin) with very high surface area and in the presence of various other components such as organic filler modifiers and compatibilisers. The general polymer degradation pathway is not substantially changed by the incorporation of the nanofiller, but the degradation is usually accelerated. This trend is found for oxidation during processing and long-term heat ageing and is even more severe for UV exposure and photooxidation. Explanation for this behaviour might be found in the nature of the nanofiller itself, the weakness of the modifier/compatibilisers used and adsorption/desorption mechanisms at the nanofiller surface. The latter are responsible for the fact that stabilisers such as antioxidants or light stabilisers often show reduced performance compared to unfilled polymers. However, through careful selection of the stabiliser system in the presence of filler deactivators, and through adjustment of stabiliser composition and loading, it is possible to achieve protection of the polymer nanocomposite against oxidation and photooxidation to an extent equivalent to unfilled materials and sufficient for industrial applications.

Apart from the practical solutions now available for the stabilisation of selected polymer nanocomposites, there is still room for fundamental investigations into the behaviour of these materials in the presence of stabilisers during processing and under long-term ageing and weathering. The available data simulating use time of nanocomposites are still limited to a few polymer substrates and a few nanofiller types. Extension of the currently available data to other polymers of relevance, to various nanofillers and also to different filler dimensions under comparable conditions can be seen as a natural step to increase the stabilisation knowledge in the field of polymer nanocomposites and to achieve a general understanding of the involved mechanisms.

Acknowledgement

The author thanks Dr. Hendrik Wermter for his contributions to the developments of stabiliser systems for nanocomposites at Ciba Lampertheim GmbH.

List of stabilisers

<u>AO-1</u>: Ciba® Irganox® 1010, Pentaerythritol tetrakis(3-(3,5-di-tert-butyl-4-hydroxyphenyl)propionate

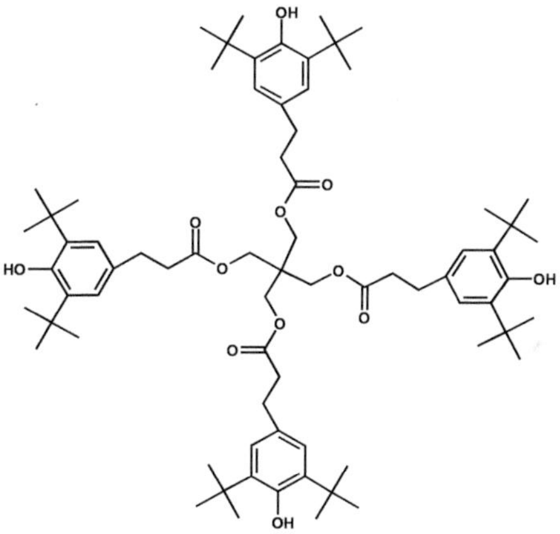

<u>AO-2</u>: Ciba® Irgatec® NC 66, proprietary composition

<u>LS-1</u>: Ciba® Tinuvin® 770, Bis(2,2,6,6-teramethyl-4-piperidyl)sebacate

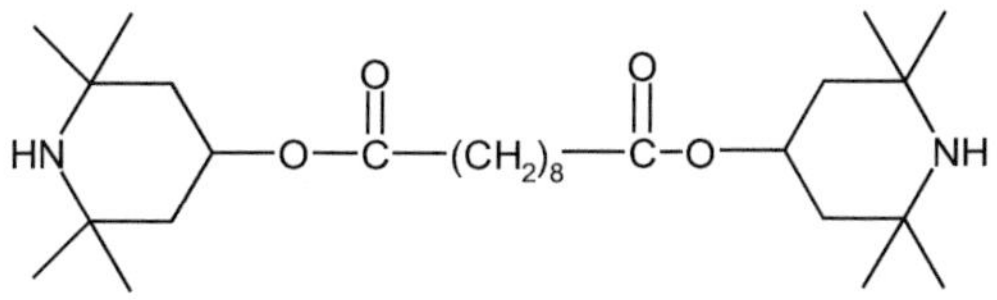

<u>LS-2</u>: Ciba® Chimassorb® 944: Poly[[6-[(1,1,3,3-tetramethylbutyl)amino]-1,3,5-triazine-2,4-diyl][(2,2,6,6-tetramethyl-4-piperidinyl)imino]-1,6-hexanediyl-[(2,2,6,6-tetramethyl- 4-piperidinyl)imino]]

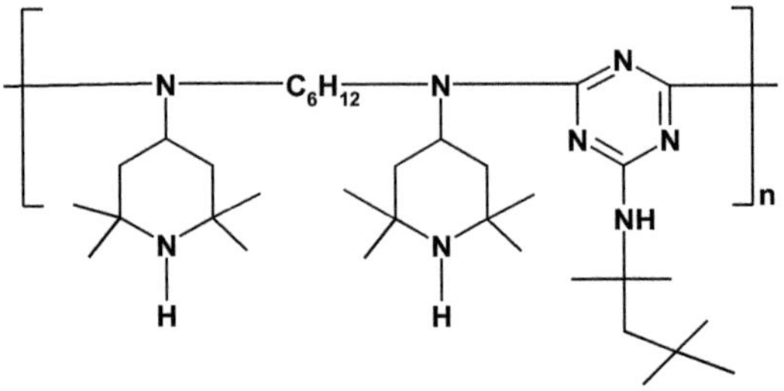

PS-1: Ciba® Irgafos® 168: Tris(2,4-ditert-butylphenyl)phosphite

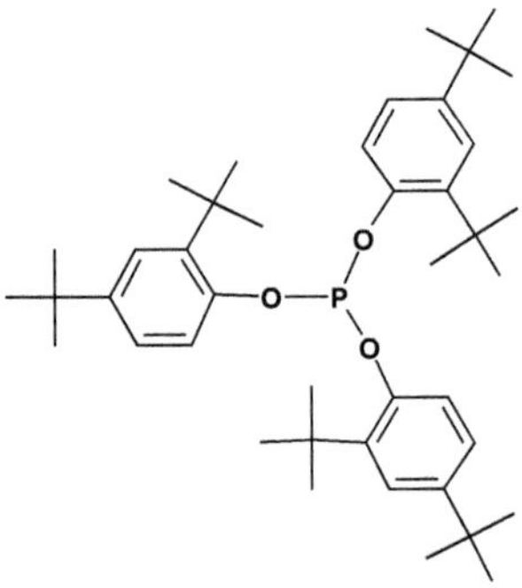

MD-1: Ciba®Irganox® MD-1024: 2',3-bis[[3-[3,5-di-tert-butyl-4-hydroxyphenyl] propionyl]] propionohydrazide

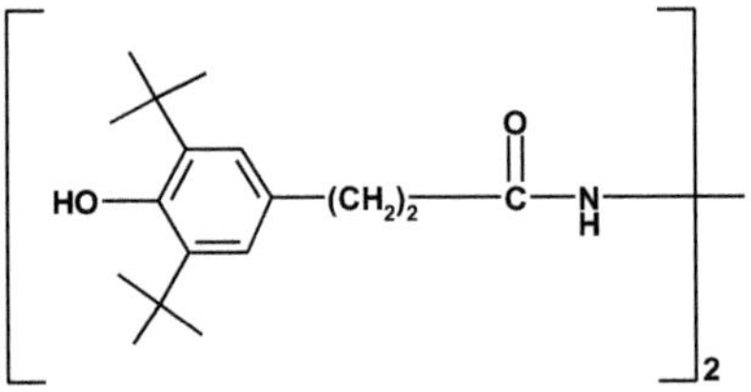

UV-1: Ciba®Tinuvin® 328: 2-(2H-benzotriazol-2-yl)-4,6-ditertpentylphenol

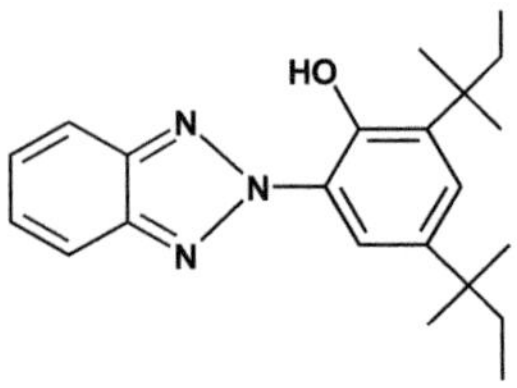

<u>UV-2</u>: Ciba®Tinuvin® 1577: 2-(4,6-Diphenyl-1,3,5-triazin-2-yl)-5-hexyloxy-phenol

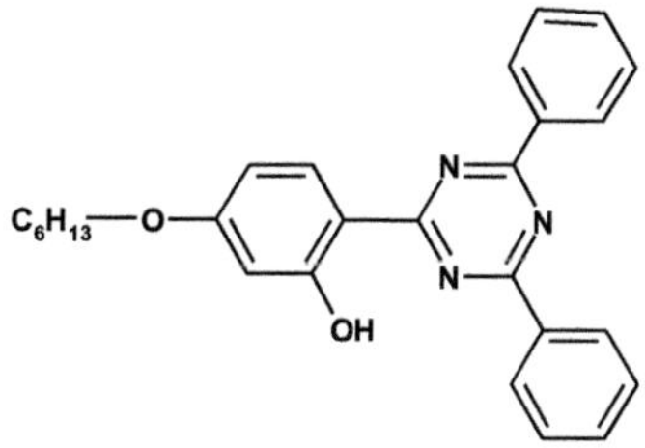

REFERENCES

1   Zweifel, H. *Plastics Additives Handbook*, 5th edition, ISBN 3-446-19579-3, Hanser Publishers, Munich, Germany, **2001**

2   Hohenberger, W. Fillers and reinforcements / coupling agents, in: *Plastics Additives Handbook*, Zweifel, H. ed., 5th edition, ISBN 3-446-19579-3, Hanser Publishers, Munich, Germany, **2001**, p. 902

3   Allen N.S.; Edge, M.; Corrales, T.; Childs, A.; Liauw, C.M.; Catalina, F.; Peinado, C.; Minihan, A.; Aldcroft, D. Ageing and stabilisation of filled polymers: an overview. *Polymer Degradation and Stability*, **1998**, 61, 183-199

4   Hu, X.; Xu, H.; Zhang, Z. Influence of fillers on the effectiveness of stabilizers. *Polymer Degradation and Stability*, **1994**, 43, 225-228

5   Rysavy, D.; Tkadleckova, H. Thermooxidative Stabilität von gefülltem Polypropylen. *Plaste und Kautschuk*, **1991**, 38, 273-275

6   Guerrica-Echevarria G.; Eguiazabal, J.I.; Nazabal, J. Effects of reprocessing conditions on the properties of unfilled and talc-filled polypropylene. *Polymer Degradation and Stability*, **1996**, 53, 1-8

7   Pan, J.; Xu H.; Qi, J.; Cen, J.; Ma, Z. Study of $CaCO_3$ filled polypropylene composite with long use life time. *Polymer Degradation and Stability*, **1991**, 33, 67-75

8   Todesco, R.; Pauquet, J.R.; Klingert, B. Stabilized filled polyolefins, US patent 5.880.186, **1999**

9   Schwarzenbach, K.; Gilg, B.; Müller, D.; Knobloch, G.; Pauquet, J.R.; Rota-Graziosi, P.; Schmitter, A.; Zingg, J.; Kramer, E. Antioxidants, in: *Plastics Additives Handbook*, Zweifel, H. ed., 5th edition, ISBN 3-446-19579-3, Hanser Publishers, Munich, Germany, **2001**, p. 57-60

10   Chinelatto M.A.; Agnelli J.A.M. Effect of the chemical stabilization system on the photooxidation of polypropylene containing talc and carbon black. *Polymer Degradation and Stability*, **1995**, 50, 13-19

11   Wypych, G. Handbook of fillers, 2nd edition, ISBN 1-895198-19-4, ChemTec Publishing, Toronto, Canada, **2000**, pages 539-557

12   Monte, S.J. Neoalkoxy titanate and zirconate coupling agent additives in thermoplastics. *Polymers and Polymer Composites*, **2002**, 10, 1-32

13   Ray S.S.; Okamoto M. Polymer/layered silicate nanocomposites: a review from preparation to processing. *Progress in Polymer Science*, **2003**, 28, 1539-1641

14   Kato M.; Usuki A.; Okada A. Synthesis of polypropylene oligomer-clay intercalation compounds. *Journal of Applied Polymer Science*, **1997**, 66, 1781-1785

15   Hasegawa N.; Kawasumi M.; Kato M.; Usuki, A.; Okada A. Preparation and mechanical properties of polypropylene-clay hybrids using maleic anhydride-modified polypropylene oligomer. *Journal of Applied Polymer Science*, **1998**, 67, 87-92

16   Hasegawa N.; Okamoto, H; Kato M.; Usuki, A. Preparation and mechanical properties of polypropylene-clay hybrids based on modified polypropylene and organophilic clay. *Journal of Applied Polymer Science*, **2000**, 78, 1918-1922

17   Borden, K.A.; Weil, R.C.; Manganaro, C.R. Optimizing properties of talc-filled polypropylene systems with a polymeric coupling agent. *ANTEC '94 conference proceedings*, Society of Plastics Engineers, **1994**, 52, 2761-2765

18   Adur, M.A.; Flynn, S.R. New coupling agent for talc-filled polypropylene. *Plastics Engineering*, **1987**, 43, 41-43

19   Pfaendner, R.; Hoffmann, K.; Meyer, F.; Rotzinger, B. Synthetic polymers comprising additive blends with enhanced effect, US patent 6.610.765, **2003**

20   Rotzinger, B. Talc-filled PP: a new concept to maintain long term heat stability. *Polymer Degradation and Stability,* **2006**, 91, 2884-2887

21   Xie, W.; Gao, Z.; Pan, W.P.; Hunter D.; Singh, A.; Vaia R. Thermal degradation chemistry of alkyl quaternary ammonium montmorillonite. *Chemistry of Materials*, **2001**, 13, 2979-2990

22   Leszczynska, A.; Njuguna J.; Pielichowski K.; Banerjee J.R. Polymer/montmorillonite nanocomposites with improved thermal properties. Part I. Factors influencing thermal stability and mechanisms of thermal stability improvement. *Thermochimica Acta,* **2007**, 453, 75-96

23   Moad, G.; Dean, K.; Edmond, L.; Kukaleva, N.; Li, G.; Mayadunne, R.T.A.; Pfaendner, R.; Schneider, A.; Simon, G.; Wermter H. Novel copolymers as dispersants/intercalants/exfoliants of polypropylene-clay nanocomposites. *Macromolecular Symposia,* **2006**, 233, 170-179

24   Gong, F.; Feng, M.; Zhao, C.; Zhang, S.; Yang, M. Thermal properties of poly(vinyl chloride)/montmorillonite nanocomposites. *Polymer Degradation and Stability,* **2004**, 84, 289-294

25   Leszczynska, A.; Njuguna J.; Pielichowski K.; Banerjee J.R. Polymer/montmorillonite nanocomposites with improved thermal properties. Part II. Thermal stability of montmorillonite nanocomposites based on different polymeric matrixes. *Thermochimica Acta,* **2007**, 454, 1-22

26   Qiu, L.; Wei C.; Qu B. Morphology and thermal stabilization mechanism of LLDPE/MMT and LLDPE/LDH nanocomposites. *Polymer,* **2006**, 47, 922-930

27   Wermter H.; Pfaendner R. Stabilization of thermoplastic nanocomposites. EP patent application 1592741, **2004**

28   Pandey J.K.; Reddy, K.R.; Kumar, A.P.; Singh, R.P. An overview on the degradability of polymer nanocomposites, *Polymer Degradation and Stability,* **2005**, 88, 234-250

29   Morlat-Therias S.; Mailhot, B.; Gonzalez, D.; Gardette J.L. Photooxidation of polypropylene/montmorillonite nanocomposites. 2. Interaction with antioxidants. *Chemistry of Materials,* **2005**, 17, 1072-1078

30   Qin, H.; Zhao, C.; Zhang, C.; Chen, G.; Yang, M. Photo-oxidative degradation of polyethylene/montmorillonite nanocomposite. *Polymer Degradation and Stability,* **2003**, 81, 497-500

31   Tidjani A.; Wilkie C.A. Photo-oxidation of polymeric-inorganic nanocomposites: chemical, thermal stability and fire retardancy investigations. *Polymer Degradation and Stability,* **2001**, 74, 33-37

32 Morlat-Therias S.; Mailhot, B.; Gardette, J.L.; Da Silva C.; Haidar B.; Vidal A. Photooxidation of ethylene-propylene-diene/montmorillonite nanocomposites. *Polymer Degradation and Stability,* **2005**, 90, 78-85

33 Kumar, B.; Rana, S.; Singh, R.P. Photo-oxidation of EPDM/layered double hydroxides composites: Influence of layered hydroxides and stabilizers. *eXPRESS Polymer Letters,* **2007**, 1, 748-754

34 Chmela, S.; Kleinova, A.; Fiedlerova, A.; Borsig, E.; Kaempfer D.; Thomann, R; Mülhaupt R. Photo-oxidation of sPP/organoclay nanocomposites. *Journal of Macromolecular Science. Part A: Pure and Appl. Chem.,* **2005**, 42, 821-829

35 Morlat S.; Mailhot, B.; Gonzalez D.; Gardette, J.L. Photo-oxidation of polypropylene/montmorillonite nanocomposites. 1. Influence of nanoclay and compatibilizing agent. *Chemistry of Materials,* **2004**, 16, 377-383

36 Chmela, S.; Fiedlerova, A.; Borsig, E.; Erler, J.; Mülhaupt R. Photo-oxidation and stabilization of sPP and iPP/boehmite-disperal nanocomposites. *Journal of Macromolecular Science. Part A: Pure and Appl. Chem.,* **2007**, 44, 1027-1034

37 La Mantia, F.P.; Tzankova Dintcheva N.; Malatesta, V.; Pagani, F. Improvement of photo-stability of LLDPE-based nanocomposites. *Polymer Degradation and Stability,* **2006**, 91, 3208-3213

38 Sloan J.M.; Patterson P.; Hsieh A. Mechanisms of photo degradation for layered silicate-polycarbonate nanocomposites. *Polymeric Materials Science and Engineering,* **2003**, 88, 354-355

39 Mailhot B.; Morlat S.; Gardette J.L.; Boucard S.; Duchet J.; Gerard J.F. Photodegradation of polypropylene nanocomposites. *Polymer Degradation and Stability,* **2003**, 82, 163-167

40 Chin H.; Solera, P.S.; Horsey, D.W.; Kaprinidis, N.; Sitzmann, E.V. Synergistic combinations of nano-scalled fillers and hindered amine light stabilizers. US patent 7.084.197, **2006**

41 Camenzind H.; Herbst, H.; Wunderlich W. Additive functionalized organophilic nano-scalled fillers. European patent application EP 1401948, **2003**

42 Giesenberg, T.; Mühlebach A.; Jung, T.; Rime, F.; Michau, L.; Müller, M.; Bauer D. Functionalized nanoparticles. European patent application EP 1805255, **2006**

Chapter 6
Toxicology of nanoparticles relevant to nanocomposites

Paul Borm

Centre of Expertise Life Sciences (CEL), Hogeschool Zuyd, Heerlen, Netherlands

6.1   Introduction

Nanoscience and nanotechnologies are expected to change industrial production and economics over the next decades. Moreover, nanotechnology and nanosciences are exciting since they sweep away the traditional barriers between mono-disciplines such as chemistry, physics and biology. Nanotechnology requires different thinking in management, collaboration, value chain propositions, education and research funding. Apart from benefits and challenges nanotechnologies also produce uncertainties and risks. Nanomaterials are introduced in the market on their claimed benefits and their chemical identity is pegged to already existing legislation regardless of some of their unique characteristics. Data are now emerging on the distribution of some nanomaterials and their toxicity, and need careful interpretation before generalizing findings to other nanomaterials. Regulation is unable to keep up with the pace of development, and treats nanoproducts as a slippery customer. Consumers are uncertain how they feel about nanotechnology, for example whether the products' claims by some proponents represent objective reality, and even if a product claiming to be associated with nanotechnology is legitimately a nanoproduct.

Much of the discussion of nanotechnology is driven by the potential hazards of nanoparticles, which are only part of the toolbox of nanoscientists. Nanoparticles have been shown by many studies to have a greater toxicological potential compared to the same chemical materials when used in a larger (micrometre) form [1,2]. Changing the material properties by reducing their size obviously also has consequences for uptake, distribution and adverse effects in the human body.

Nanoparticles, especially **organoclays, metal oxides** and **carbon nanotubes,** are important components in the construction of nanocomposites. Consequently humans and the environment may be exposed to these species when airborne in the workplace, and after spread into the environment resulting from use and wear of the materials. Data on the distribution of these nanomaterials and their toxicity are now emerging and need careful interpretation before generalizing findings to other nanomaterials. It is the purpose of this chapter to put the nanoparticle discussion into the general context of particle toxicology. Subsequently specific toxicity issue of organoclays and carbon nanotubes will be discussed. Finally, the need for unifying concepts will be discussed.

6.2   Toxicological effects of nanoparticles

6.2.1   Particle definitions

Particle toxicology has developed along the lines of industrial development and has experience with particles from micrometre size to particles down to the nanosize range [3]. This evolution from classical particles such as generated in mining and causing silicosis, through ambient particles (PM_{10} also known as particulate matter smaller than 10 micrometres) to more recent use of ultrafine and nanoparticles is shown graphically in Figure 6.1. Decades of human exposure to nanoparticles in the form of diesel soot and bulk nanoparticles from industrial processes have preceded the current emerging wave of engineered nanoparticles. Toxicological research has produced a database that is quite extensive for certain nanoparticles, such as diesel soot and carbon black, but less so for others such as fumed silica and alumina. The similarity and differences between these 'traditional' nanoparticles and the new engineered nanoparticles is important since we have some concept of the hazard that these provide and so they may act as 'benchmarks'. This database on combustion-derived nanoparticles (previously ultrafine particles) demonstrates that exposure to these particles is associated with pulmonary inflammation [4], immune adjuvant effects [5] and adverse systemic effects including blood coagulation and cardiovascular (cv) effects [6,7]. The meeting of the worlds of nanoscience, with its engineered nanoparticles, and that of toxicology, with its know-how of ultrafine particles, has led to an impressive series of workshops and meetings over the past several years. However, at the time of writing in 2008 little exchange of methods and concepts has occurred although both disciplines are expected to benefit from an exchange of methods and know-how. The field of nanocomposites is no exception to this observation, and this book chapter is therefore probably the first effort to describe the potential risks of these nanomaterials.

Until 2002 nanoparticles were referred to by the toxicology community as 'ultrafine particles' (UFP) which were defined as the fraction of ambient particulate matter which is smaller than 100 nm. Nowadays the use of the term **nanoparticles** (sometimes abbreviated to NP) is more frequent, and in this chapter we use the term nanoparticles for particles being intentionally produced with the same cut-off size of 100 nm. Nanoparticles therefore include both bulk products such as carbon blacks and TiO_2, but also particles that are engineered and produced in low quantities such as quantum dots and polymer carriers for imaging and drug delivery. For clarity, in Table 6.1 a brief summary is given of the above definitions in toxicology.

Table 6.1			
Definitions of toxicologically relevant particles with source/application and most important effects			
Particle Type	**Examples**	**Who is exposed?**	**Effects**
Crystalline silica	Coal mine dust, quartz flours, Kieselguhr	Miners	Silicosis, lung cancer
Asbestos	Crocidolite, chrysotile	Insulators, shipyard workers	Asbestosis, mesothelioma, lung cancer
Ambient particulate matter (PM)	Fly-ashes, diesel, sea salt, road dust	Everyone	Increase mortality from cardiovascular and respiratory causes
Man made vitreous fibres (MMVF)	Rockwool, ceramics	Occupational	Lung fibrosis and cancer
Nanoparticles	**Products**	**Application**	**Hazards**
Combustion-derived (not intentional)	Diesel soot	Everyone	Pulmonary and CV effects in humans
Various engineered	Carbon black	Occupational	Fibrosis and lung cancer in rats
	Amorphous silica	Occupational	Fibrosis in rats
	alumina	Occupational	Lung fibrosis in man
Metal Oxides	TiO_2	Cosmetics	Ongoing discussion
	ZnO	Cosmetics	Ongoing discussion
	Fe_xO_y	Medical imaging	Generally safe
	CeO	Catalysis	Discussion started
Carbon Nanotubes	MWCNT	Optical	Lung granumolas, Mesothelioma-mice.
	SWCNT	Electrical	Lung granulomas-rats mice

6.2.2   Effects of nanoparticles upon inhalation

The largest database on the potential toxicity of nanoparticles has originated from the literature on epidemiological and clinical effects of ultrafine particles in the ambient air. From this database the 'nanoparticle-hypothesis' has been proven to be a powerful force for research. Therefore we consider it relevant to discuss this evidence in the expectation that it will shed light on the toxicity of engineered nanoparticles. The idea that combustion-derived nanoparticles are an important component that drives the adverse effects of environmental particulate air pollution or PM_{10} comes from several sources:

1) Much of the mass of in ambient particulate matter (PM_{10}) is considered to be non-toxic and so there has arisen the idea that there is a component(s) of PM_{10} that actually drives the pro-inflammatory effects and combustion derived nanoparticles (CDNP) seem a likely candidate.

2) CDNP are the dominant particle type by number suggesting that they may be important and their small size means that they have a large surface area per unit mass. Particle toxicology suggests that, in general, more particle surface equals more toxicity.

3) Substantial toxicological data and limited data from epidemiological sources support the contention that nanoparticles in ambient particulate matter are important drivers of adverse effects.

Toxicological studies with inhaled nanoparticles have suggested several major mechanisms by which this CDNP component of PM may cause responses that explain mortality in those with existing pulmonary and cardiovascular diseases [6,8,9].

1) First, inhalation exposure to nanoparticles can invoke broncho-alveolar inflammation in susceptible individuals and exacerbate existing airways disease such as chronic obstructive pulmonary diseases (COPD) leading to death.

2) Broncho-alveolar inflammation may also cause the release of inflammatory mediators that can trigger systemic hypercoagulability of the blood, thereby increasing the risk for cardiovascular events such as those observed when ambient particle levels increase.

3) A second mechanism is the progression and destabilization of atheromatous plaques by inhalation of ambient particles. Nanoparticles properties are likely to promote plaque destabilization through their documented effects such as inflammation, and oxidative stress causing low-density lipoprotein (LDL) oxidation by lipid peroxidation.

4)   A third mechanism is provided by recent observations that upon inhalation nanoparticles deposit to a large extent in the nose, and may pass through the olfactory epithelium in the upper nose and reach the brain. Little is known about the effects of the uptake, but one could imagine disturbance in the central nervous system (CNS) control of the autonomic nervous system as a possible consequence.

5)   Finally, nanoparticles may translocate from various deposition sites (lung, gastro-intestinal (GI) tract, skin) to the systemic circulation, and distribute to secondary organs (vascular wall, heart, liver, spleen) directly affecting organ function. However, quantitative estimates of translocation differ too much to be reliable at the moment.

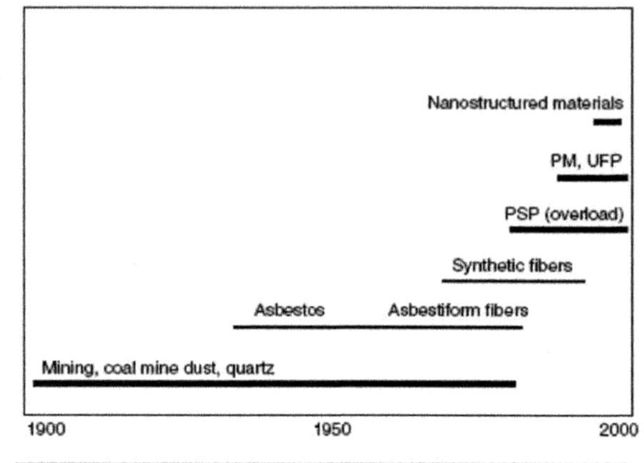

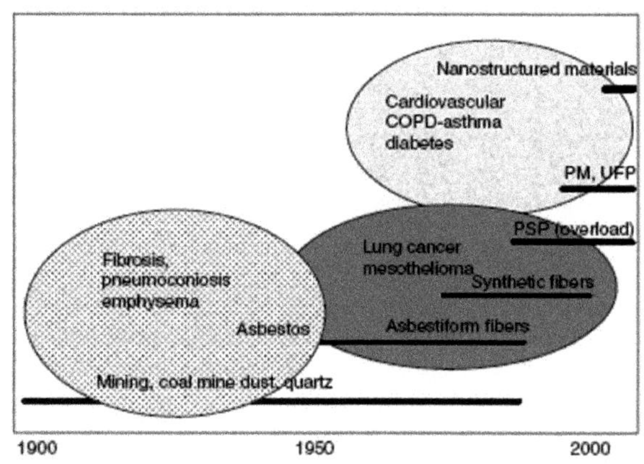

Figure 6.1

Historical development of particle toxicology along with technologies and (major) toxicological products emerging from these technologies
Panel A: time frame of particles driving particle toxicology
Panel B: particles along with the major outcomes that were studied
(PM = particulate matter, UFP = ultrafine particles, PSP = poorly soluble particles)
Reprinted with permission from Donaldson, K. and Borm, P.J.A., Particle Toxicology, Copyright 2007, Informa Healthcare, USA

6.3   Nanoparticles used in nanocomposites

Nanocomposites have been used commercially since Toyota introduced the first polymer/clay auto parts in the 1980s. Recently, advances in the ability to characterize, produce and manipulate nanometre-scale materials have led to their increased use as fillers in new types of nanocomposites. In addition it is important to recognize that nanoparticulate/fibrous loading confers significant property improvements with very low loading levels, in comparison with traditional microparticle additives requiring much higher loading levels to achieve similar performance. This in turn can result in significant weight reductions for similar performance (of obvious importance for various military and aerospace applications), greater strength for similar structural dimensions and increased barrier performance for similar material thickness.

Nanocomposites are introduced in the market on their claimed benefits and their chemical identity is pegged to already existing legislation regardless of some of their unique characteristics. Global consumption of nanocomposites has increased rapidly, exceeding 10 ktonnes in 2005, with an estimated value of $ 252 million. By 2011, it is expected to reach almost 43 ktonnes, worth some $857 million. Manufacturers now mix nanoparticulate metals, oxides and additional materials with polymers and other matrix materials to optimize the properties of composites with respect to colour and transparency, conductivity, flame retardancy, barrier properties, magnetic properties and anticorrosive properties, in addition to tensile strength, modulus and heat distortion temperature. These composites offer users significantly enhanced properties compared to conventional composite and non-composite materials [10].

Clay nanocomposites accounted for nearly one quarter (24%) of total nanocomposite consumption by value in 2005, followed by metal and metal oxide nanocomposites (19%) and carbon nanotube composites (15%). By 2011, clay nanocomposites are projected to increase their market share to 44%. Between 2005 and 2011, other market share gainers include metal/metal oxide nanocomposites and ceramic nanocomposites which are projected to attain market shares of 20% and 11.5% respectively [10]. Montmorillonite, saponite and synthetic mica have been successfully exploited for their improved stiffness, strength, toughness and stability in automotive applications of nanocomposites. One of the amazing feats of using nanocomposites in automotive industry is shown by the General Motors which produces 245 tons/year of montmorillonite-reinforced nylon-6 for use as engine covers [10]. Montmorillonite is an aluminosilicate comprising one central aluminium layer between two silicate layers which exfoliates at 1 nm thick platelets and a weight reduction of 7% is achieved. The above example represents the acceptance of nanotechnology in the automotive industry. The aerospace, defence and biomedical industries are all potential beneficiaries of these materials.

6.3.1 Carbon nanotubes

Carbon nanotubes (CNT) are comprised entirely of carbon. They can be viewed as being formed by rolling a single sheet of graphite into a seamless cylinder. There are two classes of carbon nanotubes: single-walled (SWCNT) and multi-walled (MWCNT). Multi-walled nanotubes are larger and consist of many single-walled tubes stacked one inside the other. Single-walled tubes can be as much as 10 times as strong as steel and 1.2 times as stiff as diamond, so they are used or investigated for many applications in industry such as reinforcing materials. They have a needle-like shape and are commonly a few nanometres wide but can be several millimetres in length. Therefore they have some commonality with asbestos and other fibres used in industry. The harmful effects of fibres are driven by three properties: by length, thinness and biopersistence (i.e. resistance to breakdown in the lungs). A typical long, thin and biopersistent fibre that causes lung disease is amphibole asbestos. The experience with asbestos-related lung disease suggests that MWCNTs should be treated as a special case, and that their toxicity testing and also their regulations should be benchmarked against asbestos and combustion nanoparticles.

Of course the risk of working with nanotubes is the end-product of the hazard and the exposure. So far only one study [11] has been published in which a laboratory environment was used to investigate the physical nature of the aerosol formed during handling of single-wall carbon nanotubes (SWCNT). The authors found that sufficient agitation of unrefined SWCNT released fine particles, but the concentrations generated were very low (< 53 $\mu g/m^3$). The laboratory study also demonstrated that different production methods for CNT produce different aerosols. However, other unpublished works suggest that the surface modification of the CNT is crucial in the generation of nanoparticles in the air. In addition, some agglomerates of CNT may break up into single particles.

In spite of the persistent formation of agglomerate, clusters and ropes in aqueous suspensions, now several studies have been done with CNT instilled into the lung of rodents [12-16]. These studies showed that SWCNT cause similar classical lesions in the lung as crystalline silica, including granuloma, inflammation and interstitial fibrosis. Recent work by the laboratory of Ken Donaldson in Edinburgh (UK) suggest that a dose of 50 µg of rigid, long MWCNT nanotubes with length exceeding 15 µm can cause similar lesions in the peritoneum of mice as those caused by long amosite asbestos [17] and sets the stage for the future discussion on CNT as a class of agents that may be harmful upon workplace exposure. A concomitant publication by Takagi et al [18] in a more sensitive mouse model has shown that the same type of MWCNT is able to cause tumours of the mesothelium in the peritoneum after intraperitoneal injection of 3 mg per mouse of MWCNT. These findings have caused quite a stir among users and producers of carbon nanotubes and more research is needed with appropriate inhalation models to confirm these findings. In the meantime it is recommended to refrain from applications with long (> 15 µm) MWCNT.

Just as in the case of asbestos, the mechanical handling of composites containing CNT may cause exposure to polymer-embedded CNT, which may become active after degradation of the polymer in the body. Although it seems that normal handling operations may not generate high concentrations of CNT at the workplace, the degree of CNT toxicity is still uncertain and workers should therefore be systematically screened for health effects. Interesting indicators to screen include lung inflammation markers, allergy test outcomes or cardiovascular effects (lung, skin, heart and potentially mucosa). This type of study should extend over long time periods in order to pick up long term effects and include type, level and duration of CNT exposure. Toxicological studies reviewed suggest that CNT will produce a toxic response in a time and dose dependent manner once they reach the lungs in sufficient quantity.

The studies suggest that CNT are in general more toxic than carbon black and quartz, and therefore Lam et al [19] have recommended that the occupational exposure limit for CNT should be no greater than 0.1 mg/m^3, the same as the U.S. Occupational Safety & Health Administration's permissible exposure limit (OSHA PEL) for quartz. However, considering the unique shape and aspect ratio of CNT, it remains questionable whether a mass value based exposure limit will be sufficient.

At this time there are approximately 50-60 studies published on the effects of CNTs on human health. A few of those were discussed above and are summarized in Table 6.2. However the fate of CNT in environmental compartments such as water and soil has been investigated poorly. This is surprising since CNTs have been advocated for use in environmental remediation and for bactericidal application. The fate of CNTs in environmental compartments may differ depending on their specific properties, surface chemistry and oxidative potential. Studies have shown that the state of aggregation and surface function are likely to influence the behaviour in water and soil. With regard to applications in nanocomposites it would be useful to characterize the CNT types used, and how they may be released into the environment during the product life cycle. Several studies have shown that once CNTs are released into the environment, they can be biopersistent and bioavailable to organisms. One can therefore assume that CNTs may accumulate along the food chain, and there is a need for long-term studies on ecotoxicity and bioaccumulation [20].

Precautionary measures are therefore also recommended in other stages of the CNT life cycle. Exposure may be reduced in occupational settings through redesigning the production processes or changes in the handling and transportation procedures. During product design it is important to consider whether the product function can be obtained through alternatives to CNT which are less uncertain (i.e. more is known about their capacity to cause harm, their novel effects, persistency and mobility). If not, can CNT be integrated into the product in a way that makes it possible to control the size and number of CNT containing particles released into the surroundings due to wear, tear and corrosion .

Table 6.2
Important findings on the biological activity of carbon nanotubes (CNT) in the lung using in vivo and in vitro models

Exposure + model	material	outcome	Reference
Intratracheal instillation, guinea pigs (12.5 mg)	NanoLab CNT	Granuloma,Fibrosis (lung)	Huczko, 2005 [16]
Intratracheal instillation, (0.25 and 1.25 mg/rat)	SWCNT	Inflammation, Multiple granuloma	Warheit et al, 2004 [14]
Intratracheal instillation, mice (0.1, 0.5 mg/mouse)	SWCNT	Granuloma, Inflammation> carbon black	Lam et al, 2004 [15]
Intratracheal instillation, rats (0.5- 5 mg/rat)	MWCNT	Inflammation, Fibrosis	Muller et al, 2005 [13]
Pharyngeal aspiration (10- 40 ug/mouse)	SWCNT	Progressive fibrosis, Granulomas	Shvedova et al, 2005 [12]
Intraperitoneal injection, mice	SWCNT, MWCNT, asbestos	Granuloma formation with long MWCNT and long fibrous asbestos	Poland et al, 2008 [17]
Intraperitoneal injection, mice	MWCNT, asbestos	Mesothelioma with MWCNT and asbestos	Takagi et al, 2008 [18]

6.3.2 Metal oxide particles

Metal oxides are probably the currently mostly applied nanoparticle in applications such as pigments, coatings, paints, curing agents and thermosensitive glues. Among these are titanium dioxide, iron oxides, zinc oxide, cerium oxide and alumina oxide. Typically, a number of these oxides at the micrometre level have always been considered as inert or so-called "nuisance dust" in occupational exposure limits. In such a way TiO_2, iron oxides and alumina oxides have been used as negative control particles in many classical particle studies. Even at the nano-range, iron oxides and alumina oxides are still considered as relatively non-toxic. For titanium dioxide (TiO_2) this perception has changed considerable for several reasons:

1. The original experiments elucidating the role of nanoparticles in lung inflammation were done by comparing micro- and nanosized TiO_2. The data showed that nanosized TiO_2 caused lung inflammation at much lower gravimetric levels then microsized TiO_2.

2. Some forms of TiO2, especially anatase, have a photocatalytic effect, which means that they can form reactive molecules when exposed to UV-light.

Due to its inertness and ability to trap UV-light, TiO_2 nanoparticles (and also ZnO) are used increasingly in many different products, especially sun creams where they are receiving the most attention. Uptake of nanoparticles from these products through the skin would cause a considerable burden of nanoparticles in the body and it is therefore critical whether the nanoparticles remain on the skin or can reach systemic circulation and target organs. While there is little evidence that nanoparticles in sun creams cross the stratum corneum into the dermis and from there migrate elsewhere, there remains concern about entry via damaged skin [21,22]. In sunscreens nanosized TiO_2 particles are often used as nanomirrors to partly reflect the sunlight from the skin. To reduce potential photo-catalytic adverse effects, the titanium dioxide is often coated. Surface modified TiO_2 has been the subject of considerable toxicological investigation and has shown that the hydrophobic coatings usually tend to lower the inflammatory response after inhalation or instillation. TiO_2 and other metal oxide nanoparticles are also used in many different nanocomposites with the potential for release from these products into the environments, and the effects on bacteria, fish and aquatic environments are the subject of investigation [23].

Iron oxide nanoparticles are commercially available in all sizes between 5 and 5000 nm, and are used for applications related to their red colour (in pigments), their (para)magnetic properties (magnetic trapping, and also in heat-activated 'reversible' glues) and imaging of tumours due to the increased uptake of these particles. Iron oxide may be safe in most applications, and it is even used in medical preparation for imaging and drug delivery. However, as stated before, it is usually the surface coating which determines potential toxicity and iron oxide nanoparticles are no exception to this rule. Upon inhalation or intratracheal injection iron oxide particles show the usual lesions induced by nuisance dusts, such as lung fibrosis at high dose, or effects associated with iron overload. Iron overload is caused by overwhelming the body's storage capacity for iron, which is determined by a number of proteins such as transferrine. Iron overload can lead to severe metabolic disorders.

Cerium oxide nanoparticles are being added to diesel as a catalyst for the reduction of toxic exhaust emission gases and particulate emissions in diesel vehicles. This product, Envirox™, is a fuel-borne cerium oxide catalyst supplied by Oxonica. The cerium oxide acts as a chemically active component, removing oxygen by interaction with oxidizing species, and releasing it to promote more efficient combustion. It has been claimed that Envirox™ can produce fuel savings of up to 10% and the product has been successfully trialled in buses in Hong Kong, and adopted by one of the largest bus operators in the UK. However, it is not yet clear to what extent the emission of cerium oxide will influence the current ambient exposure to nanoparticles and its potential hazards. The first toxicological screening of the micron- and nanosized cerium oxide was recently published [24] and shows little effect in *in vitro* toxicity tests. However a recent paper suggests the indication of oxidative stress in pulmonary target cells [25]. Whilst unrelated to polymeric nanocomposites, this is nonetheless interesting as a typical example of an existing engineered nanoparticle demonstrating commercial success in a new application. The result is a totally different exposed population and means of exposure, and more toxicological work in different models is needed to fully estimate potential toxicity.

6.3.3   Silica and organoclays

To fully understand the toxicology of organoclays one needs to appreciate that these are organically modified clays, typically consisting of montmorillonite, which is modified with quaternary amines. This means there are an organic component (usually at the surface) and an inorganic silica carrier with potential contamination of crystalline silica, such as quartz or cristobalite. Therefore, its potential toxicology is driven by a number of components:

        1)   the absorbed organic components such as quaternary amines
        2)   the mineral component or contaminant and surface area
        3)   the ability to release absorbed components into biological tissue.

Although it is beyond the scope of this chapter to discuss all the details of the above components, a few words are necessary on each. The mineral montmorillonite (Al_2O_3•$4SiO_2$•xH_2O) is one of the major components of bentonite and Fuller's earth. Like other secondary silicates (kaolin, mica) raw products are usually contaminated with talc and other crystalline **silica**. The latter is a well-known toxic particle known to induce lung fibrosis and lung cancer in those with occupational exposure to this material [26]. In keeping with current views of the effects of insoluble mineral particles on the lung, it is the surface of the quartz that is of prime importance in determining biological effects, since this surface makes contact with biological molecules and cell surfaces. The biological effects of silica are likely to be related to the surface reactivity of the particles, since the surface makes contact with body fluid and cells. There are several ways in which the quartz surface can generate reactive species in its vicinity when particles are present in the lung. Silanol groups (Si-OH) and ionised silanol groups (Si-O–) on the surface are considered to play a major role in interaction with membranes [27]. Common contaminating metals such as iron and aluminium may ameliorate the toxicity of quartz [27,28] but Fenton chemistry-derived hydroxyl radical may also be generated, adding to the oxidative stress. These concepts are illustrated in the graph in Figure 6.2, and may also be applied to nanoparticles.

Bentonites are frequently used in cases of poisoning after swallowing toxic chemicals because of their ability to adsorb organic species and metals. In a recent study [29] the cellular uptake and the cytotoxic potential of respirable bentonite particles of varying quartz content and activations (alkaline, acidic, organic) were determined at a range of concentrations and exposure times in a human lung fibroblasts cell line (IMR90) The red blood cell haemolysis assay gave additional information about the cytotoxic effects of bentonites on cellular membranes. Additionally, the authors analyzed the ability of bentonites to induce apoptosis in IMR90-cells. The bentonite samples were provided by Süd-Chemie (AG Moosburg) and were industrially treated with Na_2CO_3 (3–5% bentonite sample 1403, 5% bentonite sample 1400, 3% bentonite sample 1401), with HCl and an organoclay with distearyl dimethyl ammonium chloride. A summary of the outcome of toxicity tests is shown in Table 6.3 demonstrating that the organoclay was more toxic then the alkaline- or acid-activated bentonites and even more toxic then the positive control, crystalline quartz. On the other hand bentonites showed a high dose-dependent haemolytic potential at concentrations from 1.25 to 5 mg/ml. At the highest tested concentration of 5 mg/ml, two thirds of all tested bentonite samples caused lysis of red blood cells in a range of 88% to 95%. These effects were independent of the activation and the quartz content. In comparison with other mineral dusts (DQ12, chrysotile), only chrysotile showed similar results. Induction of haemolysis by bentonites was up to 2.1-fold higher than by quartz and 190-fold higher compared to gypsum. This time the organoclay sample showed the lowest hemolysis (< 20%). Together these data show that the toxicity of bentonites varies both with the silica content and the activation of the sample. It remains to be investigated what effects other organic coatings have on bentonites and which test is more relevant for in vivo toxicity. Recent studies with quaternary ammonium compound-coated montmorillonite showed inhibition of fibroblast growth, whereas amino undecanoic acid (AUA) nanoparticles showed no toxicity [30].

Table 6.3				
Outcome of in vitro toxicity tests with different types of industrial bentonites				
Material description & activation	Silica (%)	Fe (%)	Toxicity in lung fibroblasts[a]	Haemolysis in red blood cells[b]
Distearyl dimethyl ammonium-Chloride	1-2	1.7	49 ± 2	7.6
Alkaline: Na_2CO_3(3–5%)	0.5	3.2		ND
Alkaline: Na_2CO_3(5%)	5-6	3.7		73.2
Acidic: HCl	1	0.64	93 ± 56	64.7
Acidic: HCl	4-5	1.8		23.1
Untreated	1.0	1.1	78 ± 12	72.3
Untreated	5-6	3.8		43.8
Controls				
Quartz	95		63 ± 25	15.0
Gypsum	0			ND
[a] Percent viable cells after incubation of human lung fibroblast with 50 µg/cm² particles				
[b] Percent of damaged blood cells after incubation with 1.25 mg/ml in blood during 1 hour				
Data reproduced with permission from reference 29, Springer Berlin, Geh, S.; Yücel, R.; Duffin, R.; Albrecht, C.; Borm, P. J. A.; Armbruster, L.; Raulf-Heimsoth, M.; Brüning, T.; Hoffmann, E.; Rettenmeier, A. W.; Dopp, E. Cellular uptake and cytotoxic potential of respirable bentonite particles with different quartz contents and chemical modifications in human lung fibroblasts. *Archives of Toxicology,* **2006**, 80, 98-106				

Apart from concentrating on the particles themselves it is relevant to know that tertiary or **quaternary ammonium compounds** being added to the surface of these clays have a promoting ability towards allergy and asthma. In fact many chlorinated ammonium compounds are used in the cleaning industry to facilitate dust and dirt removal, for disinfection and surface maintenance. The use of surfactants like quaternary ammonium compounds has been associated repeatedly with airway irritation and as an adjuvant in sensitization to common allergens [31]. It should therefore be realized that nanoparticles coated with such agents will potentially have these effects on different targets in the body, including skin and the respiratory tract [32,33]. For the moment the common knowledge is that organoclays are no nanoparticles in dry form, but only when they become incorporated into a polymer matrix they will become intercalated and therefore dispersed into a nanodispersion.

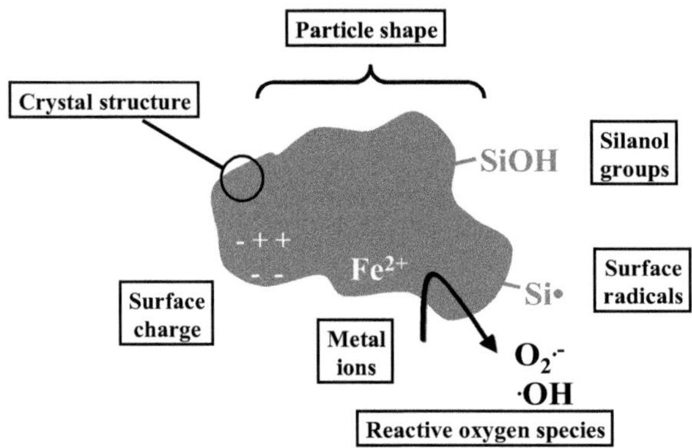

Figure 6.2
Illustration depicting the different properties that play a role in the surface reactivity of a (nano)particle, using a crystalline silica particle as an example
Reprinted with permission from Donaldson, K. and Borm, P.J.A., Particle Toxicology, Copyright 2007, Informa Healthcare, USA

6.4   Need for unifying concepts

A brief look into the near future of generating hundreds of new nanomaterials and their coatings, immediately raises the question how so many different chemical and physical entities (length, soluble toxins of different types, different surfaces) could be tested. The alternative is to develop some common form of 'harmful dose' to the machinery of the cell, based on the understanding that the ability to deliver oxidative stress is a common property of harmful particles. This oxidative stress can emanate from the particle itself and can also be a consequence of the cellular and inflammatory response to the particle [34].

The present data indicate that apart from particle size, the particle surface chemistry and particle surface charge are important parameters determining nanoparticles toxicity, translocation and subsequent effects. Toxicological studies have shown that **oxidative stress** is a central mechanism that may be responsible for the inflammatory effects of ambient particles and more specifically nanoparticles causing pulmonary responses [4,6,34]. In addition, inflammation is known to amplify the burden of reactive oxygen species to the lung, and thereby causes a persistent oxidative stress. Oxidative stress involves the generation of reactive oxygen species (ROS) and reactive nitrogen species (RNS) and the subsequent inflammatory response. The ability of particles to induce different endpoints both in lung and systemic organs seems to be associated with their ability to generate oxidants that overwhelm the endogenous anti-oxidant defence mechanisms and this is consequently called oxidative stress. Particle exposure can induce oxidative stress by a number of mechanisms, of which two are dominant:

(1) They may have oxidant generating properties themselves (i.e., acellular), and

(2) They stimulate cellular oxidant generation.

This concept is illustrated in Figure 6.2.

For several reasons nanoparticles are particularly able to generate oxidative stress when the particles deposit on the epithelium of the respiratory tract:

1) ***Nanoparticle surface area:*** The surface area of particles is known to be a major factor in driving inflammation [9,35] as well as lung carcinogenicity in experimental systems [3]. Nanoparticles of low toxicity and low solubility have a very high surface area per unit mass. They are also highly inflammogenic by virtue of this high surface area and oxidative stress, generated through mechanisms that are not well understood. Nanoparticles typically form aggregates both in the air and upon deposition that are aerodynamically larger than the classical definition of a nanoparticle, i.e. they can be larger than 100 nm. However, these nanoparticles in aggregates still demonstrate toxicity consistent with their geometric surface area and not in relation to their surface area as deduced from their aerodynamic diameter as an aggregate.

2) ***Carrier function:*** Nanoparticles may contain or adsorb components that have the ability to generate free radicals and cause oxidative stress. These components include polycyclic aromatic hydrocarbons (PAH), n-alkanes, nitro-PAHs and quinones formed by combustion or derived from unburnt fuel. Interaction of these organics with enzyme systems can results in oxidative stress via the production of oxidants such as superoxide anions. Absorbed or soluble <u>metals</u> can typically generate reactive oxygen species (ROS) in acellular and cellular systems and this is related to the inflammatory response to nanoparticles in vivo [36].

3) ***Surface modification and coatings:*** Whatever test will be used or developed, it needs to be appreciated that most suppliers apply post-production strategies to modify nanoparticles on the surface to prevent aggregation or stimulate disaggregation. In industrial applications and especially in the field of nanocomposites many different coatings are employed. There is now a body of evidence from the drug delivery and toxicological literature that surface modification as well as surface charge can have major impacts on biological response to particles, including phagocytosis, genotoxicity, and inflammation. Particle coating with polyethylene glycol (PEG) or dextran are commonly used to prevent recognition by the reticulo-endothelial system and increase the half-life of particle-conjugated drugs. A clear example, demonstrating the crucial role of particle surface, comes from work with respirable, quartz samples. Coating with aluminum lactate or the polymer polyvinyl–pyridine-N-oxide (PVNO) has a dramatic impact on the various adverse effects of the native quartz, including phagocytosis/endocytosis, oxidative DNA damage and inflammation following instillation into rat lungs [27].

REFERENCES

1. Maynard, A. D.; Aitken R. J.; Butz, T.; Colvin, V.; Donaldson, K.; Oberdörster, G.; Philbert, M. A.; Ryan, J.; Seaton, A.; Stone, V.; Tinkle, S. S.; Tran, L.; Walker, N.J.; Warheit, D. B. Safe handling of nanotechnology. *Nature,* **2006**, 444, 267-9

2. Nel, A.; Xia, T.; Madler, L.; Li, N. Toxic potential of materials at the nanolevel. *Science,* **2006**, 311, 622-627

3. Borm, P. J. A. Particle Toxicology: from coal mining to nanotechnology. *Inhalation Toxicology,* **2002**, 14, 311-324

4. Donaldson, K.; Brown, D.; Clouter, A.; Duffin, R.; MacNee, W.; Renwick, L.; Tran, L., Stone, V. The pulmonary toxicology of ultrafine particles. *Journal of Aerosol Medicine,* **2002**, 15, 213-220

5. Heinrich, J.; Wichmann, H-E. Traffic related pollutants in Europe and their effect on allergic disease. *Curent Opinion in Allergy and Clinical Immunology,* **2004** ,4, 341-348.

6. Pope, C. A. 3rd; Burnett, R. T.; Thurston, G. D.; Thun, M. J.; Calle, E. E.; Krewski, D.; Godleski, J. J. Cardiovascular mortality and long-term exposure to particulate air pollution: epidemiological evidence of general pathophysiological pathways of disease. *Circulation,* **2004**, 109, 71-77

7.   Seaton, A.; MacNee, W.; Donaldson, K.; Godden, D. Particulate air pollution and acute health effects. *The Lancet,* **1995**, 345, 176–178

8.   Oberdörster, G. Pulmonary effects of ultrafine particles. *International Archives of Occupational and Environmental Health,* **2001**, 74, 1-8

9.   Oberdörster, G.; Oberdörster, E.; Oberdörster, J. Nanotoxicology: an emerging discipline evolving from studies of ultrafine particles. *Environmental Health Perspectives,* **2005**, 113, 823-839

10. BCC Research Market report: Nanocomposites, Nanoparticles, Nanoclays, and Nanotubes, June **2006**

11. Maynard, A. D.; Baron, P. A.; Foley, M.; Shvedova, A. A.; Kisin, E. R.; Castranova, V. Exposure to carbon nanotube material: aerosol release during the handling of unrefined single-walled carbon nanotube material. *Journal of Toxicology and Environmental Health A,* **2004**, 67, 87-107

12. Shvedova, A. A.; Kisin, E. R.; Mercer,R.; Murray, A. R.; Johnson, V. J.; Potapovich, A.; Tyurina, Y. Y.; Gorelik, O.; Arepalli, S.; Schwegler-Berry, D.; Hubbs, A. F.; Antonini, J.; Evans, D. E.; Ku, B. K.; Ramsey, D.; Maynard, A.; Kagan, V. E.; Castranova, V.; Baron, P. Unusual inflammatory and fibrogenic pulmonary responses to single-walled carbon nanotubes in mice. *American Journal of Physiology- Lung Cellular & Molecular Physiology,* **2005**, 289, L 698-708

13. Muller, J.; Huaux, F.; Moreau, N.; Misson, P.; Heilier, J. F.; Delos, M.; Arras, M.; Fonseca, A.; Nagy, J. B.; Lison D. Respiratory toxicity of multi-wall carbon nanotubes. *Toxicology and Applied Pharmacology,* **2005**, 207, 221-231

14. Warheit, D. B.; Laurence, B. R.; Reed, K. L.; Roach, D. H.; Reynolds, G. A.; Webb, T.R. Comparative pulmonary toxicity assessment of single-wall carbon nanotubes in rats. *Toxicological Sciences,* **2004**, 77, 117-125

15. Lam, C. W.; James, J. T.; McCluskey, R.; Hunter, R. L. Pulmonary toxicity of single-wall carbon nanotubes in mice 7 and 90 days after intratracheal instillation. *Toxicological Sciences,* **2004**,77,126-134.

16. Huczko, A.; Lange, H.; Bysterzewsjki, M.; Baranowski, P.; Grubek-Jaworska, H.; Neijman, P. Pulmonary toxicity of 1-D carbon nanomaterials. *Fullerenes Nanotubes & Carriers in Nanotechnology,* **2005**, 13, 141-145

17. Poland, C.A.; Duffin, R.; Kinloch, I.; Maynard, A.; Wallace, W.A.H.; Seaton, A.; Stone, V.; Brown,S.; MacNee, W.; Donaldson, K. Carbon nanotubes introduced into the abdominal cavity of mice show asbestos-like pathogenicity in a pilot study. *Nature Nanotechnology,* 2008, 3, 423-428

18. Takagi, A.; Hirose, A.; Nishimura, T.; Fukumori, N.; Ogata, A.; Ohashi, N.; Kitajima, S.; Kanno J. Induction of mesothelioma in p53+/- mouse by intraperitoneal application of multi-wall carbon nanotube. *The Journal of Toxicological Sciences* 2008, 33(1), 105-116.

19. Lam, C. W.; James, J. T.; McCluskey, R.; Arepalli, S.; Hunter, R. L. A review of carbon nanotube toxicity and assessment of potential occupational and environmental health risks. *Critical Reviews in Toxicology,* **2006**, 36, 189-217

20. Helland, A.; Wick, P.; Koehler, A.; Schmid, K.; Som, C. Reviewing the environmental and human health knowledge base of carbon nanotubes. *Environmental Health Perspectives,* **2007**, 115, 1125-1131

21. Borm, P. J. A.; Robbins, D.; Haubold, S.; Kuhlbusch, T.; Fissan, H.; Donaldson, K.; Schins, R.; Stone, V.; Kreyling, W.; Lademann, J.; Krutmann, J.; Warheit, D.; Oberdorster, E. The potential risks of nanomaterials: a review carried out for ECETOC. *Particle and Fibre Toxicology,* **2006**, 3,11

22. Nohynek, G. J.; Lademann, J.; Ribaud, C.; Roberts, M.S. Grey goo on the skin? Nanotechnology, cosmetic and sunscreen safety. *Critical Reviews in Toxicology,* **2007**, 37, 251-277

23.	Adams, L. K.; Lyon, D. Y.; McIntosh, A.; Alvarez, P. J. Comparative toxicity of nano-scale TiO_2, SiO_2 and ZnO water suspensions. *Water Science and Technology,* **2006,** 54, 327-334

24.	Park, B.; Martin, P.; Harris, C.; Guest, R.; Whittingham, A.; Jenkinson, P.; Handley, J. Initial in vitro screening approach to investigate the potential health and environmental hazards of EnviroxTM - a nanoparticulate cerium oxide diesel fuel additive. *Particle and Fibre Toxicology.* **2007,** 4,12

25.	Park, E. J.; Choi, J.; Park, Y. K.; Park, K. Oxidative stress induced by cerium oxide nanoparticles in cultured BEAS-2B cells. *Toxicology,* **2008,** 245 (1-2), 90-100

26.	Gulumian, M.; Borm, P. J. A.; Vallyathan, V.; Castranova, V.; Donaldson, K.; Nelson, G.; Murray, J. Mechanistically identified suitable biomarkers of exposure, effect, and susceptibility for silicosis and coal-worker's pneumoconiosis: a comprehensive review. *Journal of Toxicology and Environmental Health- Critical Reviews.* **2006,** 9, 357-395

27.	Schins, R. P.; Duffin, R.; Höhr, D.; Knaapen, A.M.; Shi, T.; Weishaupt, C.; Stone, V.; Donaldson, K.; Borm, P. J. A. Surface modification of quartz inhibits toxicity, particle uptake, and oxidative DNA damage in human lung epithelial cells. *Chemical Research in Toxicology,* **2002,** 15, 1166-73

28.	Fubini, B.; Fenoglio, I.; Ceschino, R.; Ghiazza, M.; Martra, G.; Tomatis, M.; Borm, P. J. A.; Schins, R.; Bruch, J. Relationship between the state of the surface of four commercial quartz flours and their biological activity in vitro and in vivo. *International Journal of Hygiene and Environmental Health,* **2004,** 207, 89-104

29.	Geh, S.; Yücel, R.; Duffin, R.; Albrecht, C.; Borm, P. J. A.; Armbruster, L.; Raulf-Heimsoth, M.; Brüning, T.; Hoffmann, E.; Rettenmeier, A. W.; Dopp, E. Cellular uptake and cytotoxic potential of respirable bentonite particles with different quartz contents and chemical modifications in human lung fibroblasts. *Archives of Toxicology,* **2006,** 80, 98-106

30.	Styan, K. E.; Martin, D. J.; Poole-Warren, L. A. In vitro fibroblast response to polyurethane organosilicate nanocomposites. *Journal of Biomedical Material Research,* **2008,** 86A, 571-582

31.	Preller, L.; Doekes, G.; Heederik, D.; Vermeulen, R.; Vogelzang, P. F.; Boleij, J. S. Disinfectant use as a risk factor for atopic sensitization and symptoms consistent with asthma: an epidemiological study. *European Respiratory Journal.* **1996** 9, 1407-1413

32.	Bernstein, J. A.; Stauder, T.; Bernstein, D. I.; Bernstein, I. L. A combined respiratory and cutaneous hypersensitivity syndrome induced by work exposure to quaternary amines. *Journal of Allergy and Clinical Immunology,* **1994,** 94, 257-259

33.	Burge, P. S.; Richardson, M. N; Occupational asthma due to indirect exposure to lauryl dimethyl benzyl ammonium chloride used in a floor cleaner. *Thorax,* **1994,** 49, 842-843

34.	Donaldson, K.; Stone, V.; Borm, P. J. A.; Jimenez, L. A.; Gilmour, P. S.; Schins, R. P. F.; Knaapen, A. M.; Rahman, I.; Faux, S. P.; Brown, D. M.; MacNee, W. Oxidative stress and calcium signalling in the adverse effects of environmental particles (PM_{10}). *Free Radical Biology and Medicine,* **2002,** 34, 1369-1382

35.	Tran, C. L.; Buchanan, D.; Cullen, R. T.; Searl, A.; Jones, A. D.; Donaldson, K. Inhalation of poorly soluble particles. II. Influence of particle surface area on inflammation and clearance. *Inhalation Toxicology,* **2000,** 12, 1113-1126

36.	Schaumann, F.; Borm, P. J. A.; Herbrich, A.; Knoch, J.; Pitz, M.; Schins, R. P. F.; Luetti, B.; Hohlfeld, J. J.; Heinrich, J.; Krug, N. Metal rich particles (PM2.5) causes airway inflammation in healthy subjects. *American Journal of Respiratory and Critical Care Medicine,* **2004,** 170, 898-903

Chapter 7
Flame retardancy from nanocomposites – from research to technical products

Günter Beyer

Kabelwerk Eupen AG, Eupen, Belgium

7.1 Introduction

Fire hazards are mainly the combination of different factors including ignitability, ease of extinction, flammability of volatiles generated, amount of heat released on burning, rate of heat release, flame spread, smoke obscuration, and smoke toxicity. The most important factors are the rates of heat release, rate of smoke production, and rate of toxic gas release [1]. An early high rate of heat release causes a fast ignition and flame spread; furthermore it controls the fire intensity and is much more important than ignitability, smoke toxicity or flame spread. The time for people to escape from a fire is controlled by the heat release rate [2].

Once a fire starts in a room containing flammable materials, it will generate heat, which can heat up and ignite additional combustible materials. As a consequence the rate at which the fire progresses increases, because more and more heat is released and a progressing increase of the room temperature is observed. The radiant heat and the temperature can rise to such an extent that all materials within the room will be ignited easily, resulting in an extremely high rate of fire spread. This point in time, termed *flashover,* leads to a fully developed fire. Escape from the room will then be nearly impossible, and spread of the fire to other rooms is very likely. When a fire goes to flashover, all polymers will release roughly 20% of its weight as carbon monoxide, resulting in too much toxic smoke. Therefore, most people die in big fires, and 90% of fire deaths are the result of fires becoming too big, resulting in too much toxic smoke [3].

Fire statistics reported more than 12 million fires every year in the United States, Europe, Russia, and China. Fires kill roughly 300,000 people worldwide every year according to information from the WHO (World Health Organization); several hundreds of thousands of people are also injured every year. Most people are killed in less developed countries, but each year about 5000 people are killed by fires in Europe and more than 4000 people in the USA.

Direct property losses by fires are roughly 0.2 % of the gross domestic product and the total costs of fires are around 1 % of the gross domestic product [4]. Therefore, it is important to develop well-designed flame retardant materials to decrease both fire risks and fire hazards.

Polymers are used in more and more fields of applications, and specific mechanical, thermal, and electrical properties are required. One further important property is the flame retardant behaviour of polymers, which can be fulfilled traditionally by using intrinsically flame retardant polymers like poly(vinyl chloride) (PVC) or fluoropolymers, and flame retardants like aluminium trihydrate (ATH), magnesium dihydroxide (MDH), organic brominated compounds, or intumescent systems based on nitrogen- or phosphorus-based compounds to prevent the burning of such polymers like polyethylene (PE), polypropylene (PP), ethylene-vinyl acetate copolymer (EVA), polyamide (PA) or other polymers.

These flame retardants sometimes exhibit serious disadvantages. Use of aluminium trihydrate and magnesium dihydroxide in flame retardant cables requires a very high loading of these fillers for the applied polymers EVA, PE or PP; filling levels of more than 60 weight-% are necessary to achieve a suitable flame retardancy. Clear disadvantages of these filling levels are the high density and lack of flexibility in end products, the poor mechanical properties, and the problematic compounding and extrusion steps. In Europe, there are, at least, reservations about the general use of brominated compounds as flame retardants. Intumescent systems are relatively expensive and electrical requirements can restrict the use of these products.

A new class of materials, called **nanocomposites** overcomes the disadvantages of the traditional flame retardant systems. Generally the term *nanocomposite* describes a two-phase material with a suitable nanofiller [usually a modified layered silicate like montmorillonite (organoclay) or carbon nanotubes] dispersed in the polymer matrix at the nanometre (10^{-9} m) scale.

Compared with pure polymers, the corresponding nanocomposites show tremendous improvements; the content of nanofillers within the polymer matrix is usually between 2 weight-% and 10 weight-%.

The most important properties improved by nanocomposites are mechanical properties such as tension, compression, bending, and fracture, barrier properties like permeability, and solvent resistance, translucence, and ionic conductivity.

The review by Ray and Okamoto discusses these improvements on detail [5].

Other highly interesting properties exhibited by polymer nanocomposites concern their increased thermal stability and flame retardancy at very low filler levels [6-8]. The low filler content in nanocomposites for improved thermal stability is highly attractive for the industry because the end products can be made cheaper and easier to process.

Several research groups have reported on the preparation and flame retardant properties of EVA-based nanocomposites. EVA nanocomposites were prepared in a Brabender mixer by Camino et al. [9] and the thermal degradation was improved dramatically by reducing the influence of oxygen during thermooxidation. . Hu et al. [10] prepared intercalated EVA nanocomposites; only 5 % of filler content improved the flame retardancy of the nanocomposites. Camino et al. [11] described the synthesis and thermal behaviour of layered EVA nanocomposites; the nanofiller was a synthetic modified fluorohectorite, which is a layered silicate, and protection against thermal oxidation and mass loss was observed in air. The modified silicates accelerated the deacetylation of the polymer but reduced the thermal degradation of the deacetylated polymer due to the formation of a barrier at the surface of the polymer. Zanetti et al. [12] mixed modified fluorohectorite with EVA in an internal mixer and indicated that the acccumulation of the filler on the surface of a burning specimen created a protective barrier to heat and mass loss during combustion. There was suppressed dripping of burning particles in the vertical combustion in the case of nanocomposites, reducing the hazard of flame spread to surrounding materials. Melt intercalated and additionally gamma-irradiated PE / EVA nanocomposites were prepared by Hu et al. [13-14] based on a modified montmorillonite; increasing the clay content from 2 to 10 % was beneficial for the improvement of flammability properties. Thermogravimetric analysis (TGA) data showed that nanodispersion of the modified montmorillonite within the polymer inhibited the irradiation degradation of the PE/EVA blend, which led to nanocomposites with better irradiation-resistant properties than those of the nonfilled blend. Other authors have described the preparation of EVA-based nanocomposites in more detail. Sundararaji and Zhang [15] used a twin-screw extruder and found intercalation of modified montmorillonites with EVAs differing in melt flow index and vinyl acetate content. The use of maleated EVA obviously improved the exfoliation, probably due to chemical interaction between the maleated EVA and the filler. Camino et al. [16] studied the effect of different compounding machines on the properties of EVA nanocomposites. A discontinuous batch mixer, a single-screw extruder, a counter rotating- and a co-rotating intermeshing twin-screw extruder were used. Hu et al. [17] prepared EVA nanocomposites on a twin screw extruder and a twin-roll mill. Morgan [18] compared natural and synthetic clays to improve polymer flammability. The natural clay was a montmorillonite mined and refined in the United States; the synthetic clay was a fluorinated synthetic mica. Both clays were converted into organoclays by ion exchange with an alkyl ammonium salt and were then used to synthesize polystyrene (PS) based nanocomposites by melt blending. Both nanocomposites showed very similar reductions in the peak of heat release rate. The major differences between the natural and synthetic clay were improved colour, and a better batch-to-batch consistency but also higher costs for the synthetic clay.

Concerning the reaction mechanism of degradation and FR-behaviour of EVA nanocomposites, Wilkie et al. [19] found that in the early EVA degradation, the loss of acetic acid seemed to be catalysed by the hydroxyl groups which were present on the edges of the montmorillonite. The thermal degradation of EVA in the presence and in the absence of the modified clay showed that the formation of reaction products differed in quantity and identity. The products were formed as a result of radical recombination reactions that could occur because the degrading polymer was contained within the layers for a long enough time to permit the reactions. The formation of these new products explained the variation of heat release rates. In cases with multiple degradation pathways, the presence of the modified montmorillonite could promote one of these at the expense of another and thus led to different products and hence a different rate of volatilisation. From the investigations of PA-6, PS, poly(methyl methacrylate) (PMMA), styrene-acrylonitrile (SAN), acrylonitrile-butadiene-styrene (ABS), high impact polystyrene (HI-PS), PE and PP, Wilkie et al. [20] proposed a more general explanation of the FR properties of nanocomposites. Since the clay layers acted as a barrier to mass transport and led to superheated conditions in the condensed phase, extensive random scission of the products formed by radical recombination was an additional degradation pathway of polymers in the presence of clay. The polymers that showed good fire retardancy upon nanocomposite formation exhibited significant intermolecular reactions, such as interchain aminolysis and acidolysis, radical recombination, and hydrogen abstraction. In the case of polymers that degraded through a radical pathway, the relative stability of the radicals was the most important factor for the prediction of the effect that nanocomposite formation had on the reduction in the peak of heat release rate. The more stable was the radical produced by the polymer, the better was the fire retardancy, as measured by the reduction of the heat release rates of the polymer-clay nanocomposites.

Other nanostructured fillers have been described as flame retardants. Frache et al. [21] investigated the thermal and combustion behaviour of PE-hydrotalcite nanocomposites. Hydrotalcites were synthesized and then intercalated with stearate anions, because of the compatibility of the long alkyl chain with the PE chains. The presence of the inorganic filler shielded PE from thermal oxidation and a reduction of 55 % in the peak of heat release rate was observed. Nelson et al. [22] generated different kinds of nanocomposites using modified silica; PMMA–silica and PS–silica nanocomposites were obtained by single screw extrusion. Although these nanocomposites exhibited higher thermal stabilities and oxygen indices, they burned faster than virgin polymers according to horizontal burning tests, suggesting that nanocomposites themselves cannot be considered as flame retardant materials. In combination with traditional flame-retardant additives, flame retardancy and better mechanical properties could be achieved using less flame-retardant additives in the presence of nanofillers. Zammarano et al. [23] studied the flame retardant properties of modified layered double hydroxides (LDH) nanocomposites, which can be more effective than modified montmorillonites in the reduction of heat release rates. This may be related to the layered structure of LDHs and their hydroxyl groups and water molecules. Zammarano at al. [24] also reported on synergistic effects for LDHs in particular with ammonium polyphosphate.

In general, polymer nanocomposites exhibited low flammability when evaluated by cone calorimetry [25] with the samples in horizontal position; but they failed for other tests with samples in vertical positions, such as limiting oxygen index (LOI) [26], and Underwriters' Laboratory test (UL 94) [27]. The correlation between UL 94 or LOI and cone calorimeter is not well established as the tests run at very different fire and combustion scenarios. In a critical paper on the flammability of layered silicate polymer nanocomposites [28], there was no relevant flame behaviour enhancement for the nanocomposites in LOI and UL 94 tests except for the dripping which becomes limited. The authors suggested the interesting feature that the influence of the barrier vanished for small external heat flux (like in the cases of LOI and UL 94). They also noticed that the fire behaviour observed for nanocomposites was different in terms of slower burning velocity and hindered dripping, and postulated that viscosity was the governing parameter. In addition to this, many other authors indicated that in a typical cone calorimetry evaluation, nanocomposites tend to burn slowly and nearly completely. In terms of total heat evolved, no relevant difference between nanocomposites and reference material is generally detected proving that the charred layer slows the escaping flammable molecules and spreads out the time of combustion. Using kinetic analysis, Bourbigot et al. [29] suggested that the clay acts as a char promoter slowing the degradation and providing a transient protective barrier to the nanocomposite in combination with the aluminosilica barrier which arises from the clay. The combination of these two effects is an important factor lowering the heat release rate.

Therefore nanocomposites need further improvements to increase the ignition time which is usually less than that of the neat polymer. The first convincing explanation of this behaviour was proposed by Kashiwagi et al . [30-32]. Several of these papers showed that single- and multi wall carbon nanotubes enhanced the thermal stability of polymers without using any organic treatment or additional additives. The carbon nanotubes were at least as effective flame retardants as organoclays. PP and PMMA were investigated and the dispersion of the nanotubes within the polymer matrix was the important key parameter for good flame retardancy. Kashiwagi reported that the ideal structure of a protective surface layer (consisting of clay particles and some char) was net-like and had sufficient physical strength not to be broken or disturbed by bubbling. The protective layer should remain intact over the entire burning period. The requested formation of a continuous, network-structured protective char was easiest with high aspect ratio nanoscale particles. Kashiwagi et al. [33] pointed out that in general, a variety of highly extended carbon-based nanoparticles, such as single- and multiwall carbon nanotubes as well as carbon nanofibres, will form this kind of a network if the nanofillers formed a "jammed" network structure in the polymer matrix such that the material as a whole behaves rheologically like a gel. Also Schartel et al. [34] reported on the flame retardancy of multi wall carbon nanotubes in PA-6; again, the increased melt viscosity of the nanocomposites and the fibre-network character were the dominant factors influencing the fire performance.

Leroy et al. [35] investigated the influence of the aspect ratio of fillers on flame retardancy for the system EVA, magnesium dihydroxide (MDH), and talc. Talc particles of different lamellarity and specific surface area were tested, leading to the conclusion that for highly lamellar talc particles the fire retarding behaviour became similar to that of EVA-MDH-modified montmorillonite-based nanocomposites but with a significant intumescence. This intumescence, which occurred during the pre-ignition period in cone calorimeter tests, may to be related to three phenomena caused by the lamellar particles (modified montmorillonite or talc), which were heterogeneous bubble nucleation, increased viscosity, and charring promotion. Ferry et al. [36] described similar results on the intumescence effect of talc in EVA-MDH-zinc borate-talc. Zinc borate acted as binder in EVA-MDH-zinc borate formulations, as shown by Le Bras et al. [37].

Recently Jho et al. [38] pointed out that modified montmorillonites alone are not sufficient as flame retardants used in cable applications. Also, Wilkie [39] gave a clear statement: *"It is apparent that nanocomposite formation alone is not the solution to the fire problem, but it may be a component of the solution. We, and others, have been investigating combinations of nanocomposites with conventional fire retardants."*

It is now accepted that a nanocomposite based on organoclays alone is not sufficient to pass some of the important fire tests. Nanocomposites exhibit low flammability when tested in a cone calorimeter but fail in fire tests with vertical positions; therefore they need further improvements to increase the ignition time which is usually less than that of the virgin polymer.

Charring polymers like PA-6 and PA-6 nanocomposites were used by Bourbigot et al. [40] to improve the flame retardancy of EVA nanocomposites. The organoclay increased the efficiency of the char as a protective barrier by thermal stabilisation of a phosphorcarbonaceous structure in the intumescent char and additionally, the formation of a "ceramic" layer. Hu et al. [41] used a blend of PA-6 and EVA-nanocomposite to improve the flame retardancy of PP.

Often, combinations of nanofillers with traditional micro-sized traditional flame retardants demonstrated synergistic effects.

A halogen-free flame retardant nanocomposite was reported by Hu et al. [42] using PA-6, modified montmorillonite, MDH, and red phosphorus. This system showed higher mechanical and flame retardant properties than those of a classical flame-retarded PA-6 and therefore a synergistic effect for all the three fillers. Ferry et al. [43] partially substituted MDH in flame retardant EVA by organoclays; improvements in self-extinguishability were reported, and the main mechanism was connected to a phenomenon of intumescence leading to the formation of a foam-like structure during the pre-ignition period. Horrocks et al. [44] demonstrated that the combination of organoclays with ammonium polyphosphate or polyphosphine oxide showed synergistic effects in flame retardancy for PA-6. Whaley et al. [45] investigated blends of EVA and ethylene-co-octene copolymers with MDH and modified montmorillonites for cable compounds. The time dependence of the char layer formation in such systems suggested that optimal loadings for the montmorillonite could be different for applications with different cable jacketing thickness, and therefore the true performance of nanocomposite-based jacketing compounds need to be assessed in actual cable constructions.

Shen et al. [46-47] reported on the flame retardant improvements gained by using a filler combination of modified montmorillonites, MDH, and zinc borate in EVA. A modified montmorillonite with a smaller particle size gave a better UL 94 performance than one of a larger size, and a stronger char was formed in the presence of zinc borate having a very fine particle size.

Ristolainen et al. [48] used modified montmorillonite as a partial substitute for ATH in PP-ATH composites and observed enhanced flame retardancy with composites containing both fillers. Wilkie and Zhang [49] studied the fire behaviour of PE combined with ATH and a modified montmorillonite. The combination of PE with 2.5 % modified montmorillonite and 20 % ATH gave a 73% reduction in the peak heat release rate, which was the same as that obtained when 40% ATH was used alone. A further increase in the montmorillonite loading did not improve the fire properties. Mechanical properties such as elongation at break could be improved when comparing compounds with or without montmorillonite at the same reduction in peak heat release rate.

Cusak et al. [50] found that zinc hydroxystannate greatly enhanced the performance of an ATH-organoclay synergistic fire-retardant system in an EVA formulation, allowing reductions in the overall filler level with no or little compromise in terms of flame retardant or smoke suppressant properties.

This chapter will review in detail the properties of nanocomposites-based on organoclays or carbon nanotubes, and the synergistic effects of these fillers with micro-sized aluminium trihydrate as a traditional flame retardant for cables and technical applications will be presented.

7.2 Organoclay nanocomposites

7.2.1 Processing and structure of EVA/organoclay-based nanocomposites

Poly(ethylene-co-vinyl acetate) copolymers (EVA) with different weight-% of vinyl acetate have demonstrated their ability to promote nanocomposite formation by melt blending with organoclays [51-53]. Depending on the nature of the filler distribution within the matrix, the morphology of nanocomposites can evolve from the intercalated structure with a regular alternation of layered silicates and polymer monolayers to the exfoliated (delaminated) structure with layered silicates randomly and homogeneously distributed within the polymer matrix. The easiest and technically most attractive way to produce these types of materials is kneading a polymer in the molten state with a modified layered silicate such as montmorillonite. The native Na^+ interlayer cation within the silicate has been exchanged by a quaternary alkylammonium cation. The modified filler, called an *organoclay,* is much more compatible with the polymer matrix.

Information on the nanocomposite morphology was obtained by transmission electron microscopy (TEM) and X-ray diffraction (XRD) observation. Compounding was done on a twin-roll mill and exfoliated silicate sheets were observed together with small stacks of intercalated sheets. This structure may be described as a semi-intercalated semi-exfoliated structure that did not change principally with the vinyl acetate content of the EVA matrix, even a larger amount of stacks were observed for EVA with lower vinyl acetate contents [52]. There were no great differences within the morphology of these nanocomposites.

7.2.2 Thermal stability of EVA/organoclay-based nanocomposites

Thermogravimetric analysis (TGA) is widely used to characterize the thermal stability of polymers. The mass loss of a polymer due to volatilization of products generated by the thermal decomposition is monitored as function of a temperature ramp. Nonoxidative decomposition occurs when the material is heated under an inert gas flow like helium or nitrogen, while the use of air or oxygen allows investigation of oxidative decomposition reactions. The experimental conditions highly influenced the reaction mechanism of the degradation. EVA is known to decompose in two consecutive steps. The first step is identical in both oxidative and nonoxidative conditions, occurring between 350 °C and 400 °C and involving the loss of acetic acid. The second step involved the thermal decomposition of the obtained unsaturated backbone either by further radical scissions (non-oxidative decomposition) or by thermal combustion (oxidative decomposition), (Figure 7.1).

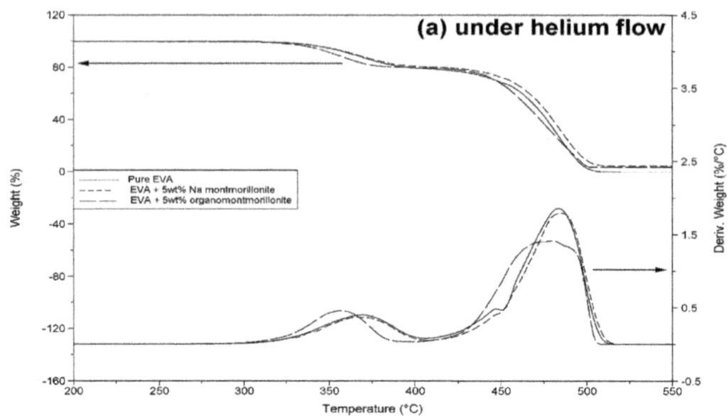

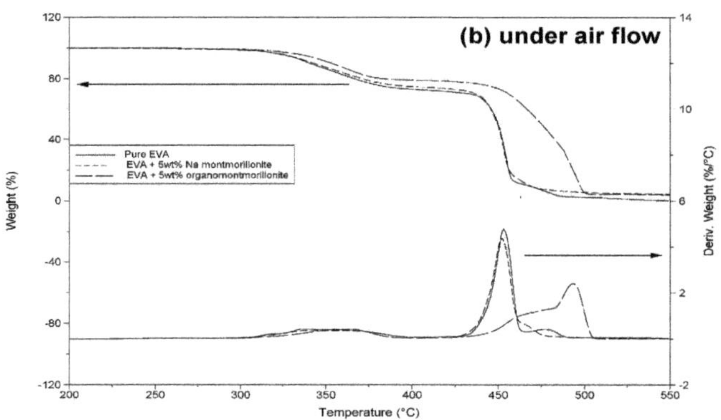

Figure 7.1

TGA of EVA, EVA microcomposite with 5 weight-% Na^+-montmorillonite and EVA nanocomposite with 5 weight-% organoclays under helium and airflow

Heating rate : 20°C/min; EVA : Escorene UL 00328 with 28 weight-% vinyl acetate content

Reproduced with permission from reference 53, Beyer,G., Flame retardant properties of EVA-nanocomposites and improvements by combination of nanofillers with aluminium trihydrate. *Fire and Materials*, **2001**, 25, 193-197, copyright 2001, John Wiley & Sons Limited.

In helium the EVA nanocomposite showed a negligible reduction in the thermal stability compared to pure EVA or EVA filled with Na-montmorillonite (microcomposite). In contrast, when decomposed in air, the same nanocomposite exhibited a rather large increase in the thermal stability because the maximum of the second degradation peak was shifted 40 °C to higher temperatures while the maximum of the first decomposition peak remained unchanged [52] (Table 7.1). In this case the explanation for the improved thermal stability was char formation occurring under oxidative conditions. The char acted as a physical barrier between the polymer and the superficial zone where the combustion of the polymer occured. The optimum thermal stabilization was obtained at an organoclay level of 2.5 - 5.0 weight-%, as indicated by the results in Table 7.1 on the maximal temperatures for the main degradation peak for EVA nanocomposites.

It is very important for technical applications that nanocomposites must be additionally stabilized as outlined in chapter 5 (Pfaender); as otherwise the end products may have greatly reduced life times, which limit the proper technical applications of nanocomposites.

Table 7.1

Maximal temperature at the main degradation peak measured by TGA under air flow at 20°C/min for EVA and EVA-based nanocomposites with different organoclay contents;

(EVA = Escorene UL 00328 with 28 weight-% vinyl acetate content, Organoclay = Nanofil 15)

Organoclay content (weight-%)	Maximal temperature at the main degradation peak (C°)
0	452.0
1	453.4
2.5	489.2
5	493.5
10	472.0
15	454.0

7.2.3 Flammability properties of EVA/organoclay-based nanocomposites

From an engineering point of view, it is important to identify the principle fire hazard associated with products so that relevant properties can be measured. Extensive research at the National Institute for Standards and Technology in the United States led to an important conclusion that allows significant simplification of the problem. The heat release rate, in particular the peak of heat release rate, is the single most important parameter in a fire and can be viewed as the "driving force" of the fire [54]. Therefore, today the universal choice of an engineering instrument to measure flame retardant properties of polymers is the cone calorimeter. The measuring principle is the oxygen depletion with a relationship between the mass of oxygen consumed from the air and the amount of heat released. In a typical cone calorimeter experiment, polymer samples placed in aluminium dishes are exposed to a defined heat flux (typically, 35 kW/m^2 or 50 kW/m^2). Simultaneously, the properties "rate of heat release (RHR)", "peak of heat release rate" (PHRR), "time to ignition", "total heat released", " mass loss rate", "mean CO yield", "mean specific extinction area" and others are measured.

The flame retardant properties of the EVA nanocomposites were determined by a cone calorimeter with a heat flux of 35 kW/m^2 (Figure 7.2). Under such conditions, simulating a developing fire scenario, the effect of the organoclay was already observed for 3 weight-%. A decrease by 47 % of the PHRR compared to virgin EVA, as well as a shift towards longer times, was detected for a nanocomposite containing 5 weight-% of the organoclay. Increasing the filler content to 10 weight-% did not significantly improve the reduction of the PHRR. As a decrease in PHRR indicates a reduction of burnable volatiles generated by the degradation of the polymer matrix, such a decrease clearly showed the flame retardant effect due to the presence of the organoclays and their "molecular" distribution throughout the matrix. Furthermore, the flame retardant properties were improved by the fact that the PHRR was spread over a much longer period. The flame retardant properties were due to the formation of a char layer during the nanocomposite combustion. This char acted as an insulating and non-burning material and reduced the emission of volatile products (fuel) into the flame area. The silicate layers of the organoclay played an active role in the formation of the char but also strengthened it and made it more resistant to ablation.

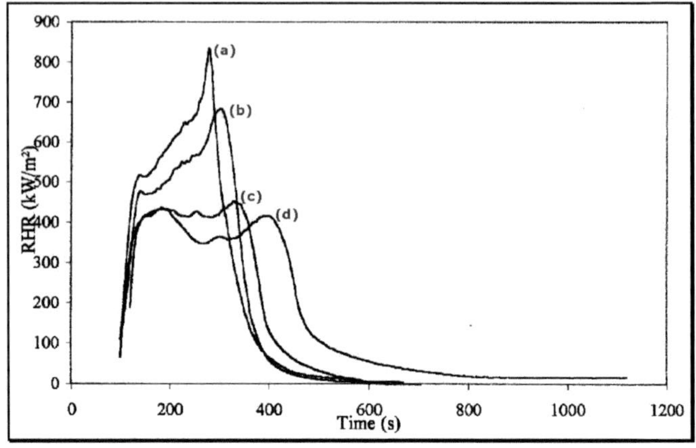

Figure 7.2
Rate of heat release at heat flux = 35 kW/m^2
for various EVA (Escorene UL 00328 with 28 weight-% vinyl acetate content) based materials
a)   Pure EVA matrix and EVA matrix with 5 weight-% of Na$^+$-montmorillonite
b)   EVA + 3 weight-% of organoclay Nanofil 15
c)   EVA + 5 weight-% of organoclay Nanofil 15
d)   EVA + 10 weight-% of organoclay Nanofil 15
Reproduced with permission from reference 53, Beyer,G., Flame retardant properties of EVA-nanocomposites and improvements by combination of nanofillers with aluminium trihydrate. *Fire and Materials*, **2001**, 25, 193-197, copyright 2001, John Wiley & Sons Limited.

Cone calorimeter experiments at a heat flux of 35 kW/m^2 showed that virgin EVA was completely burned without any residue. In contrast, a very early strong char formation was found for an EVA nanocomposite with an analogous cone calorimeter experiment; this char was stable and did not disappear by combustion (Figure 7.3). Finally, compared to a pure EVA matrix, the nanocomposite burnt without producing burning droplets (UL 94

vertical procedure) [55], a characteristic feature that furthermore limits the propagation of a fire. Burning droplets will be an important additional characteristic for cables to be classified within the Euroclasses B1ca, B2ca, Cca or Dca, defined by a draft for the European standard prEN 50399, see later in this chapter.

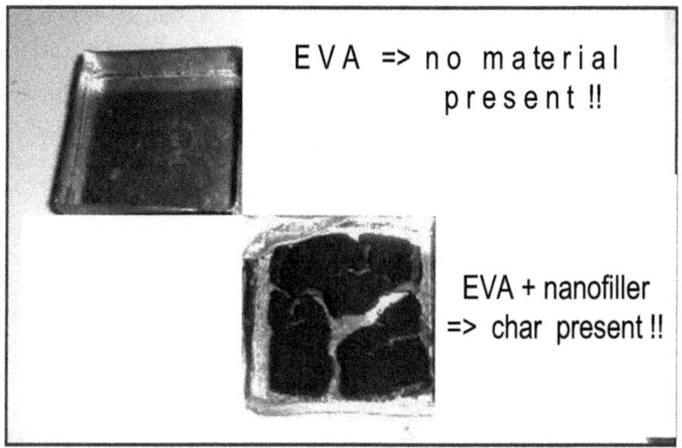

Figure 7.3

Samples of pure EVA and EVA nanocomposite with 5 weight-% organoclays
after cone calorimeter combustion for 200 sec
Heat flux = 35 kW / m^2; polymer plates of 100 x 100 x 3 mm within aluminium dishes
EVA : Escorene UL 00328 with 28 weight-% vinyl acetate content
Organoclay : Nanofil 15
Reproduced with permission from reference 53, Beyer,G., Flame retardant properties of EVA-nanocomposites
and improvements by combination of nanofillers with aluminium trihydrate. *Fire and Materials*, **2001**, 25, 193-
197, copyright 2001, John Wiley & Sons Limited.

7.2.4 NMR investigation and FR mechanism of EVA/organoclay-based nanocomposites

The degradation of EVA and EVA nanocomposites was investigated by solid phase cross-polarization magic angle spinning carbon-13 nuclear magnetic resonance spectroscopy (CP-MAS-^{13}C-NMR). Bourbigot [56] described the measurement method in detail. EVA (Exxon's EVA Escorene UL 00112 with 12 weight-% vinyl acetate content) and a nanocomposite based on the same EVA with 5 weight-% of the organoclay were degraded by irradiation in a cone calorimeter with a heat flux of 50 kW/m^2. Samples were taken out from the heat flux after 50, 100, 150, 200 and 300 seconds, and the presence of EVA and the char formation were measured by the shift (in ppm) of the signals related to the signal from the standard tetramethylsilane:
The following results were obtained [57]:

Before irradiation of EVA and EVA nanocomposite:
- 33 ppm   =>   –CH$_2$– : polymer backbone of EVA
- 75 ppm   =>   –CH$_3$:   acetate group of EVA
- 172 ppm   =>   –C=O :   acetate group (small signal) of EVA

After irradiation of EVA:
- 50 seconds :   New signals at 130 ppm (char: aromatics and polyaromatics) and 180 ppm (–C=O with start of oxidation), EVA signals still present
- 150 seconds :   no signals => no organic material present

After irradiation of EVA nanocomposite:
- 50 seconds : New signals at 130 ppm (char: aromatics and polyaromatics) and 180 ppm (–C=0 with start of oxidation), EVA signals still present
- 100 seconds : char formation and EVA signals still present
- 200 seconds : char formation and EVA signals still present
- >300 seconds : no signals => no organic material present.

Obviously, the formation of nanocomposites clearly promoted char formation and delayed the degradation of EVA.

7.2.5   Intercalation versus exfoliation in EVA/organoclay-based nanocomposites

It is often reported in the literature that exfoliation is the most effective structure for maximal enhancements of properties improved by nanocomposites. Therefore, it was of interest to shift the ratio of the mixed intercalated-exfoliated structure that was observed within EVA nanocomposites by twin-roll mill compounding [57] to the purely exfoliated structure. This was done by melt compounding EVA (Exxon's Escorene EVA UL 00328) with 5 phr (parts per hundred parts resin) of organoclay by a co-rotating twin screw extruder (27 mm screw diameter, 40 L/D). Two screw designs were used: a first screw for maximal mixing using mixing elements, and the second screw for maximal dispersion using kneading blocks. The screws were used from 300 up to 1200 rpm. TEM and XRD demonstrated that for the highest shear rate (1200 rpm) and highest friction (second screw), the mixed structures were shifted to the exfoliated structure. However, cone calorimeter data showed that there were no changes for the PHRRs for all these melt-compounded nanocomposites. Obviously, the mixed intercalated-exfoliated structures within the EVA nanocomposites already had the maximal reductions for PHRR. This result is important for companies using EVA-based nanocomposites as a flame retardant system, because it simplifies the task of organoclay dispersion as a main processing step.

7.2.6   Combination of the classical flame retardant filler ATH with organoclays

Cable compounds must be flame retardant to achieve a low flame spread defined by the widely used International Electrotechnical Commission cable fire test (IEC 60332-3-24) [58]. A combination of 65 weight-% of ATH and 35 weight-% of a high filler level accepting polymer matrix like EVA must often be used for cable outer sheaths [59].

The performances of two compounds were compared. Both compounds were prepared on a BUSS Ko-kneader (46 mm screw diameter, 11 L/D). One compound was made from 65 weight-% ATH and 35 weight-% EVA with 28 % vinyl acetate content and a second compound was made from 60 weight-% ATH, 5 weight-% of organoclays, and 35 weight-% EVA. Both compounds were investigated with TGA in air and cone calorimeter at 50 kW/m^2 heat flux. TGA in air clearly showed a delay in the degradation brought about by the small amount of organoclays (Figure 7.4).

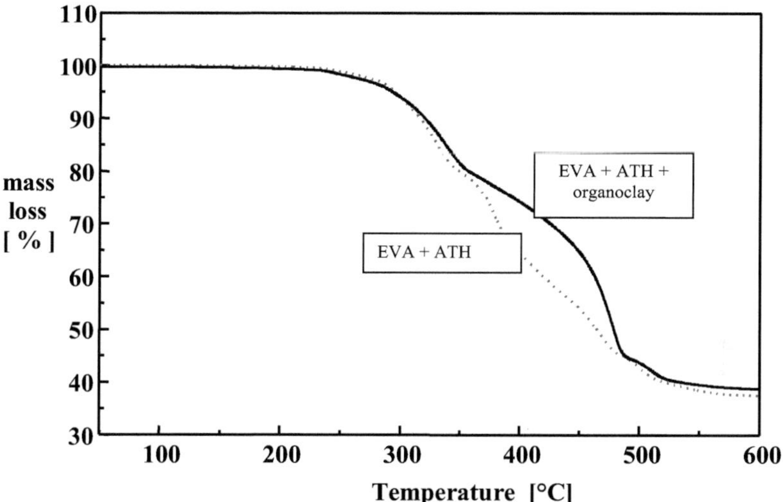

Figure 7.4
TGA in air of a compound with 35 weight-% EVA and 65 weight-% ATH compared with a nanocomposite
compound with 35 weight-% EVA, 60 weight-% ATH and 5 weight-% of organoclays
EVA : Escorene UL 00328 with 28 weight-% vinyl acetate content
Organoclay : Nanofil 15
ATH : Martinal OL 104 LE
Reproduced with permission from reference 53, Beyer,G., Flame retardant properties of EVA-nanocomposites
and improvements by combination of nanofillers with aluminium trihydrate. *Fire and Materials*, **2001**, 25, 193-
197, copyright 2001, John Wiley & Sons Limited.

The char of the EVA-ATH-organoclay compound generated by cone calorimeter was very rigid und showed
only very few small cracks; but the char of the EVA-ATH compound was much less rigid (low mechanical
strength) with many big cracks. This could explain why for the nanocomposite the PHRR was reduced to 100
kW/m^2, compared to 200 kW/m^2 for the EVA-ATH compound. To obtain the same decrease in the PHRR by the
flame retardant filler ATH only, the content of ATH must be increased to 78 weight-% within the EVA-ATH
compound.

The great improvements in flame retardancy by the organoclays also opened the possibility of decreasing the
level of ATH within the EVA polymer matrix. To maintain 200 kW/m^2 as an acceptable peak heat release level,
the content of ATH could be decreased from 65 weight-% to 45 weight-% by including just 5 weight-%
organoclay within the EVA polymer matrix. The reduction in the total amount of these fillers resulted in
improved mechanical and rheological properties of the EVA-based nanocomposite.

7.2.6.1 Coaxial cable passing UL 1666 fire test with an organoclay/ATH-based outer sheath

There are many applications for indoor cables passing the large-scale fire test UL 1666 (riser test for cables) [60]
with a 145-kW burner in a two-storey facility. This very severe fire test defines the following important points of
measurements:

1. maximal temperature of fire gases at 3.66 metres: 454 °C
2. maximal height for flames: 3,66 metres

Compounds with halogenated flame retardants are often used to pass this test, but more and more flame retardant non-halogen (FRNH) cables are requested by the market for the riser test. Cables based on nanocomposite compounds demonstrate their promising performances for this fire test.

The first example worldwide for the application of nanocomposites as a new flame retardant concept for the cable industry was reported by Kabelwerk Eupen AG; Beyer [57] described FRNH cables passing UL 1666 and the cable is shown in Figure 7.5. The outer sheath was based on a nanocomposite with an industrial EVA-ATH-organoclay composition. The analogous coaxial cable was tested with an outer sheath based on EVA-ATH. In both compounds the relation of polymer to filler was the same and Table 7.2 presents the results. The improved flame retardant properties were due to the formation of a char layer during nanocomposite combustion. This insulating and nonburning char reduced the emission of volatile products from the polymer degradation into the flame area and thus minimized the maximal temperature and height of the flames.

Figure 7.5

FRNH coaxial cable (1/2-in. diameter) with a nanocomposite-based outer sheath
passing the UL 1666 cable fire test
Reproduced with permission from reference 55, Beyer, G. Flame retardancy of nanocomposites based on organoclays and carbon nanotubes with aluminium trihydrate. *Polymers for Advanced Technologies*, **2006**, 17, 218-225, copyright 2006, John Wiley & Sons Limited.

Table 7.2
Fire performances of FRNH coaxial cables with EVA-ATH and EVA-ATH-organoclay outer sheaths

UL 1666 requirements	EVA-ATH compound	EVA-ATH-organoclay compound
maximal temperature at 3.66 m < 454 °C	1954 °C	327 °C
maximal flame height: < 3.66 m	> 3.66 m	1.83 m

7.2.6.2 Medium voltage cables with organoclay/ATH-based outer sheaths

The cable producers General Cable and Silec Cable have also reported both on the improved flame retardancy and cost reduction achieved by nanocomposites for different medium voltage cables [61].

The first cable considered was a medium voltage cable (rated up to 30 kV) used in tunnels. In fact there was no previous design because of two issues related to the outer sheath that could not be properly balanced: flame retardancy and high values for tensile strength, elongation, and abrasion. The use of a nanocomposite (mixture of ATH and organoclays) for the outer sheath demonstrated a good overall behaviour and complied with the requirements of flame propagation test on bunched cables according to IEC 60332-2-23 (category B) [62]. The cables demonstrated an improved performance (slow combustion and low smoke emission) due to the effect of char produced by the nanocomposite-based outer sheath, which acted as a fire barrier, and also had good mechanical properties like abrasion resistance. The optimum design has not yet been achieved, according to the authors but they have already obtained a significant reduction in the overall thickness of the nanocomposites-based outer sheath. This implied lower overall diameter (4.2 %) and therefore lower cable manufacturing costs.

The second cable (rated up to 30 kV) was also a medium voltage cable but designed for fixed installations such as distribution networks or industrial installations. The new nanocomposite-based compound for the outer sheath contained a combination of the fillers ATH and organoclays. This substituted the previous double layer design (ATH-based outer sheath and an additional fire barrier of mica tape) by a single flame retardant outer sheath. Again in this case the authors indicated that the char formation effect is responsible for the improved flame retardancy. They also stated that for this cable type the purely ATH-based outer sheath (in combination with the mica tape fire barrier) also passed all flame tests with a low and smooth combustion. However, the crucial achievement was to dispense with the mica tape fire barrier, resulting in a 5.4 % reduction in the overall diameter for the nanocomposite-based cable. This fact resulted in simplification of the manufacturing process, with the use of only one nanocomposite-based compound and only one extruder. One can also speculate that the production line speed has been increased because the production step involved in applying the mica tape is a low speed process compared to the extrusion of the outer sheath. The authors stated their views that regarding the fire reaction issues affecting electrical cables, nanocomposites will be one of the main solutions in the near future.

7.2.6.3 Energy cables passing prEN 50399 with an organoclay ATH-based outer sheath

For the new fire tests in Europe, nanocomposites will have a good chance to be used regularly by the industry in the near future.

The European Commission's Fire Regulators Group, consisting of representatives from each member country, have reviewed various proposals to create a new fire performance hierarchy for products like cables installed in buildings within the governing document known as the *Construction Products Directive* (CPD). Input was received from its own sponsored studies as well as from investigations by industry groups. The European Commission's FIPEC study (Fire Performance of Electrical Cables) presented a fire hazard assessment methodology for both communications and energy cables in buildings based primarily on modifications to the international cable fire test protocol according to different categories defined in IEC 60332-3 [58,62]. The modifications enabled measurements of heat release rate by oxygen consumption technique, smoke release rate and the possibility of measuring toxic gas production rates by FTIR instrumentation. Parameters are flame spread (FS), peak heat release rate (PHRR), total heat released (THR), fire index growth rate (FIGRA), smoke production rate (SPR), and total smoke production (TSP). The possible ranking of cables is summarized in Table 7.3, and the tests are described in the proposal prEN 50399, which is currently (November 2008) under consultation circulation and is soon expected to be accepted as a formal EN standard. This test set-up will allow testing of cables in one of the most advanced test arrangements available in Europe. Highest flame retardant cables are in class B and cables with no flame retardancy are in class F.

<table>
<tr><td colspan="4" align="center">Table 7.3
Classes of reaction to fire performance for electric cables</td></tr>
<tr><td>Class</td><td>Test method(s)</td><td>Classification criteria</td><td>Additional classification</td></tr>
<tr><td>A_{ca}</td><td>EN ISO 1716</td><td>PCS $\leq$ 2,0 MJ/kg (1)</td><td></td></tr>
<tr><td rowspan="2">$B1_{ca}$</td><td>$FIPEC_{20}$ Scen 2 (5)
and</td><td>FS $\leq$ 1.75 m and
$THR_{1200s} \leq$ 10 MJ and
Peak HRR $\leq$ 20 kW and
FIGRA $\leq$ 120 Ws⁻¹</td><td rowspan="2">Smoke production (2, 6) and
Flaming droplets/particles (3) and
Acidity (4)</td></tr>
<tr><td>EN 50265-2-1</td><td>Height (H) $\leq$ 425 mm</td></tr>
</table>

$B2_{ca}$	FIPEC$_{20}$ Scen 1 (5) *and*	FS $\leq$ 1.5 m; *and* THR$_{1200s}$ $\leq$ 15 MJ; *and* Peak HRR $\leq$ 30 kW; *and*	Smoke production (2, 7) and Flaming droplets/particles (3) and Acidity (4)
	EN 50265-2-1	FIGRA $\leq$ 150 Ws^{-1} H $\leq$ 425 mm	
C_{ca}	FIPEC$_{20}$ Scen 1 (5) *and*	FS $\leq$ 2.0 m; *and* THR$_{1200s}$ $\leq$ 30 MJ; *and* Peak HRR $\leq$ 60 kW; *and* FIGRA $\leq$ 300 Ws^{-1}	Smoke production (2, 7) and Flaming droplets/particles (3) and Acidity (4)
	EN 50265-2-1	H $\leq$ 425 mm	
D_{ca}	FIPEC$_{20}$ Scen 1 (5) *and*	THR$_{1200s}$ $\leq$ 70 MJ; *and* Peak HRR $\leq$ 400 kW; *and* FIGRA $\leq$ 1300 Ws^{-1}	Smoke production (2, 7) and Flaming droplets/particles (3) and Acidity (4)
	EN 50265-2-1	H $\leq$ 425 mm	
E_{ca}	EN 50265-2-1	H $\leq$ 425 mm	
F_{ca}	No performance determined		

(1) For the product as a whole, excluding metallic materials, and for any external component (i.e. sheath) of the product.

(2) **s1** = TSP$_{1200}$ $\leq$ 50 m^2 *and* Peak SPR $\leq$ 0.25 m^2/s

s1a = **s1** and transmittance in accordance with EN 50268-2 $\geq$ 80%

s1b = **s1** and transmittance in accordance with EN 50268-2 $\geq$ 60% < 80%

s2 = TSP$_{1200}$ $\leq$ 400 m^2 *and* Peak SPR $\leq$ 1.5 m^2/s

s3 = not s1 or s2

(3) For FIPEC$_{20}$ Scenarios 1 and 2: **d0** = No flaming droplets/particles within 1200 s; **d1** = No flaming droplets/particles persisting longer than 10 s within 1200 s; **d2** = not d0 or d1.

(4) EN 50267-2-3: **a1** = conductivity < 2.5 µS/mm *and* pH > 4.3; **a2** = conductivity < 10 µS/mm *and* pH > 4.3; **a3** = not a1 or a2. No declaration = No Performance Determined.

(5) Air flow into chamber shall be set to 8000 ± 800 l/min.

FIPEC$_{20}$ Scenario 1 = prEN 50399-2-1 with mounting and fixing as below

FIPEC$_{20}$ Scenario 2 = prEN 50399-2-2 with mounting and fixing as below

(6) The smoke class declared for class B1$_{ca}$ cables must originate from the FIPEC$_{20}$ Scen 2 test.

(7) The smoke class declared for class B2$_{ca}$, C$_{ca}$, D$_{ca}$ cables must originate from the FIPEC$_{20}$ Scen 1 test.

The value FIGRA is defined by the ratio of PHRR/time to PHHR and it represents the fire growth rate, which may work as an indication of the propensity to cause a quickly growing fire. It is reported from cone calorimeter investigations that nanocomposites showed impressive reductions of peak of heat release rates and FIGRA, especially for higher external heat fluxes [63].

Two test methods were developed: scenario 1 according to prEN 50399-2-1, and scenario 2 according to prEN 50399-2-2, the former being slightly more severe than IEC 60332-3-24, and the latter being much more severe and suitable for high hazard installations. The conclusions reached by the FIPEC study were that current tests (e.g. IEC 60332-3-24, category C) are not sensitive enough to differentiate the enhanced fire performances needed for high density telecommunication cable installations, and the parameter that has the most effect on the test results is the method of mounting cables on the ladder.

Cables with the following specification were tested according the FIPEC scenarios 1 and 2 (designation: NHXMH-J, 4 x 16 mm^2; insulation: crosslinked polyethylene; bedding: a flame retardant non halogen compound; outer sheath: a flame retardant non-halogen EVA compound either based classically on ATH only or based on a nanocomposite by combination of organoclays and ATH [57]). The values listed in Table 7.4 clearly demonstrate the improvements seen in both for flame spread and FIGRA due to the nanocomposite-based outer sheath. During the fire tests (Table 7.3) there was no dripping of burning polymer from the cable with the nanocomposite-based outer sheath; this is an additional important requirement within the CPD regulation.

Table 7.4

Flame retardant properties of cables by FIPEC scenarios 1 and 2according to prEN 50399

	NHXMH-J 4 x 16 mm^2 Classical purely ATH-based outer sheath	NHXMH-J 4 x 16 mm^2 Nanocomposite-based outer sheath
FIPEC Scenario 1	Flame spread = 0.49 m	Flame spread = 0.48 m

20.5 kW flame 20 minutes burning time	PHRR = 27.1 kW FIGRA = 63.2 W s^{-1}	PHRR = 22.9 kW FIGRA = 20.3 W s^{-1}
FIPEC Scenario 2 30 kW flame & plate 30 minutes burning time	Flame spread = 1.85 m PHRR = 58.6 kW FIGRA = 53.5 W s^{-1}	Flame spread = 1.21 m PHRR = 55.8 kW FIGRA = 47.9 W s^{-1}

7.2.7 Synergistic effects with halogenated flame retardants

Beside the synergistic effect of organoclays with ATH or MDH, halogenated flame retardants also demonstrated a very similar synergistic effect with organoclays in technical products.

The bromine-containing flame retardant decabromodiphenyl ether (DBDPO) was used by the filler producer Nanocor to study the synergistic effects obtained with organoclays in PP. Both organoclays (Nanomer 1.44PA) and organoclays concentrates (Nanomer C.44PA) were incorporated using a twin screw extruder. The organoclay concentrate could also be incorporated into the formulation using a single screw extruder equipped with a mixing screw such as the "Nano-Mixer™" co-developed by New Castle Industries and Nanocor, Inc.

The flame retardant characteristics were quite similar to the EVA-MDH nanocomposites: namely reduced burning rate, minimal dripping and sparkling. UL 94 test results (Table 7.5) indicated good synergy between DBDPO/Sb_2O_3 with the organoclays. 6 weight-% of organoclay I.44PA addition reduced the DBDPO/Sb_2O_3 use by 7 weight-% and maintained the same UL 94 rating. In addition, the incorporation of the organoclays in the DBDPO/Sb_2O_3 system increased the mechanical properties and lowered the FR agent migration. This enables formulators to fine tune the composition to meet different application requirements.

Table 7.5
PP-DBDPO-Nanomer compositions and UL 94 FR 1/8" Rating

Homo-PP (wt%)	73.3	80	77	74	74	68
DBDPO (wt%)	20	15	15	15	15	15
Sb_2O_3 (wt%)	6.7	5.0	5.0	5.0	5.0	5.0
Nanomer I.44 P (wt%)	0	0	3.0	6.0	0 0	-- 0
NanoMax-PP (wt%)					6	12
UL 94 rating	V-0	Fail	V-2	V-0	V-2	V-0

NanoMax-PP is a masterbatch of Nanomer I.44P in a polymer (1/1 ratio of organoclay/polymer)

7.2.8 Commercial examples of nanocomposite-based compounds

Nanocor is already supplying Nanomer® products to the market. Nanocomposites for flame retardancy applications are also available from Gitto/Global Corporation, Lunenburg, MA, USA. One example of a FR application is in heavy-duty electrical enclosures. These enclosures, typically injection molded polypropylene, vary in size from around 30cm^3 to 1 m^3. Because they house electrical items, flammability is a central concern, but weight is also important since many enclosures must also be portable. In addition, electrical components can themselves be quite heavy, making enclosure strength a key element. By switching to a nanocomposite, the level of the FR additive package can be significantly reduced, taking the enclosure's specific gravity down from 1.35 to 1.16 and reducing overall weight by 18%. Both flexural and tensile moduli increase 25% on average without loss of the Izod impact. The enclosures maintain their original UL 94 V-0 rating, yet they are both stronger and lighter.

The second commercial polymer nanocomposite with additional flame retardants rated for regulatory tests is a class of materials made by PolyOne Corporation under the trade name "Maxxam FR". These materials, according to the company's website, are PE- and PP-nanocomposites with a variety of fire classification according to UL 94 available. The product line is the result of a strategic alliance between PolyOne and Nanocor. This suggests that the clays used in these systems are organoclays made via masterbatch approaches since Nanocor only supplies modified montmorillonites as organoclays for nanocomposite applications and is strongly promoting the masterbatch route for nanocomposite preparation. Halogenated and non-halogenated flame retardant versions are available in this product line, but the precise details of the technology are currently unavailable [64].

7.3   Carbon nanotube composites

7.3.1   General properties of carbon nanotubes

Carbon nanotubes (CNT) are tubular derivatives of fullerenes. They were first observed in arc discharge experiments and exhibit properties which are quite different from those of the closed cage fullerenes such as C_{60}, C_{70} and C_{76} etc. Special topologies are responsible for the unique and interesting properties of CNTs. As novel carbon materials the CNTs are of great interest within the field of materials science research. Due to their high mechanical strength, capillary properties, and remarkable electronic structures, a wide range of potential uses is reported. Typical applications for CNTs are supports for metals in the field of heterogeneous catalysis, materials for hydrogen storage, as composite materials in polymer science [65-66] and for immobilization of proteins and enzymes. Several techniques including arc discharge, laser ablation and catalytic methods and others have been developed for the production of CNTs [66]. Many materials science researchers and also companies (Nanocyl, Hyperion Catalysis International, Bayer MaterialScience, Arkema) are working on development of methods for a large-scale production of CNTs to realize their speculated applications.

CNTs can consist of one (singlewall carbon nanotubes, SWCNT) or more (multiwall carbon nanotubes, MWCNT) cylindrical shells of graphitic sheets. Each carbon is completely bonded to three neighbouring carbon atoms through sp^2 hybridisation to form a seamless shell. CNTs can have a very high aspect ratio, above 1000. It is reported [67] that polymer degradation can be retarded by CNTs as indicated by thermogravimetric analysis. In PVOH a loading level of 20 weight-% MWCNTs shifts the beginning of the polymer degradation and the single degradation peak to higher temperatures.

7.3.2   Synthesis and purification of CNTs

Crude MWCNTs and SWCNTs have been produced by catalytic decomposition of acetylene on Co- and/or Fe catalysts on different supports [68]. The CNTs contained used catalysts and other by-products. The catalysts and support contents of different MWCNT samples are shown in Table 7.6.

Crude MWCNTs contained Co, Fe and alumina. Purified MWCNTs were synthesized from crude MWCNTs by dissolution of the catalyst support alumina in concentrated NaOH, dissolution of the metal catalyst Co in concentrated HCl, drying at 120 °C in an air oven and additionally drying at 500 °C under vacuum.

Crude SWCNTs contained Co and MgO. Purified SWCNTs were synthesized from crude SWCNTs by dissolution of the catalyst support MgO in concentrated HCl, purification by air oxidation at 300°C and then drying at 120 °C in an air oven.

<table>
<tr><td colspan="7" align="center">Table 7.6
Properties of crude and purified MWCNTs</td></tr>
<tr><td rowspan="3">Sample</td><td colspan="2">Nanotubes</td><td colspan="2">Catalyst</td><td colspan="2">Support</td></tr>
<tr><td>Length</td><td>Diameter</td><td>Co</td><td>Fe</td><td>Al_2O_3</td><td>MgO</td></tr>
<tr><td>(µm)</td><td>(nm)</td><td>(wt.%)</td><td>(wt.%)</td><td>(wt.%)</td><td>(wt.%)</td></tr>
<tr><td>Crude MWCNTs</td><td>ca. 50</td><td>5-15</td><td>0.3</td><td>0.3</td><td>19</td><td>-</td></tr>
<tr><td>Purified MWCNTs</td><td>ca. 50</td><td>5-15</td><td>0.2</td><td>0.3</td><td>0.2</td><td>-</td></tr>
</table>

7.3.3   Flammability of EVA/MWCNT compounds and EVA/MWCNT/organoclay compounds

It was now possible for the first time to investigate the flame retardant properties of carbon nanotube compounds using a cone calorimeter at 35 kW/m^2 [68,69]. All compounds were melt-blended in a Brabender mixing chamber. It is evident from the results in Table 7.7 that all the filled polymers had improved flame retardant properties. For EVA and the EVA-based nanocomposites containing 2.5 phr of filler, the PHRR decreased as follows: EVA > organoclays ~ purified MWCNTs. For EVA and EVA-based composites containing 5.0 phr of fillers, the PHRR decreased as follows: EVA > organoclays > purified MWCNTs = crude MWCNTs. Crude MWCNTs were as effective in the reduction of PHRR as purified MWCNTs! Increasing the filler content from 2.5 to 5.0 phr caused an additional flame retardant effect and this improvement was most significant when purified or crude MWCNTs were used.

A synergistic effect for flame retardancy between MWCNTs and organoclays was observed for the nanocomposite containing 2.5 phr of purified MWCNTs and 2.5 phr of organoclays (Figure 7.6). The latter sample was found to be the best flame retardant compound.

The variation of the screw velocity from 45 rpm to 120 rpm did not change the flame retardant properties for composites containing 5.0 phr of crude MWCNTs (Table 7.7, samples 6a and 6b).
There was also no reduction in time to ignition for the MWCNT-based EVA composite, in contrast to the only organoclay-based EVA composite, which undergoes an early thermal degradation of the quaternary ammonium compound within the galleries of the organoclay [70].

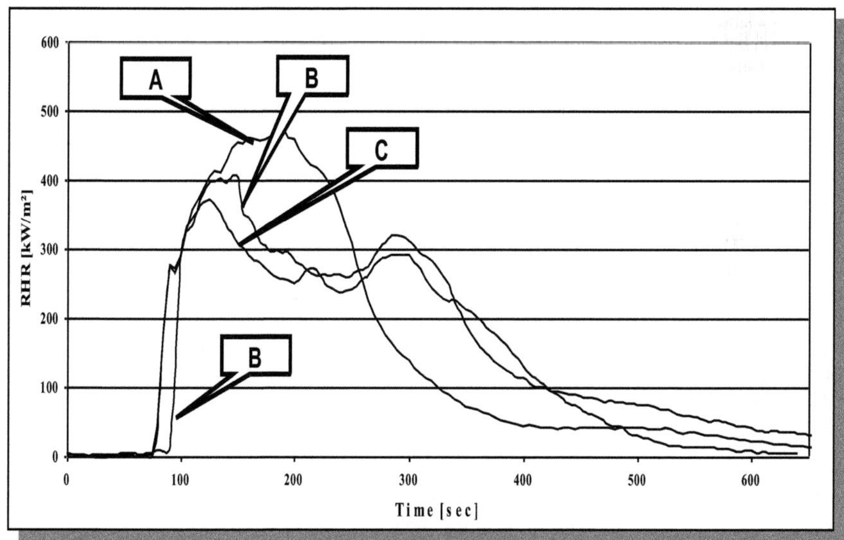

Figure 7.6
Rates of heat release at heat flux = 35 kW/m²
for various EVA-based materials :
EVA + 5,0 phr organoclays (A)
EVA + 5,0 phr pure MWCNTs (B)
EVA + 2,5 phr organoclays + 2,5 phr pure MWCNTs (C)
EVA : Escorene UL 00328 with 28 weight-% vinyl acetate content
Organoclay : Nanofil 15
Reproduced with permission from reference 55, Beyer, G. Flame retardancy of nanocomposites based on organoclays and carbon nanotubes with aluminium trihydrate. *Polymers for Advanced Technologies*, **2006**, 17, 218-225, copyright 2006, John Wiley & Sons Limited.

Table 7.7
Peak of heat release rates at heat flux = 35 kW/m²for various compounds with organoclays and multiwall carbon nanotubes
(EVA = Escorene UL 00328 with 28 weight-% vinyl acetate content, Organoclay = Nanofil 15)

Sample	EVA	MWCNT (phr)		Organolay	PHRR
	(parts resin)	Purif.	Crude	(phr)	(kW/m²)
EVA	100.0	-	-	-	580

1 *	100.0	2.5	-	-	520
2 *	100.0	5.0	-	-	405
3 *	100.0	-	-	2.5	530
4 *	100.0	-	-	5.0	470
5 *,**	100.0	2.5	-	2.5	370
6a *	100.0	-	5.0	-	403
6b ***	100.0	-	5.0	-	405

*: The screw velocity was 45 rpm and the mass temperature was 136 °C.
**: The nanotubes and the organoclay were premixed before their addition.
***: The screw velocity was 120 rpm and the mass temperature was 142 °C.

7.3.4 Crack density and surface results of charred MWCNT compounds

For flame retardant EVA-based composites containing 5 phr of fillers (Table 7.7), the crack density increased in the order purified MWCNTs < organoclays. A very important synergistic effect reducing the crack density to a very low level was observed for the nanocomposite containing both 2.5 phr purified MWCNTs and 2.5 phr organoclays (Figures 7.7a-c). The synergistic effect for improved flame retardancy by the filler combination MWCNTs and organoclays can be explained by the improved (less cracked) surface of the char. The char acted as an insulating and nonburning material with reduction of emission of volatile products (fuel) into the flame area. The fewer cracks present within the char, the better was the reduction of emission of fuel and therefore the reduction of PHRR. The fillers played an active role in the formation of this char, but obviously the MWCNTs, with their long aspect ratio, could strengthen it and made it more resistant to mechanical cracking.

Figure 7.7a
EVA with 5 phr organoclays after combustion

EVA : Escorene UL 00328 with 28 weight-% vinyl acetate content
Organoclay : Nanofil 15
Reproduced with permission from reference 55, Beyer, G. Flame retardancy of nanocomposites based on organoclays and carbon nanotubes with aluminium trihydrate. *Polymers for Advanced Technologies*, **2006**, 17, 218-225, copyright 2006, John Wiley & Sons Limited.

Figure 7.7b
EVA with 5 phr pure MWCNTs after combustion
EVA : Escorene UL 00328 with 28 weight-% vinyl acetate content
Reproduced with permission from reference 55, Beyer, G. Flame retardancy of nanocomposites based on organoclays and carbon nanotubes with aluminium trihydrate. *Polymers for Advanced Technologies*, **2006**, 17, 218-225, copyright 2006, John Wiley & Sons Limited.

Figure 7.7c
EVA with 2,5 phr pure MWCNTs and 2,5 phr organoclays after combustion
EVA : Escorene UL 00328 with 28 weight-% vinyl acetate content
Organoclay: Nanofil 15

Reproduced with permission from reference 55, Beyer, G. Flame retardancy of nanocomposites based on organoclays and carbon nanotubes with aluminium trihydrate. *Polymers for Advanced Technologies*, **2006**, 17, 218-225, copyright 2006, John Wiley & Sons Limited.

7.3.5   Flammability of LDPE/CNT compounds

Compounds of SWCNTs and MWCNTs in low density polyethylene (LDPE, BPD 8063 from INEOS) were melt blended in a Brabender mixing chamber according to the formulations indicated in Table 7.8 and Table 7.9. The corresponding cone calorimeter measurements are shown in Figures 7.8 and 7.9.

The results from the cone measurements of different carbon nanotubes in LDPE demonstrated that SWCNTs did not act as flame retardants in LDPE. MWCNTs acted as flame retardants in LDPE with no reductions on time to ignition (in contrast to organoclays), and crude MWCNTs showed similar reductions for the PHRR as purified MWCNTs.

Table 7.8 SWCNTs in LDPE			
Sample description	LDPE	SWCNT (weight-%)	
	(weight-%)	Purified	Crude
BPD 8063	100.0	-	-
5 SWCNT	95.0	5	-
10 SWCNT	90.0	10	-
5 crude SWCNT	95.0	-	5
10 crude SWCNT	90.0	-	10

Table 7.9 MWCNTs in LDPE			
Sample description	LDPE	MWCNT (weight-%)	
	(weight-%)	Purified	Crude
BPD 8063	100.0	-	-
5 MWCNT	95.0	5	-
10 MWCNT	90.0	10	-
5 crude MWCNT	95.0	-	5
10 crude MWCNT	90.0	-	10

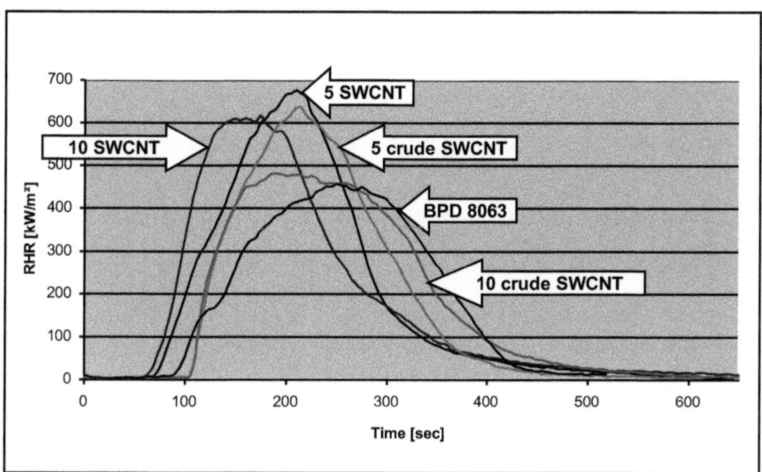

Figure 7.8
Rates of heat release for SWCNTs in LDPE
Heat flux = 35 kW/m²
Reproduced with permission from reference 71, Beyer, G. Filler blend of carbon nanotubes and organoclays
with improved char as a new flame retardant system for polymers and cable application. *Fire and Materials,*
2005, 29, 61-69, copyright 2005, John Wiley & Sons Limited.

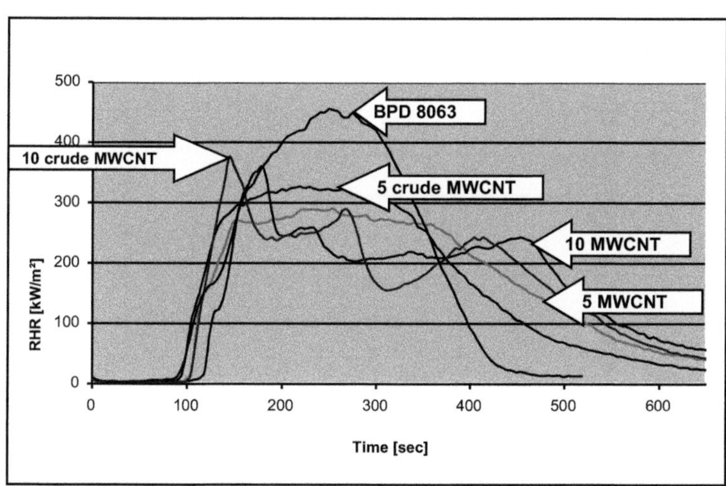

Figure 7.9
Rates of heat relase for MWCNTs in LDPE
Heat flux = 35 kW/m²

Reproduced with permission from reference 71, Beyer, G. Filler blend of carbon nanotubes and organoclays with improved char as a new flame retardant system for polymers and cable application. *Fire and Materials*, **2005**, 29, 61-69, copyright 2005, John Wiley & Sons Limited.

7.3.6   Cable with the new fire retardant system MWCNT/organoclay/ATH

It was of interest to apply the results seen for the synergistic filler-based FR-system based on MWCNTs and organoclays to real products [71]. Consequently, the Belgian carbon nanotube supplier Nanocyl SA synthesized 1.5 kg of MWCNTs. The aim of the experiment was to investigate whether the filler blend of MWCNTs and organoclays could be used in a real cable compound without causing processing problems. [71]. To produce a flame-retardant insulated wire using a real production cable extruder, a minimum of approximately 60 kg of compound was needed to fill the extruder and produce a small sample.

A well running purely organoclay-based cable compound designated compound 1 was selected, and the weight ratio between the organoclay and the MWCNTs was changed stepwise. The sum of both fillers always remained constant (Table 7.10) within the cable compounds. Compounding was carried out on a twin-roll mill, and the reductions of PHRRs for the three compounds were measured (Figure 7.10). The results clearly indicated that for the filler blend and the completely MWCNT-based compounds the first PHRR was reduced maximally. The second PHRR was observed at the longest times for the two compounds 2A and 2B, indicating that the chars were less cracked (more stable over time). Therefore a 1:1 blend of MWCNTs and organoclays was used (compound 2A, see Table 7.10 for the cable compound). This allowed production of the required quantity of the cable compound. Formulation 2A was compounded using an 11 L/D BUSS Ko-kneader with a 46 mm screw diameter, and 60 kg were produced without any processing problems. Processing on the BUSS Ko-kneader improved the PHRRs compared to those with the corresponding twin-roll mill compounding (Figure 7.11).

<table>
<tr><td colspan="2" align="center">Table 7.10
Blends of MWCNTs and organoclays in a cable compound formulations</td></tr>
<tr><td align="center">Compound</td><td align="center">Composition</td></tr>
<tr><td align="center">1</td><td align="center">Technical cable compound
(EVA-PE-ATH-organoclay / processing additives)</td></tr>
<tr><td align="center">2A</td><td align="center">Same formulation as compound 1, but 50 % organoclay substituted by the same amount of MWCNTs</td></tr>
<tr><td align="center">2B</td><td align="center">Same formulation as compound 1, but 100 % organoclay substituted by the same amount of MWCNTs</td></tr>
</table>

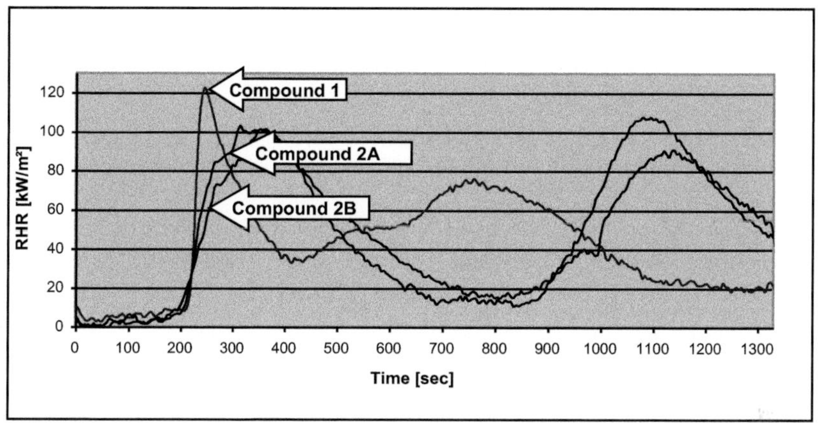

Figure 7.10

Rates of heat release for compounds made by twin-roll mill with different filler blends "MWCNTs / organoclays"
Heat flux = 35 kW/m²
Reproduced with permission from reference 55, Beyer, G. Flame retardancy of nanocomposites based on organoclays and carbon nanotubes with aluminium trihydrate. *Polymers for Advanced Technologies*, **2006**, 17, 218-225, copyright 2006, John Wiley & Sons Limited.

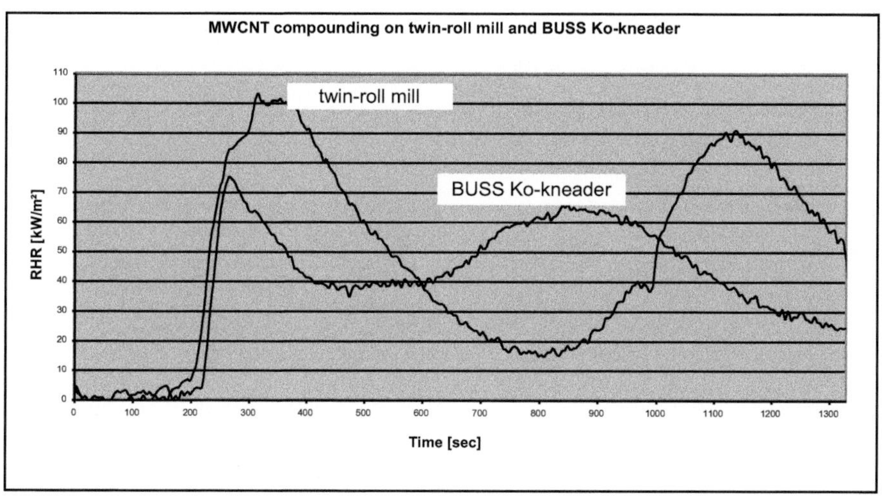

Figure 7.11

Rates of heat release for cable compound 2A with the filler blend "MWCNTs / organoclays"
by twin-roll mill and BUSS Ko-kneader
Heat flux = 35 kW/m²
Reproduced with permission from reference 55, Beyer, G. Flame retardancy of nanocomposites based on organoclays and carbon nanotubes with aluminium trihydrate. *Polymers for Advanced Technologies*, **2006**, 17, 218-225, copyright 2006, John Wiley & Sons Limited.

Two insulated wires with identical geometric parameters were produced on a 20 L/D single-screw cable extruder with an 80 mm screw diameter. The diameter of the copper conductor was 1.78 mm, and the wall thickness for the insulation was 0.8 mm. For one wire the insulation was compound 1 (filler combination based on ATH-organoclay) and for the other wire the optimised compound 2A (filler combination of ATH-organoclay-MWCNTs) was used as insulation.

The MWCNT-based compound 2A showed a remarkable increase in viscosity over that of the standard nanocomposite compound 1, as indicated by reduced revolutions per minute of the screw and increased power take-up by the extruder motor. A high pressure capillary viscosimeter showed a higher viscosity for the compound 2A for all shear rates up to 3000 s^{-1}.

Results of a small scale fire test according to IEC 60332-1 (the insulation was exposed to a Bunsen burner ignition source) were very similar for both insulated wires. No dripping of burning polymer was noted, and the charred lengths were identical. However the char of the insulation made with the compound 2A was much less cracked than the char generated from the compound 1. This may be the result of the strengthening effect of the MWCNT, with its very great L/D ratio; a proposal of such a mechanism for the function of carbon nanotubes was published recently [72].

Heat release rates and times to ignition of the two insulated wires were measured using a cone calorimeter. The wires were cut in samples of 10 cm, and the standard cone sample holder was filled up with the cable pieces. Twenty-six wires were mounted by building up a single layer of wires with no gap between them [64] and the ends of the wires were not sealed; this mounting was called a *single layer design*. Also, four cut wires with no sealing of the ends, simulating an unjacketed cable [73] were put together. An aramid fibre binder was used to maintain the integrity of the bundles; this mounting with 24 wires/layer and two layers in total was called a *bundle design* (Figure 7.12).

Both mounting designs demonstrated quite different cone calorimeter results (Figures 7.13 and 7.14).

For the single layer design the cone tests were finished within 20 minutes. This is the flame application time for cables as defined in the new European proposal for flame tests of cables (prEN 50399). Wires insulated with the compound 1 had a higher PHRR within the first 5 minutes. For the bundle design, the duration of the cone calorimeter tests was very long, due to the insulation effect of the first charred layer on those located underneath in the second layer; therefore, the tests were stopped after 20 minutes. This mounting design is representative of many end-product applications. Wires insulated with the filler-blended compound 2A did not show any increase in the PHRR within the first 10 minutes in contrast with the wires insulated with the purely organoclay-based compound 1.

Figure 7.12
Different insulated wire mounting designs for cone calorimeter tests
Reproduced with permission from reference 55, Beyer, G. Flame retardancy of nanocomposites based on organoclays and carbon nanotubes with aluminium trihydrate. *Polymers for Advanced Technologies*, **2006**, 17, 218-225, copyright 2006, John Wiley & Sons Limited.

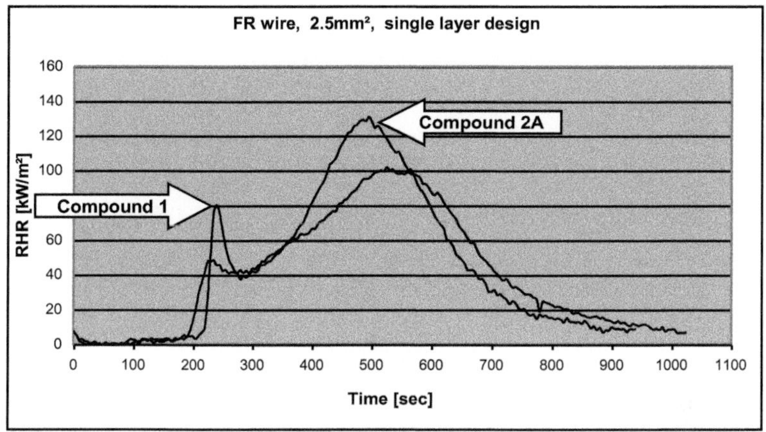

Figure 7.13
Rates of heat release for the single layer design
Heat flux = 35 kW/m²
Reproduced with permission from reference 55, Beyer, G. Flame retardancy of nanocomposites based on organoclays and carbon nanotubes with aluminium trihydrate. *Polymers for Advanced Technologies*, **2006**, 17, 218-225, copyright 2006, John Wiley & Sons Limited.

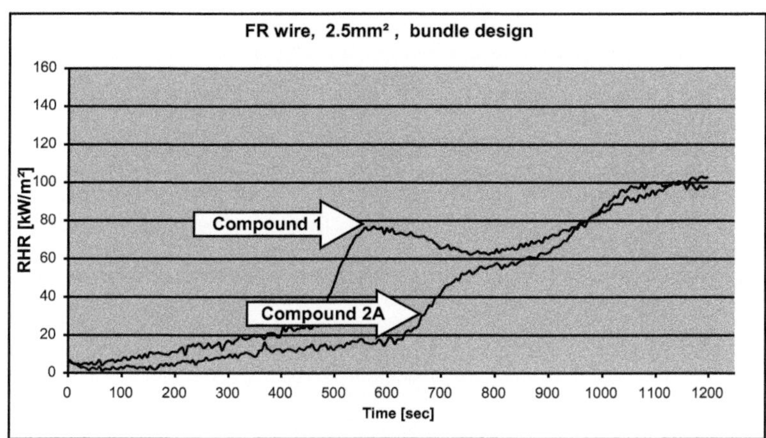

Figure 7.14
Rates of heat release for the bundle design
Heat flux = 35 kW/m²
Reproduced with permission from reference 55, Beyer, G. Flame retardancy of nanocomposites based on organoclays and carbon nanotubes with aluminium trihydrate. *Polymers for Advanced Technologies*, **2006**, 17, 218-225, copyright 2006, John Wiley & Sons Limited.

7.4 Outlook

The use of nanocomposites based on organoclays in combination with metal hydroxides is now regularly seen in the cable industry. Kabelwerk EUPEN AG from Belgium, the inventor of this technology, has used it over a number of years for their range of flame retardant non-halogen cables. The company also has worldwide patents on this technology. Other companies will follow this way for their flame retardant product ranges. In Europe one can expect a wider application of the nanocomposite-based flame retardant technology driven by the enhanced requirements of the Construction Products Directive. It is also likely that countries including China and Japan which are following the European trend towards non-halogenated flame-retardant products will adopt this technology in the near future.

However, and despite all of the success to date, there are still areas which need technological breakthroughs or improved understanding to develop the flame retardant nanocomposites. In all of the reported examples, the organic treatment of the montmorillonite clay was a quaternary alkyl ammonium cation. While technical success was seen in all of these systems, the quaternary alkyl ammonium cation can show some problems, most notably a lack of thermal stability above 200 °C due to a Hoffman degradation at this temperature [74] If the organic treatment decomposes, the clay will lose its interface with the polymer, and the polymer will de-intercalate, forming a microcomposite, which in turn results in a loss of flammability properties [70]. Possible solutions have been published to improve the thermal stability of the organic treatment, including phosphonium [75] and imidazolium treatments [76–78].

7.5 Summary

Nanocomposites made by melt-blending poly(ethylene-co-vinyl acetate, EVA) with modified layered silicates are used for halogen-free flame-retardant cables. Combinations of organoclays and traditional flame retardants based on metal hydroxides like ATH or MDH must be used for proper flame retardancy, as outlined for a coaxial cable fulfilling the UL 1666 riser test and medium-voltage cables passing the IEC 60332-3 fire tests. For the new European fire tests, which will be regulated by the Construction Products Directive, nanocomposite-based cables demonstrate benefits.

MWCNTs act as very efficient flame retardants at low filler contents in EVA. An optimized formulation for flame retardant insulated wires was developed based on the filler blends of 'multiwall carbon nanotubes and organoclays'. Small scale fire tests according to IEC 60332-1, with flames of a Bunsen burner attacking the insulation of the wire, showed that the char was strengthened by the long L/D ratio of the MWCNTs. The improved char results in better flame retardant performance of the wires.

Much research is currently underway around the world to investigate the combinations of organoclays and other nanostructured fillers in combination with traditional non-halogenated or halogenated traditional flame retardants for technical applications, to enhance the worldwide markets for flame retardant polymers.

REFERENCES

1   Hirschler, M.M. Fire performance of organic polymers, thermal, decomposition, and chemical decomposition. *Polymeric Materials: Science and Engineering*, Vol. 83, pages 79-80, ACS Meeting, Washington, DC August **2000**

2   Babrauskas, V. The generation of CO in bench scale fire tests and the prediction for real scale fires. *Fire and Materials*, **1995**, 19, 205-213

3   Hirschler, M.M. Fire safety, smoke toxicity and halogenated materials. Commentary in: *Flame Retardancy News*, April **2005**, Business Communications Co., Norwalk, CT, USA

4   Stevens, G.C. Countervailing risks and benefits in the use of flame retardants. *Proceedings of the Flame Retardants 2000 Conference*, pages 131-145, Interscience Communications, UK, 8-9 February **2000**

5   Ray, S.S.; Okamoto, M. Polymer/layered silicate nanocomposites: A review from preparation to processing. *Progress in Polymer Science*, **2003**, 28, 1539-1641

6   Beyer, G. Nanocomposites offer new way forward for flame retardants. *Plastics Additives & Compounding*, **2005**, September / October issue, 7, 32-35

7   Gilman,J.W.; Kashiwagi, T.; Giannelis, E.P.; Manias ,E.; Lomakin, S.; Lichtenhan, J.D.; Jones, P. Nanocomposites: Radiative gasification and vinyl polymer flammability. *Fire Retardancy of Polymers: The use of Intumescence*, Le Bras, M.; Camino, G.; Bourbigot, S.; Delobel, R. (editors), ISBN 0-85404-738-7,Royal Society of Chemistry, Cambridge, **1998**, pages 203-221

8   Gilman, J.W.; Kashiwagi, T.; Lichtenhan, J.D. A revolutionary new flame retardant approach. *SAMPE J.,* **1997**, 33, 40-46

9   Camino, G.; Zanetti, M.; Riva, A.; Braglia, M.; Falqui, L. Thermal degradation and rheological behaviour of EVA/montmorillonite nanocomposites. *Polymer Degradation and Stability*, **2002**, 77, 299-304

10   Hu, Y.; Tang, Y.; Wang, S.; Gui, Z.; Fan, W., Chen Z., Preparation and flammability of ethylene-vinyl acetate copolymer/montmorillonite nanocomposites. *Polymer Degradation and Stability*, **2002**, 78, 555-559

11   Camino, G.; Mülhaupt, R.; Zanetti, M.; Thomann; R. Synthesis and thermal behaviour of layered silicate–EVA nanocomposites. *Polymer*, **2001**, 42, 4501-4507

12   Zanetti, M.; Camino, G.; Mülhaupt, R. Combustion behaviour of EVA/fluorohectorite nanocomposites. *Polymer Degradation and Stability*, **2001**, 74, 413-417

13   Hu, Y.; Lu, H.; Kong, Q.; Cai, Y.; Chen, Z.; Fan, W. Influence of gamma irradiation on high density polyethylene/ethylene-vinyl acetate/clay nanocomposites. *Polymers for Advanced Technologies*, **2004**, 15, 601-605

14   Hu, Y.; Lu, H.; Kong, Q.; Chen, Z.; Fan, W. Gamma irradiation of high density poly(ethylene)/ethylene-vinyl acetate/clay nanocomposites: possible mechanism of the influence of clay on irradiated nanocomposites. *Polymers for Advanced Technologies*, **2005**, 16, 688-692

15   Sundararaj, U.; Zhang, F. Nanocomposites of ethylene-vinyl acetate copolymer (EVA) and organoclay prepared by twin-screw melt extrusion. *Polymer Composites*, **2004**, 25, 535-542

16   Camino, G.; Gianelli, W.; Dintcheva, N.T.; Verso, S.L.; La Mantia, F.P. EVA-Montmorillonite Nanocomposites: Effect of Processing Conditions. *Macromolecular Materials and Engineering*, **2004**, 289, 238-244

17   Hu, Y.; Tang, Y.; Wang, J.; Gui, Z.; Chen, Z.; Zhuang, Y.; Fan, W.; Zong, R. Influence of organophilic clay and preparation methods on EVA/montmorillonite nanocomposites. *Journal of Applied Polymer Science*, **2004**, 91, 2416-2421

18   Morgan, A.B.; Chu, L.; Harris, J.D. A flammability performance comparison between synthetic and natural clays in polystyrene nanocomposites. *Fire and Materials*, **2005**, 29, 213-229

19   Wilkie, C.A..; Costache, M.C.; Jiang, D.D. Thermal degradation of ethylene–vinyl acetate copolymer nanocomposites. *Polymer*, **2005**, 46, 6947-6958

20   Wilkie, C.A..; Costache, M.C.; Jang, B.N. The relationship between thermal degradation behavior of polymer and the fire retardancy of polymer/clay nanocomposites. *Polymer*, **2005**, 46, 10678-10687

21   Frache, A.; Constantino, U.; Gallipoli, A.; Nchetti, M; Camino, G.; Bellucci, F. New nanocomposites constituted of polyethylene and organically modified ZnAl-hydrotalcites. *Polymer Degradation and Stability*, **2005**, 90, 586-590

22   Nelson, G.L.; Yngard, R.; Yang, F. Flammability of polymer-Clay and polymer-silica nanocomposites. *Journal of Fire Sciences*, **2005**, 23, 209-226

23   Zammarano, M.; Franceschi, M.; Bellayer, S.; Gilman, J.W.; Meriani, S. Preparation and flame resistance properties of revolutionary self-extinguishing epoxy nanocomposites based on layered double hydroxides. *Polymer*, **2005**, 46, 9314-9328

24   Zammarano, M.; Franceschi, M.; Mantovani, F.; Minigher, A.; Celotto, M.; Meriani, S. Flame resistance properties of layered-double-hydroxides/epoxy nanocomposites; *Proceedings of the 9th European Meeting on Fire Retardancy and Protection of Materials*, M. Le Bras et al. (editors), USTL Pub., Villeneuve D'Ascq. France, 17-19 September **2003**

25   Babrauskas, V. Development of the cone calorimeter: A bench scale heat release rate apparatus based on oxygen consumption, *Fire and Materials*, **1984**, 8, 81-95

26   Limiting Oxygen Index, Standard test method for measuring minimum oxygen concentration to support candle-like combustion of plastics, ASTM standard D2863, 2008, http://www.astm.org.

27   UL 94, Test for flammability of plastic materials for parts in devices and appliances,1966-10-00, Underwriters Laboratories Inc., http://ulstandardsinfonet.ul.com.

28   Bartholmai, M; Schartel, B. Layered silicate polymer nanocomposites: New approach or illusion for fire retardancy ? Investigations of the potentials and the tasks using a model system. *Polymers for Advanced Technologies*, **2004**, 15, 355-364

29   Bourbigot, S; Gilman, J. W.; Wilkie, C.A. Kinetic analysis of the thermal degradation of polystyrene-montmorillonite nanocomposite. *Polymer Degradation and Stability*, **2004**, 84, 483-492

30   Kashiwagi, T.; Grulke, E.; Hilding, J.; Harris, R.; Awad, W.; Douglas. J. Thermal degradation and flammability properties of poly(propylene)/carbon nanotube composites. *Macromolecular Rapid Communications*, **2002**, 23, 761-765

31   Kashiwagi, T.; Grulke, E.; Hilding, J.; Groth K.; Harris, R.; Butler, K.; Shields, J.; Kharchenko, S.; Douglas, J. Thermal and flammability properties of polypropylene/carbon nanotube nanocomposites. *Polymer,* **2004**, 45, 4227-4239

32   Kashiwagi, T.; Du, F.; Winey, K.; Groth, K.; Shields, J.; Bellayer, S.; Kim, H.; Douglas, J. Flammability properties of polymer nanocomposites with single-walled carbon nanotubes: effects of nanotube dispersion and concentration. *Polymer,* **2005**, 46, 471-481

33   Kashiwagi, T.; Du, F.; Douglas, J.; Winey, K.I.; Harris, R.; Shields, J. Nanoparticle networks reduce the flammability of polymer nanocomposites. *Nature Materials,* **2005**, 4, 928-933

34   Schartel, B.; Pötschke, P.; Knoll, U.; Abdel-Goad, M. Fire behaviour of polyamide 6/multiwall carbon nanotube nanocomposites. *European Polymer Journal,* **2005**, 41, 1061-1070

35   Leroy, E.; Lopez-Cuesta, J.-M.; Clerc, L. Influence of talc physical properties on the fire retarding behaviour of (ethylene–vinyl acetate copolymer/magnesium hydroxide/talc) composites. *Polymer Degradation and Stability,* **2005**, 88, 504-511

36   Ferry, L.; Durin-France, A.; Lopez-Cuesta, J.-M.; Crespy, A. Magnesium hydroxide/zinc borate/talc compositions as flame-retardants in EVA copolymer. *Polymer International,* **2000**, 49, 1101-1105

37   Le Bras, M.; Bourbigot, S.; Carpentier, F.; Delobel, R. Rheological investigations in fire retardancy: application to ethylene-vinyl-acetate copolymer-magnesium hydroxide/zinc borate formulations. *Polymer International,* **2000**, 49, 1216-1221

38   Jho, J.Y.; Hong, C.H.; Lee, Y.B.; Bae, J.W.; Nam, B.U.; Nam, G.J.; Lee, K. J. Tensile and flammability properties of polypropylene-based RTPO/clay nanocomposites for cable insulating material. *Journal of Applied Polymer Science,* **2005**, 97, 2375-2381

39   Wilkie, C.A.. Nanocomposite formation as a component of a fire retardant system. *Proceedings of the 10th European Meeting on Fire Retardancy and Protection of Materials,* FRMP '05, BAM/EUROLAB, Berlin, Germany, 07- 09 September **2005**

40   Bourbigot, S.; Le Bras, M.; Dabrowski, F.; Gilman, J. W.; Kashiwagi, T. PA-6 clay nanocomposite hybrid as char forming agent in intumescent formulations. *Fire and Materials,* **2000**, 24, 201-208

41   Hu, Y.; Tang, Y.; Xiao, J.; Wang, J.; Song, L.; Fan, W. PA-6 and EVA alloy/clay nanocomposites as char forming agents in poly(propylene) intumescent formulations. *Polymers for Advanced Technologies,* **2005**, 16, 338-343

42   Hu, Y.; Song, L.; Lin, Z.; Xuan, S.; Wang, S.; Chen, Z.; Fan, W. Preparation and properties of halogen-free flame-retarded polyamide 6/organoclay nanocomposite. *Polymer Degradation and Stability,* **2004**, 86, 535-540

43   Ferry, L.; Gaudon, P.; Leroy, E.; Lopez Cuesta, J-M. Intumescence in ethylene-vinyl acetate copolymer filled with magnesium hydroxide and organoclays. *Fire Retardancy of Polymers: new applications of mineral fillers,* pages 302-312, Royal Society of Chemistry, **2005**, ISBN 0-85404-582-1

44   Horrocks, A.R.; Kandola, B.K.; Padbury, S. A. Interactions between nanoclays and flame retardant additives in polyamide 6, and polyamide 6,6 films. *Fire Retardancy of Polymers: new applications of mineral fillers,* pages 223-238 Royal Society of Chemistry, **2005**, ISBN 0-85404-582-1

45   Whaley, P.D.; Cogen, J.M.; Lin, Th. S.; Bolz, K.A. Nanocomposite flame retardant performance: laboratory testing methodology. *Proceedings of the 53rd International Wire & Cable Symposium,* www.iwcs.org, November **2004**, pages 605-611, Philadelphia, Pennsylvania, USA,

46   Shen, K.K.; Olsen, E. Borates as FR in halogen-free polymers. *Proceedings of the 15th Annual BCC Conference on Flame Retardancy,* Stamford, CT, USA, 06-09 June **2004**

47   Shen, K.K.; Olsen, E. Recent advances on the use of borates as fire retardants in halogen-free systems. *Proceedings of the 16th Annual BCC Conference on Flame Retardancy,* Stamford,CT, USA, 22-25 May **2005**

48   Ristolainen, N.; Hippi, U.; Seppälä, J.; Nykänen, A.; Ruokolainen, J. Properties of polypropylene/aluminum trihydroxide composites containing nanosized organoclay. *Polymer Engineering & Science,* **2005**, 45, 1568- 1575

49   Wilkie, C.A..; Zhang, J. Fire retardancy of polyethylene-alumina trihydrate containing clay as a synergist. *Polymers for Advanced Technologies,* **2005**, 16, 549-553

50   Cusack, P.A.; Cross, M.S.; Hornsby, P.R. Effects of tin additives on the flammability and smoke emission characteristics of halogen-free ethylene-vinyl acetate copolymer. *Polymer Degradation and Stability*, **2003**, 79, 309-318

51   Beyer, G.; Alexandre, M.; Henrist, C.; Cloots, R.; Rulmont, A.; Jérôme, R.; Dubois, Ph.Poster-presentation: Preparation, morphology, mechanical and flame retardant properties of EVA/layered silicate Nanocomposites. *World Polymer Congress, IUPAC Macro 2000, 38th International Symposium on Macromolecules*, Warsaw, **2000**

52   Beyer, G.; Alexandre, M.; Henrist, C.; Cloots, R.; Rulmont, A.; Jérôme, R.; Dubois, Ph. Preparation and properties of layered silicate nanocomposites based on ethylene vinyl acetate Copolymers *Macromolecular Rapid Communications*, **2001**, 22, 643-646

53   Beyer,G., Flame retardant properties of EVA-nanocomposites and improvements by combination of nanofillers with aluminium trihydrate. *Fire and Materials*, **2001**, 25, 193-197

54   Babrauskas, V.; Peacock, R.D. Heat release rate: the single most important variable in fire hazard,.*Fire Safety Journal*, **1992**, 18, 255-272

55   Beyer, G. Flame retardancy of nanocomposites based on organoclays and carbon nanotubes with aluminium trihydrate. *Polymers for Advanced Technologies*, **2006**, 17, 218-225

56   Bourbigot, S.; Le Bras, M.; Leeuwendal, R.; Shen, K.K.; Schubert, D. Recent advances in the use of zinc borates in flame retardancy of EVA. *Polymer Degradation and Stability*, **1999**, 64, 419-425

57   Beyer, G. Flame retardancy of nanocomposites – from research to technical products. *Journal of Fire Sciences*, **2005**, 23, 75-87

58   IEC 60332-3-24, Tests on electrical cables under fire conditions - Part 3-24: Test for vertical flame spread of vertically-mounted bunched wires or cables; Category C, 2000-10-00, International Electrotechnical Commission, http://www.iec.ch.

59   Herbert, M.J.; Brown, S.C. New developments in ATH technology and applications. *Proceedings of the Flame Retardants '92 Conference*, UK Plastics and Rubber Institute, Elsevier Applied Science, pages 100-119, London, UK, 12-13 January, **1992**

60   L 1666: UL standard for safety test for flame propagation height of electrical and optical-fiber cables installed vertically in shafts, http://ulstandardsinfonet.ul.com.

61   Calveras,D.; Grabolosa, J.; Canerot, I. Use of nanofillers in high performance of MV cables under fire conditions. Presentation at the 7th JICABLE Conference (international conference on insulated power cables) Société de l'Électricité, de l'Électronique & des Technologies de l'Information et de la Communication, Versailles, France, 24 to 28 June, **2007**

62   IEC 60332-3-23, Tests on electrical cables under fire conditions - Part 3-24: Test for vertical flame spread of vertically-mounted bunched wires or cables; Category B, 2000-10-00, International Electrotechnical Commission, http://www.iec.ch.

63   Schartel, B.; Hartwig, A.; Pütz, D.; Bartholmai, M.; Wendschuh-Josties, M. Combustion behaviour of epoxide based nanocomposites with ammonium and phosphonium bentonites. *Molecular Chemistry and Physics*, **2003**, 204, 2247-2257

64   Morgan, A. Flame retarded polymer layered silicate nanocomposites: a review of commercial and open literature systems. *Polymers for Advanced Technologies*, **2006**, 17, 206-217

65   Breuer, O.; Sundararaj, U. Big returns from small fibers: a review of polymer/carbon nanotube composites. *Polymer Composites*, **2004**, 25, 630-645

66   Tang, B.Z..; Xu, H. Preparation, alignment, and optical properties of soluble poly(phenylacetylene)-wrapped carbon nanotubes. *Macromolecules*, **1999**, 32, 2569-2576

67 Shaffer, M. Carbon nanotube modified polymers. *ERO Workshop Nanostructures in Polymer Matrices*, Risley Hall, Derbyshire, UK, 10-13 September, **2001**

68   Beyer, G. Carbon nanotubes as flame retardants for polymers. *Fire and Materials*, **2002**, 26, 291-293

69   Beyer, G. Improvements of the fire performance of nanocomposites. *Proceedings of the 13th Annual BCC Conference on Flame Retardancy*, Stamford,CT, USA, 03-06 June **2002**

70   Gilman, J.W.; Jackson, C.L.; Morgan, A.B.; Harris, R.; Manias, E.; Giannelis, E.P.; Wuthenow, M.; Hilton, D.; Philipps, H. Flammability properties of polymer-layered-silicate nanocomposites. Polypropylene and polystyrene nanocomposites. *Chemistry of Materials*, **2000**, 12, 1866-1873

71   Beyer, G. Filler blend of carbon nanotubes and organoclays with improved char as a new flame retardant system for polymers and cable application. *Fire and Materials*, **2005**, 29, 61-69

72   Beyer, G.; Gao, F.; Yuan, Q. A mechanistic study of fire retardancy of carbon nanotube/ethylene vinyl acetate copolymers and their clay composites. *Polymer Degradation and Stability*, **2005**, 89, 559-564

73   Elliot, P.J.; Whiteley, R.H. A cone calorimeter test for the measurement of flammability properties of insulated wire. *Polymer Degradation and Stability*, **1999**, 64, 577-584

74   Xie, W.; Gao Z.; Pan, W. P.; Hunter, D.; Singh, A.; Vaia, R. Thermal degradation of alkyl quaternary ammonium montmorillonite. *Chemistry of Materials*, **2001**, 13, 2979–2990

75   Xie, W.; Xie, R.; Pan, W. P.; Hunter, D.; Koene, B.; Tan L. S.; Vaia, R. Thermal stability of quaternary phosphonium modified montmorillonites. *Chemistry of Materials*, **2002**, 14, 4837–4845

76   Gilman, J. W.; Awad; W.H.; Davis, R.D.; Shields, J.; Harris, R.H.; Davis, C.; Morgan, A.B.; Sutto, T.E.; Callahan, J.; Trulove, P.C.; DeLong, H. C. Polymer/layered silicate nanocomposites from thermally stable trialkylimidazolium-treated montmorillonite. *Chemistry of Materials*, **2002**, 14, 3776-3785

77   Bottino, F.A.; Fabbri, E.; Fragala, I.L.; Malandrino, G.; Orestano, A.; Pilati, F.; Pollicino, A. Polystyrene-clay nanocomposites prepared with polymerizable imidazolium surfactants. *Macromolecular Rapid Communications*, **2003**, 24, 1079–1084

78   Wang, Z.M.; Chung, T.C.; Gilman, J.W.; Manias, E. Melt-processable syndiotactic polystyrene/montmorillonite, nanocomposites. *Journal Polymer Science* B, **2003**, 41, 3173–3187

Chapter 8
Polyhedral oligomeric silsesquioxane flame retardancy
Joseph Lichtenhan,
Hybrid Plastics Inc., Hattiesburg, Mississippi, USA

8.1   Introduction

This chapter explores the use of another class of nanoscale materials, the polyhedral oligomeric silsesquioxanes in retarding the flammability of polymers. Polyhedral oligomeric silsesquioxanes are a very broad family of chemicals sold under the trade name POSS®. They are properly categorized as nanostructured chemicals and commercially utilized as monomers, grafting agents, processing aids, catalysts, and surface modification agents [1].

8.2   POSS chemical technology and unique features

POSS [1] are defined by the following features: They are *single molecules with an exact chemical formula weight*. They have rigid *polyhedral geometries* with well-defined three-dimensional cage-like shapes. They have specific *nanoscopic sizes* which range between 0.7 nm to 5 nm in diameter, depending on the central cage size and type of organic substituents. They have *systematic chemistries* that enable precise control over size, stereochemistry, reactivity, physical properties and chemical compatibility. The topology of POSS cages controls their physical state and consequently they exist as solids, waxes, and liquids. POSS are manufactured in multi-ton scale from commodity silanes via a one step self-assembly process. In some instances the cage core can be derived directly from silicates [2]. The inorganic silicon oxygen core is hollow, and both thermally and chemically robust. The central core is externally functionalized with organic groups to provide compatibility with man-made or biological systems (Figure 8.1).

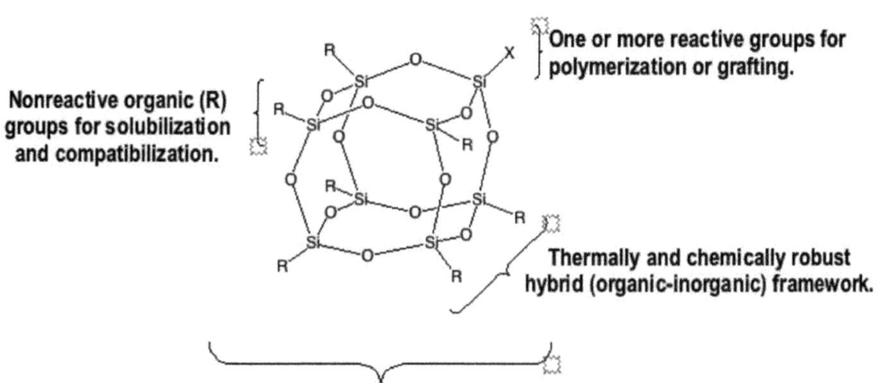

Figure 8.1

Anatomy of a POSS Chemical

The chemical nature of POSS enables its incorporation into polymers to be thermodynamically governed by the Gibbs free energy of mixing equation ($\Delta G = \Delta H - T\Delta S$). The nature of the R group and the ability of reactive groups on the cage to interact with polymers and surfaces contribute to a favourable enthalpic (ΔH) term while the entropic term (ΔS) is significant due to the molecular nature and modest molar mass of the cage. Despite their high molar mass, of approximately 1000 atomic mass units (amu), POSS densities are typically in the range of 1.12 g/ml. This is because the cage core occupies only 5 % of the overall cage volume. Most of the volume is occupied by the R groups, which lie on the outside of the cage. Matching the chemical potential between a POSS cage and the desired polymer system enables POSS to become dispersed at the 1-3 nm level. Consequently POSS can contribute approximately 3600 m^2/g of surface area and this is an important consideration for flammability control, surface modification, and thermomechanical properties.

The incorporation of microscale and nano-scale particles as fillers into polymers has been scientifically well explored with respect to fire retardancy. The conclusions for these studies are (1) macroscopic and nanoscopic particles can serve as a heat shield to limit thermal conduction into the bulk sample (2) macroscopic and nanoscopic particles can limit the evolution of gas "fuel" due to formation of a char layer, and (3) macroscopic and nanoscopic particles can reduce surface regeneration due to an increase in melt viscosity [3]. The inability to quickly regenerate a surface during combustion reduces the rate of heat released while burning.

Systematic studies on the effect of decreasing particle sizes to the 1 nm level (or at least below 10 nm), where the increase of interface and surface area is unprecedented, do not exist. Similarly there are no reports on the fire retardancy of POSS copolymers, or POSS surface-modified fillers. Literature on the fire retardancy of POSS systems has primarily involved blending individual POSS cages and silsesquioxane resins with polymers. These findings reveal that flammability performance of polymer/POSS blends depends on the type of polymer matrix, its compatibility with POSS, and the composition and structure of the POSS. The inorganic-organic composition of POSS is intriguing with respect to char formation, crosslinking, and reduction of volatiles at the burning polymer surface primarily because of the molecule's resemblance to a silica particle, and the favourable impact of nanoparticles toward improving fire retardance as discussed above.[4] The ability to incorporate metals into the cage is also intriguing as a method to control solid-state chemistry as an additional fire retarding mechanism [5]. In light of known fire resistant organic polymer compositions, and previous findings for nanoparticulates, it would seem reasonable to assume that POSS cages containing R groups such as phenyl and vinyl may improve fire performance via crosslinking and charring. Yet it should be realized that the central core of an octameric Si_8O_{12} POSS molecule is able to contribute only 416 amu of $SiO_{1.5}$ to a char layer. Therefore the amount of POSS loading needed to realize a uniform surface SiO_2 char would dissuade its singular use as a fire retardant. Secondly the temperature at which a POSS cage will react to form a char needs to be matched to the decomposition temperature of the polymer. Thirdly symmetrically substituted octameric POSS cages bearing phenyl and vinyl groups are notoriously insoluble and disperse poorly in polymers. Consequently use of these systems has not been productive for attaining the desired fire performance.

The mechanism for thermochemical decomposition for POSS cages has been well studied [6]. POSS cages bearing R groups such as methyl ($MeSiO_{1.5}$)$_8$ or isobutyl ($iBuSiO_{1.5}$)$_8$ are well known to sublime at 200 °C despite their respective molar masses of 537 amu and 874 amu. Interestingly even the octavinyl POSS cage (Vinyl-$SiO_{1.5}$)$_8$ is known to have approximately 100 mTorr vapour pressure at 150 °C while undergoing simultaneous sublimation and charring at 200 °C (reaching 1100 mTorr vapour pressure) [7]. Many POSS cages however undergo homolytic cleavage of their organic groups at approximately 360 °C in air and these are lost as volatile (and combustible) byproducts. For certain POSS systems the oxidation process can be so efficient that white powders resembling silica result [8]. The remaining cage core simultaneously undergoes solid-state crosslinking reactions to produce a glassy surface char layer. In most instances the cage structure is retained up to 550 °C where upon further heating an amorphous silicon-oxy-carbide char is formed. POSS cages bearing phenyl and imide groups show particularly interesting thermochemistry. For example, octaphenyl POSS $(PhSiO_{1.5})_8$ and dodecaphenyl POSS $(PhSiO_{1.5})_{12}$ undergo competitive homolytic bond cleavage and crosslinking beginning at approximately 500 °C and the process is complete around 625 °C after a 20 % total mass loss. The 80 % char yields are higher than what could be expected from the Si_8O_{12} cage contribution alone. The glassy black nature of the char and gravimetric data indicate that charring of phenyl groups and their incorporation into black glass occurs for these materials. While this analysis may seem useful it is only likely to be applicable to high temperature materials such as phenolics, polyimides, and bismaleimides, which have decomposition temperatures in the same range as aromatic POSS systems. Phenyl-containing POSS have been successfully introduced into phenolic resins marketed by Thermalguard Technology LLC [9-10]. For phenolics the POSS cage is utilized to improve the oxidation resistance and thermal insulating characteristics of phenolic char layers.

Incorporation of high temperature POSS into engineering plastics would not be expected to provide significant fire resistance because polymer depolymerization would occur well before POSS decomposition. One notable exception is the use of octatetramethylammonium POSS (octa-TMA) in melt blended polystyrene. Despite the inherent incompatibility of octa-TMA POSS with polystyrene, the peak heat release rate was reduced by 54.5 % and the average heat release rate was reduced by 41.5 % relative to polystyrene. The intriguing aspects of this report include the synergistic use of octa-TMA POSS to release water during combustion and to simultaneously form a netlike surface char (Figure 8.2).

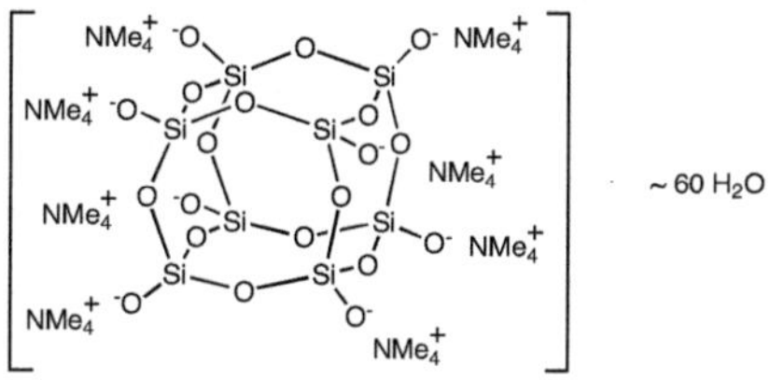

Figure 8.2
Chemical structure of octatetramethylammonium (octa-TMA) POSS hydrate

8.3   Successful use of POSS as a fire retardant

In addition to the POSS phenolic systems mentioned previously, the use of POSS to achieve fire retardancy in polyphenylene ether (PPE) [11] and polycarbonate (PC) [12] and to simultaneously improve their processability has been reported. Ikeda's work [11 (a-c)] provides an excellent opportunity to examine the impact of cage structure, composition and chemical-level dispersion on flammability. Using a twin screw compounder, Ikeda prepared 5 weight-% POSS / PPE blends from four related POSS chemicals (Figure 8.3). Each of the compounds is capable of reacting with PPE and thereby achieving an acceptable level of dispersion. The data indicate that amine-containing systems were most effective at reducing the PPE flammability rating. This result is surprising and counter intuitive given that phenyl silsesquioxane resin (B) shows a 90 weight-% char yield while the diaminosiloxane (A) and POSS systems (C, D) produce less than a 10 weight-% char yield as measured by TGA.

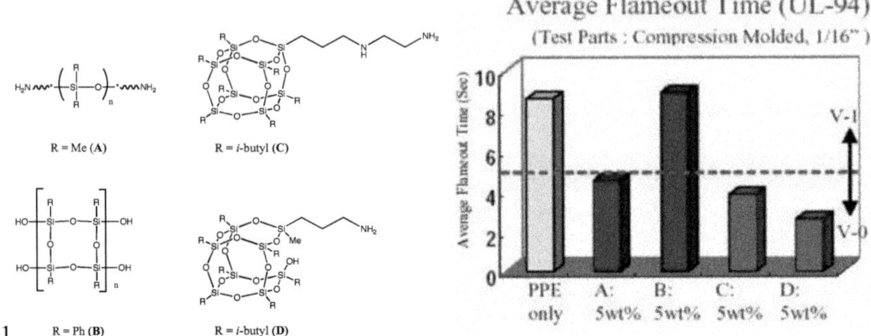

Figure 8.3

Chemical comparisons (A-D) and resulting impact on UL-94 properties at 5 weight-% loading into PPE. Reproduced with permission from reference 11(a); Ikeda, M., *Utilization of POSS in Industrial Applications* at the Nanostructured Chemicals Workshop, Huntington Beach, California Sept 7-8, **2002,** Paper No. 17, pp 1-11, published by Hybrid Plastics Inc.

These flammability results are perhaps best understood by the fact that diaminosiloxane (A) and phenyl silsesquioxane resin (B) are much less compatible with PPE than the fully dispersible POSS systems C and D. The dispersion differences are not unexpected given that polymer blends tend to disperse into micro sized domains. Of further importance may be that both POSS C and D begin to undergo a slow degradation at 225 °C which is well before the TGA degradation point for PPE (425 °C). Diaminosiloxane improves the flameout time to some extent but is less effective than the less volatile and more readily dispersible POSS systems. Both POSS systems significantly reduce the flammability of PPE with POSS system D reaching the desired V=O level. The flammability performance difference between POSS C and POSS D is difficult to rationalize from a compositional perspective. However it is known that incompletely condensed POSS cages are more easily crosslinked than completely condensed cages, and the propensity of POSS cages to crosslink and char during combustion was both desired and visually observed by Ikeda (Figure 8.4).

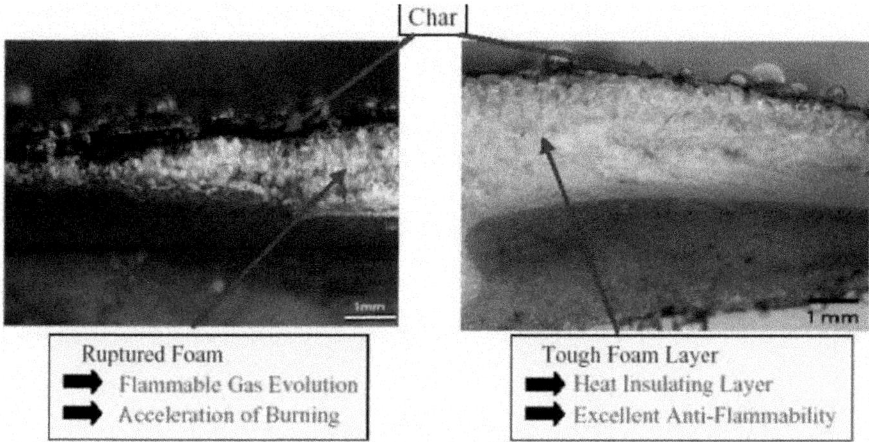

Figure 8.4

Comparisons of 100 % PPE (left picture) and of 5 weight-% POSS (D) / 95 weight-% PPE (right picture) after UL-94 testing

Reproduced with permission from reference 11(a); Ikeda, M., *Utilization of POSS in Industrial Applications* at the Nanostructured Chemicals Workshop, Huntington Beach, California Sept 7-8, **2002,** Paper No. 17, pp 1-11, published by Hybrid Plastics Inc.

The photographs in Figure 8.4 show that the char formed on PPE is significantly more foamy and likely to be capable of continued volatilization. The PPE/POSS system however produces a physically tougher char layer that remains intact and will be more likely to provide a thermally insulating barrier. Ikeda's findings suggest that use of incompletely condensed POSS cages should be further studied as fire retardant additives.

8.4   Conventional fire retardants and POSS

Brominated derivates of phenyl- and vinyl substituted POSS cages not only afford desirable compatibility with most engineering polymers, they have also been shown to be self-extinguishing fire retardants in polypropylene and polyethylene, when used at 15 weight-% loadings in combination with antimony trioxide at a 5 weight-% loading [13]. This method is advantageous because it affords an alternate (not wholly organic) form of halogenated fire retardant to control gas phase chemistry and results in a heavy surface char.

Brominated aliphatic POSS such as octadibromoethyl POSS (Figure 8.5) are thermally stable up to 275 °C and afford 40 weight-% char yields as measured by TGA. Consequently octadibromoethyl POSS is best suited for incorporation into olefinic engineering polymers. In contrast, brominated aromatic POSS such as bromophenyl POSS and dibromophenyl POSS (Figure 8.5) show a much higher thermal stability (425 °C) and are suitable for compounding into polymers such as polyamides, polyacetals, and cyclic olefin copolymers.

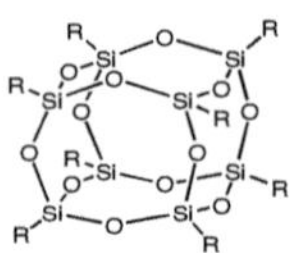

R = CH₂BrCH₂Br
Octadibromoethyl POSS
Mwt: 1919.57
C 10.01%, H 1.68% ,
Br 66.6%, O 10%, Si 11.7%

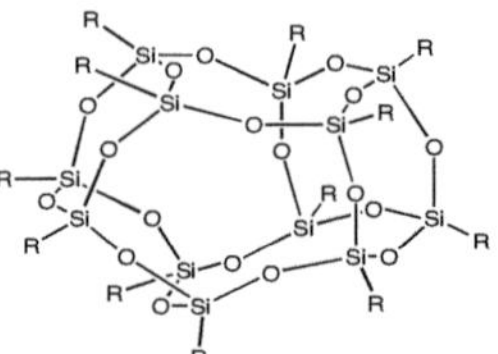

R = C₆H₄Br
Bromophenyl POSS
Mwt: 2483.03
C 34.63%, H 1.94%
Br 38.4%, O 11.53%, Si 13.5%

R = C₆H₃Br₂
Dibromophenyl POSS
Mwt: 3443.77
C 25.11%, H 1.05%
Br 55.69%, O 8.36%, Si 9.79%

Figure 8.5

Brominated POSS as fire retardants

There are only two reports on the use of POSS as a synergist with conventional fire retardants. This is somewhat surprising given the propensity of POSS to aid dispersion [14] (a-c) as well as improve processing and mechanical properties. This is a likely area for future development as it holds the promise of more effective use of fire retardant packages while reducing their processing and mechanical property drawbacks.

Chigwada et al. [15] investigated the synergy between vinyl ester/POSS nanocomposites and phosphorous-based fire retardants. Their work showed a decrease in the total heat released of 40-50 % when both additives were present with little negative impact on mechanical performance. Despite this achievement the goal of attaining an equivalent level of fire retardancy to that via use of halogenated fire retardants was not realized.

A synergy between octamethyl POSS and a commercial phosphate fire retardant in recycled polyethylene terephthalate (PET) has been identified. An increase in limiting oxygen index (LOI) values and reduction in both the peak heat release and in CO_2 emissions, as measured by cone calorimetry, were attained despite very poor compatibility and dispersion of octamethyl POSS in PET [16]. The result is surprising in view of the fact that octamethyl POSS should sublime out of the PET during processing. The findings suggest that a solid-state reaction between octamethyl POSS and the phosphate is probably occurring and in turn produces a phospho-silicate char during combustion. The synergistic effect should be investigated further with use of trisilanol phenyl POSS or trisilanol iso-octyl POSS which are both highly dispersible in PET.

8.5   POSS and fire-retardant coatings for textiles

The literature contains only one reported use of POSS as a fire retardant in textile fibre [17]. The work reported the incorporation of vinyl POSS into multifilament polypropylene (PP) yarns at 10 weight-% loading in which the POSS provided some benefit by delaying the time to ignition, however the overall flammability as measured by heat release rate was not reduced (Figure 8.6).

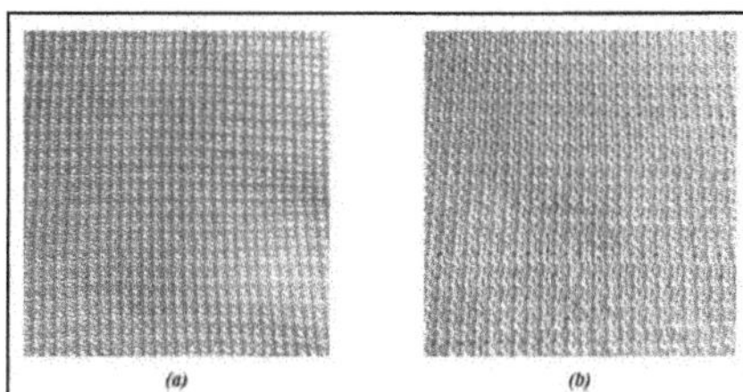 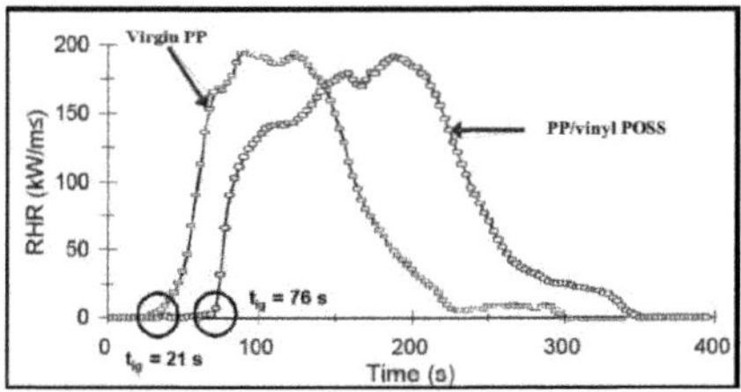

Figure 8.6
Fabric photo and plot showing the heat release rate (RHR) for
(a) knitted PP; (b) PP/Vinyl POSS fabrics at 35 kW/m^2
Reproduced with permission from reference 17; The Royal Society of Chemistry, Bourbigot, S.; Le Bras, M.; Flambard, X.; Rochery, M.; Devaux, E.; Lichtenhan, J. Polyhedral Oligomeric Silsesquioxanes – Application to Flame Retardant Textiles, *Fire Retardancy of Polymers: New Applications of Mineral Fillers,* Edited by M. Le Bras, S. Bourbigot, S. Duquesne, C. Jama and C.A. Wilkie, **2005,** pp. 189-201.

The same vinyl POSS was incorporated into a polyurethane coating that was subsequently applied to woven PET textiles and afforded both an increase in time to ignition and a 50 % decrease in peak heat release rate and rate of heat release. Again the mechanism of action was attributed to a more robust char layer at the fibre surface [18].

8.6   Commercial applications

Each month, on average, there are 5-10 patents and 20-25 publications published describing POSS usage. As of 2008 there were over 700 patents and 2700 academic papers in total which reflect a high level of academic and commercial interest in POSS. In most cases, the public declaration of commercial use of POSS is frowned upon by technical and business management because it invites unwanted competition and reduces economic advantages for products in competitive markets. Nevertheless, commercial use of POSS in products and processes is growing rapidly and covers over fifty different market segments. Commercial strengths for POSS in thermoplastics involve its use as a flow aid for difficult-to-process resins and optical resins, where the size and highly dispersed nature of POSS cages do not degrade properties. For thermoset resins POSS is utilized to provide oxidation resistance, elevated use temperature, and solvent resistance primarily in adhesives. POSS is also utilized as a dispersion aid for fillers such as TiO$_2$, Gd$_2$O$_3$, SiO$_2$ and other metal oxides, where a high level of homogeneity and increased filler-polymer interface is desirable for both economic and performance advantages.

POSS is also commercially utilized to achieve improvements in the fire retardancy of PPE, cyclic olefin copolymers (COC) and phenolic resin systems. In these cases, POSS usage is successful because it affords additional property enhancements along with fire retardancy improvement. For example the processing and moulding of fire retarded polymers is often difficult due to the raised viscosity and decreased mechanical properties of filled systems. Very often the combination of improved processing and fire retardant performance can result in greater productivity and reduced materials usage, and hence more commercially viable products.

8.7   Conclusions

Although a significant amount of work has been carried out on POSS chemicals as blendable fire retardants, it is clear that additional areas for research are worthy of exploration. These include synergistic relationships between POSS and conventional fire retardants, antioxidants, and processing aids. Whilst it would already seem that the chemical composition of POSS should make it an ideal additive to fire retardant packages, the complex nature of POSS and its interaction with polymer matrices still requires further research, in order to be fully understood and commercially developed.

REFERENCES

1   Nanostructured® (2,610,806), Nanoreinforced® (2,628,735), POMS® (3,138,678) and POSS® (2,548,048) are registered US Trademarks of Hybrid Plastics Incorporated.

2   Asuncion, M.Z.; Hasegawa, I; Kampf, J.W.; Laine, R.M. The selective dissolution of rice hull ash to form $[OSiO_{1.5}]_8[R_4N]_8$ (R= Me, CH_2CH_2OH) octasilicates. Basic nanobuilding blocks and possible models of intermediates formed during biosilicification processes. *Journal of Materials Chemistry*, **2005**, 15, 2114-2121.

3   Flame Retardant Mechanism of the Nanotube-based Nanocomposites. *Final Report*. Kashiwagi, T NIST Department of Fire Protection Engineering, University of Maryland, College Park, MD *20742 NIST GCR 07-912* September **2007.**

4   Lichtenhan, J. D.; Gilman, J. W. Preceramic Additives as Fire Retardants for Plastics. *United States Patent*, 6,362,279 **2002.**

5   Fina, A.; Abbenhuis, H. Cl.; Tabuani, D.; Camino, G.   Metal Functionalized POSS as Fire Retardants in Polypropylene. *Polymer Degradation and Stability,* **2006**, 91, 2275-2281.

6   Mantz, R.A.; Lichtenhan, J.D.; Jones, P.F.; Chaffee, K.P.; Gilman, J.W.; Ismial, I. M. K.; Burmeister, M.J. Thermolysis of Polyhedral Oligomeric Silsesquioxane (POSS) Macromers and POSS-Siloxane Copolymers. *Chemistry of Materials,* **1996**, 8, 1250-1259.

7   Simkovic,V. Novel Low Dielectric Constant Thin Film Materials by Chemical Vapor Deposition. Thesis submitted to the Faculty of the Virginia Polytechnic Institute and State University, in partial fulfillment of the requirements for the degree of Master of Science In Materials Science and Engineering,July 16, **1999**, Blacksburg, Virginia.

8   Zeng, J.; Bennett, C.; Jarrett, W. L.; Iyer S.; Kumar, S.; Mathias, L. J.; Schiraldi, D. A.   Structural Changes in Trisilanol POSS During Nanocomposite Melt Processing. *Composite Interfaces,* **2005**, 11, 673–685.

9(a)   TG-1001™ Laminating Resin: www.thermalguard-technology.com/TG%201001%20properties%20-%202.pdf.

9(b)   TG-1009™ Laminating Resin: www.thermalguard-technology.com/TG%201009%20properties%20-%202.pdf.

10   POSS® Resorcinol see: Nanoreinforced® Thermoset Resins Catalog, www.hybridplastics.com.

11(a)   Ikeda, M., *Utilization of POSS in Industrial Applications*, at the Nanostructured Chemicals Workshop, Huntington Beach, California Sept 7-8, **2002** Proceedings published electronically by Hybrid Plastics Inc. Paper No. 17, pp 1-11.

11(b)   Saito, H.; Ikeda, M. Polyphenylene ether resin composition containing silicon compounds, US Patent Application US 2004/0138355, July 15, 2004.

11(c)   Ikeda, M., Saito, H. Improvement of Polymer Performance by Cube-Oligosilsesquioxane, *Reactive & Functional Polymers,* **2007**, 67, 1148-1156

12   Song, L.; He, Q.; Hu, Y. Chen, H. Liu, L. Study on Thermal Degradation and Combustion Behaviors of PC/POSS Hybrids, *Polymer Degradation and Stability,* **2008**, 93, 627-639.

13   Hait, S.; Wheeler, P.; Gilman, J. W.; Lichtenhan, J. D. Hybrid Plastics and NIST/BFRL unpublished findings.

14(a) Misra, R.; Fu, B. X.; Morgan, S. E. Surface Energetics, Dispersion, and Nanotribomechanical Behavior of POSS/PP Hybrid Nanocomposites. *Journal of Polymer Science, Part B: Polymer Physics,* **2007**, 45, 2441–2455.

14(b) Wheeler, P.A.; Misra, R.; Cook, R. D.; Morgan, S.E. Polyhedral Oligomeric Silsesquioxanes Trisilanols as Dispersants for Titanium Oxide Nanopowder. *Journal of Applied Polymer Science,* **2008**, 108, 2503-2508.

14(c) See: POMS® and Nanopowders on catalog at www.hybridplastics.com.

15   Chigwada G.; Jash P.; Jiang D.D.; Wilkie C.A. Fire Retardancy of Vinyl Ester Nanocomposites: Synergy with Phosphorus Based Fire Retardants. *Polymer Degradation and Stability,* **2005,** 89, 85-100.

16   Vannier, A.; Duquesne, S.; Bourbigot, S.; Castrovinci, A.; Camino, G.; Delobel, R. The use of POSS as Synergist in Intumescent Recycled Poly(ethylene terephthalate). *Polymer Degradation and Stability,* **2008**, 93, 818-826.

17   Bourbigot, S.; Le Bras, M.; Flambard, X.; Rochery, M.; Devaux, E.; Lichtenhan, J. Polyhedral Oligomeric Silsesquioxanes – Application to Flame Retardant Textiles, *Fire Retardancy of Polymers: New Applications of Mineral Fillers",* Edited by M. Le Bras, S. Bourbigot, S. Duquesne, C. Jama and C.A. Wilkie, Published by Royal Society of Chemistry, **2005,** pp. 189-201.

18   Devaux, E.; Rochery, M.; Bourbigot, S. Polyurethane/Clay and Polyurethane/POSS Nanocomposites as Flame Retarded Coating for Polyester and Cotton Fabrics, *Fire and Materials,* **2002**, 26, 149-154.

Chapter 9

Barrier property enhancement by nanocomposites

Tie Lan and Ying Liang

Nanocor, Inc., Hoffman Estates, USA

9.1   Introduction

Polymer nanocomposite materials have been the hot topic in research and development for the last 10 years in the plastics industry. The dimension of nano-sized particles allows intimate interaction between filler and polymer matrix phases. This interaction and the nature of most of the nanostructured fillers provide novel property enhancements at very low loading levels which cannot be achieved with traditional filler materials. Properties such as stiffness, dimensional and heat stability, flame retardancy and barrier enhancement are the most recognized areas for performance enhancement. Nanostructured particles with platy geometry are particularly interesting in the barrier enhancement area. Modified montmorillonite clays (organoclays) have been the major fillers used in achieving barrier enhancement for plastic materials, due to their rich surface chemistry and the high aspect ratio of individual montmorillonite aluminum silicate layers. However, it is essential to fully disperse the modified montmorillonite clays into the plastic materials to realize their high aspect ratio, products and technologies have been developed to incorporate organoclays into plastics via in-situ polymerization and melt compounding. For packaging applications, particularly for food packaging, it is important to have regulatory approval status for commercialization.

Organoclay

Montmorillonite clay is the major component of bentonite, which has been used historically as a thickener, sealant, binder, lubricant and absorption agent. The development of the petrochemical industries was also dependent on the use of clay as a drilling agent and cracking catalyst. The deposit of bentonite is quite abundant around the world. According to a marketing report in 2006 by Vicente Flynn International it is believed that the total deposit of bentonite is more than 9 billion tonnes, which could give more than 700 years of bentonite supply based on the current usage level. In detail, the world bentonite production in 2007 is estimated at around 12.8 million tonnes, of which US production is estimated at 4.1 million tonnes. Not all bentonite deposits are suitable as plastic additives due to the variation of compositions and purity levels, since the plastics industry requires higher purity than traditional industries such as paints and coatings. Proper choice of clay mines and purification processes are needed prior to the organophilic modification of the clay by an exchange reaction. Traditional bentonite refinement processes use gravity separation and water washing after soda ash treatment of the crude bentonite. Nanocor, a subsidiary of Amcol, has developed a patented processing technology to extract montmorillonite clay from bentonite [1].

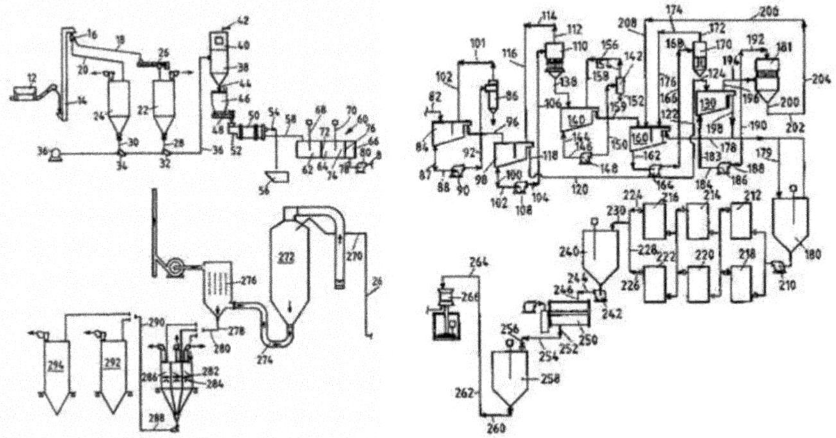

Figure 9.1

Flow chart of Nanocor's patented organoclay purification process

The process of removing impurities, illustrated in Figure 9.1, includes the following steps:

- separating the clay from rocks and other large non-clay impurities,

- dispersing the clay and smaller impurities in water to provide a clay slurry,

- passing the clay slurry through a series of hydrocyclones to remove the larger particles while retaining montmorillonite,

- use of an ion exchange column, wherein the multivalent cations are replaced by monovalent cations, such as sodium, lithium and/or hydrogen and

- centrifuging the clay to extract a majority of the particles having a size in the range of about 20 μm. This process ensures that the resulting montmorillonite has excellent consistency and high quality.

Cationic exchange reactions are used to modify the surface nature of bentonite to produce fillers generally called organoclays. Quaternary ammonium surfactants are the most common surface treatment agents. Then polymer chains can enter the modified clay interlayer space to form nanocomposite materials.

It can be debated whether modified montmorillonites for the paint and coatings industry can be called organoclays, simply due to their hydrophobic nature. However, as stated before, the requirements for modified montmorillonites for the plastic industry are higher than the requirements on these fillers for the paint and coating industries. This is due to the plastic processing conditions at elevated temperatures, and the requirements of regulatory approval for food packaging applications, which limit the controlled use of quaternary ammonium chemicals. A minimum amount of surface modification is required to eliminate release of volatiles during compounding and migration to the surface to contaminate the packaged foodstuff [2]. The existence of byproducts within the montmorillonite clay is quite harmful in the clay chemistry. Kaolin clay, chlorite, feldspar, quartz, amorphous silica and chalk are the common inorganics co-existing with montmorillonite clays. Their presence compromises the precision in the amount of organic modifier added. In addition, it creates defects in the final compound due to having a generally larger particle size than the dispersed organoclay platelets.

9.2   Formation of polymer-clay nanocomposites

Polymer-clay nanocomposites are prepared on a commercial scale via two major routes, namely melt compounding and in-situ polymerization. The melt compounding process is suitable for most polymer systems. It is an extrusion process, which utilizes the hydrophobicity of organoclays to allow molten polymer segments to enter the clay gallery. Furthermore, the shear force in the extrusion will disrupt the clay layer stacking order to disperse the organoclay in the polymer matrix.

In-situ polymerization involves pre-dispersion of organoclays in monomer systems, and it is suitable for most ring opening polymerization and free radical polymerization reactions. To date full commercial success has been achieved in polyamide-6 (PA-6) and epoxy systems, The requirement of in-situ polymerization is to allow monomer to enter the clay gallery, then start the polymerization within the gallery resin and grow polymer chain from the clay gallery. Further migration into the gallery resin will expand the clay layer to form well dispersed nanocomposites. In situ polymerization offers better nanoclay efficiency than compounding and potential cost benefits. Consequently extensive research is ongoing on in situ polymerization of polyolefin systems such as polyethylene and polypropylene, although compounding has already been demonstrated successfully.

Nanocomposites based on polyamide-6 have been successfully produced using the in-situ approach. The organoclay Nanomer® I.24TL was developed for in-situ polymerization of polyamide-6, which contains 12-aminododecanoic acid (ADA) as surface modifier. ADA effectively participates in the polyamide-6 polymerization to become a part of the polymer network (Figure 9.2).

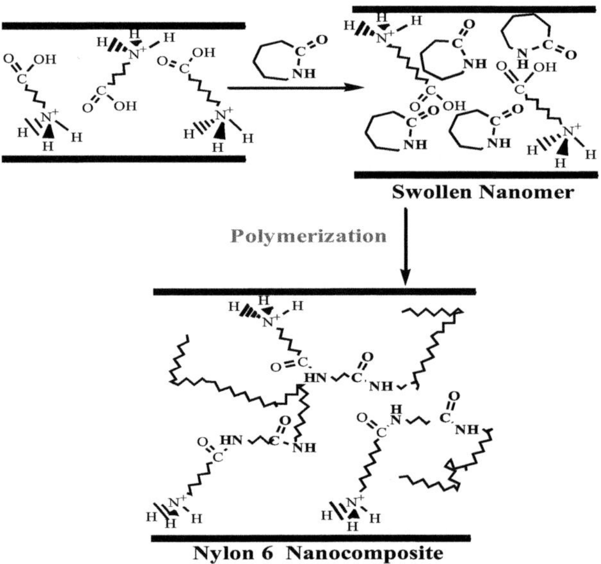

Figure 9.2

Polyamide-6 nanocomposite synthesis by in-situ polymerization with Nanomer® I.24TL and caprolactam

This reaction, described as a tethering effect, forms covalent bonds between the surface modifier and the polymer matrix. Tethering results in the surface modifier becoming part of the polyamide molecule itself, thus eliminating the possibility of ADA migration out of polyamide-6. The US Food and Drug Administration (FDA) has approved the use of in-situ polyamide-6 nanocomposites containing I.24TL for direct food contact applications.

Extrusion processes have been used to prepare a variety of plastic-clay nanocomposites. For low polarity polymers like polyethylene (PE) and polypropylene (PP), low polarity surface modifiers such as dimethyl dialkyl ammonium surfactants and compatibilizers are required. The compatibilizers are grafted PE or PP, such as maleic anhydride grafted PE or PP. Organoclay masterbatches have also been developed for the production of PP- and PE nanocomposites. These masterbatch formulations already contain the compatibilizers and simplify compounding and material sourcing.

For polymers with medium and high polarity, alternative clay surface modifiers should be used. For example, in an ethylene-vinyl alcohol system (EVOH), a highly polar surface modifier containing several ethylene oxide segments has been used to make the organoclay disperse into the EVOH resins [3]. Polar surface treatment is also required for polyamide systems [4]. In a nanocomposite system with the polymer PA-6 MXD6 (produced by Mitsubishi Gas Chemical), the surface modifier is not only functioning as compatibilizer, it also provides control of the polyamide crystallinity, giving the final nanocomposites an excellent processing capability in stretch blow moulding applications.

Organoclay products are typically supplied in the form of agglomerates. For example, an organoclay particle with a particle size of 20 micron may contain around one million clay plates, and for dispersion of the clay into the polymer system to take place the kinetics should be considered. The kinetics involve the wetting of the agglomerate surface with the molten polymer chains and diffusion of polymer chains into the clay gallery, followed by disassociation of the clay layer stacking order. High shear extrusion energy should be applied in the polymer compounding process. One way to increase the shear force is to apply high rpm (revolution per minute) extrusion with tight gaps for the polymer to flow through. This approach has been proven to be quite effective for certain polymer like polyolefins. For most heat and shear sensitive polymers, high shear stress with low shear rate at lower processing temperature and lower rpm is desired. Suitable technology has been developed by Japan Steel Works, Ltd. particularly by applying their

NIC (special kneading cylinder) and TKD (twist kneading rotor) compounding technologies [5].

9.3 Nano-effects in barrier enhancement

Polymer-based layered silicate nanocomposites are plastics containing low levels of dispersed platy minerals with at least one dimension in the nanometer range. The most common minerals are modified montmorillonite clay. Its aspect ratio exceeds 300, giving rise to enhanced barrier and mechanical properties. In general, every 1 weight-% of these organoclays creates a 10 % property improvement. Their interactions with polymer molecules alter the morphology and crystallinity of the polymer matrix, leading to improved processability in addition to the improvements in barrier properties, strength and stability. In general, plastic materials are classified as three categories of oxygen barrier materials. Polyolefin resins, like PP and PE are low barrier resins. PA-6, PA-66, and polyethylene terephthalate (PET) are medium barrier resins. EVOH, PA-6 MXD6 and polyvinylidene chloride (PVDC) resins are high barrier resins. It has been demonstrated that organoclays can enhance the barrier properties of different classes of barrier materials.

For polyolefins, the masterbatch route is the preferred way to synthesize nanocomposites. Masterbatches are typically compounded in high shear extruders with the base polymer, a compatibilizer and the organoclay. Even through the organoclay in the masterbatch is only intercalated; the whole internal surface has been wetted by the polymer matrix. Final nanocomposites can be generated by diluting the masterbatch with additional polymers. A polypropylene nanocomposite (nano-PP) with a PP homopolymer was made with 7 weight-% Nanomer® I.31PS via a masterbatch process. Nanomer I.31PS is a montmorillonite clay modified by octadecyl ammonium in the clay interlayer gallery and silane coupling on the OH-group on the edges of the clay layers. Films were made via cast or blown film processes. Oxygen transmittance rates (OTR) were measured in an oxygen permeation tester from Mocon and data are listed in Table 9.1. It is interesting to notice that the barrier improvement factor for the blown film is higher than that of the cast film. This could be caused by the enhanced orientation of the organoclays in the blown film. The blown film has a high stretch ratio and promotes a better alignment of organoclays along the film surface.

Table 9.1

OTR properties of homo-PP nanocomposite blown and cast films

Processing	Nanomer grade	Loading level (weight-%)	OTR (cm^3 mm/m^2 day atm)	Improvement of OTR
Blown film	Control	0	992	/
Blown film	I.31PS	7	401	2.5x
Cast film	Control	0	718	/
Cast film	I.31PS	7	501	1.4x

PP copolymers have also been converted into nanocomposites. Blown films were made from standard blown film line. Mechanical and barrier properties were measured and listed in Table 9.2. These nanocomposite-based films also showed improvements in barrier properties compared with the neat polymer. Film stiffness also increased for the nanocomposites, which could provide down-gauging opportunities for packaging design.

Table 9.2

OTR properties of PP copolymer nanocomposite blown films

Nanomer grade	Loading level (weight-%)	Tensile modulus (MPa)	Improvement of tensile modulus (%)	OTR (cm^3 mm/m^2 day atm)	Improvement of OTR
Control	0	909		205	/
I.30P	6	1751	93	137	1.5 x
I.31PS	6	1495	64	108	2 x

As stated previously, PA-6 resins can be made by an in situ polymerization route. A series of PA-6 nanocomposites containing Nanomer® I.24TL were prepared via in-situ polymerization. The dependence of mechanical and barrier properties on the loading level of organoclay were evaluated for nanocomposites containing 2, 4, 6 and 8 weight-% organoclay I.24TL. Mechanical properties and heat deflection temperature (ASTM 648) are listed in Table 9.3.

Table 9.3			
Mechanical properties of PA-6 nanocomposites			
Organoclay I.24TL (weight-%)	Flexural modulus (MPa)	Tensile modulus (MPa)	Heat distortion temperature under load (°C)
0	2836	2961	56
2 (% improvement)	4326 (53%)	4403 (49%)	125 (123%)
4 (% improvement)	4578 (61%)	4897 (65%)	131 (134%)
6 (% improvement)	5388 (90%)	5875 (98%)	136 (143%)
8 (% improvement)	6127 (116%)	6370 (115%)	154 (175%)

The high aspect ratio of organoclays and the interaction between polymer chains and nano-dispersed silicate layers create an increase of more than 110 % in flexural and tensile moduli, and a 175% increase in heat distortion temperature under load (DTUL), at a loading level of 8 weight-% ADA modified montmorillonite. In addition, smooth and transparent films were successfully cast using standard techniques and equipment. These films were tested for gas permeation at 65 % relative humidity. Oxygen transmission rates (OTR) improved as I.24TL addition levels increased. At 8 weight-% addition level, OTR reduction was 80 %. This makes PA-6 nanocomposites particularly appropriate for packaging applications requiring improved barrier performance. Water vapour transmittance rates (WVTR) of these nanocomposite samples were also reduced by the addition of Nanomer I.24TL. The WVTR- and OTR improvements versus Nanomer loading are illustrated in Figure 9.3.

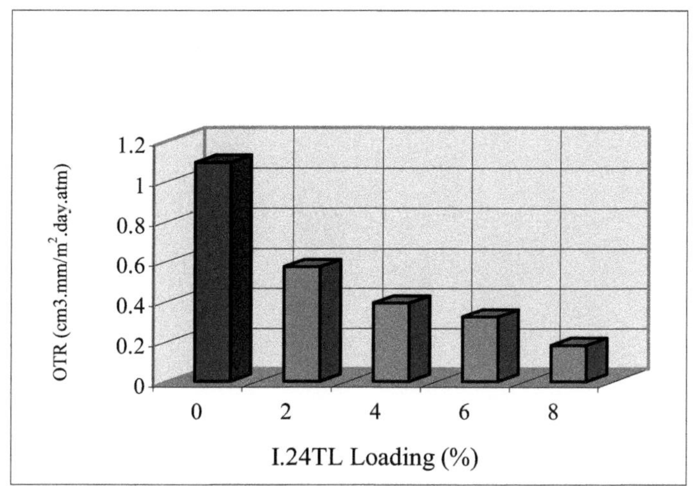

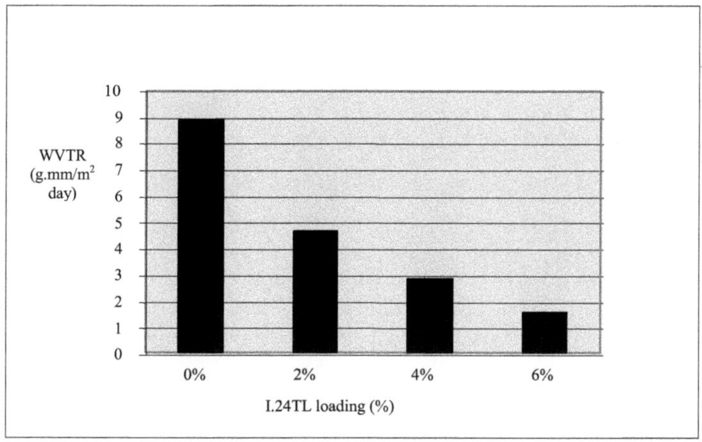

Figure 9.3

OTR and WVTR of PA-6 nanocomposites made by in situ polymerization

Oxygen barrier properties of the PA-6 nanocomposite films were also tested under different relative humidity levels (Figure 9.4) and they clearly demonstrate that the barrier effect of nanocomposites is functioning at various relative humidity (RH) levels. This is typical for a passive barrier enhancement mechanism.

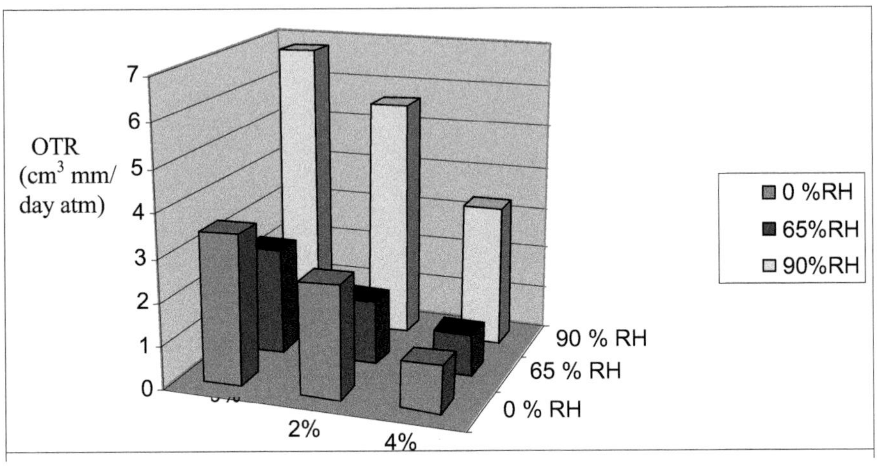

Figure 9.4

OTR of PA-6 nanocomposites by in-situ polymerization at different relative humidity levels

Commercially produced PA-6 nanocomposite was made into cast film and its mechanical properties were tested. Nanocomposite films show significantly enhanced film stiffness and strength. At the same time, the films maintain nearly the same strain-at-break as the neat PA-6 (Table 9.4). These data are quite important for the application of PA-6 nanocomposites. Combinations of stiffness, strength and barrier properties allow new film structure designs with 50% thickness reduction, which is advantageous for multilayer film structures, as the final multilayer film should have the same barrier and mechanical properties. Alternatively, if one keeps the same thickness of the PA-6 nanocomposite as neat PA-6, the new package should have twice the shelf-life of the equivalent PA-6 version. The nanocomposite has also enhanced stiffness and heat resistance.

PET has been the main polymer used for beverage and food containers in the last 20 years, and many researchers are interested in incorporating organoclays into PET to enhance its barrier performance and heat distortion temperature. In-situ polymerization as well as melt compounding methods has been used. However, technological advances and commercial success have been limited to date. This could be caused by the high processing temperature of PET and its very high sensitivity to moisture. Current organoclay technology limits the maximum processing temperature to 270 °C, in order to minimize degradation of the surface modifier. The current solution for PET barrier containers is to make multilayer structures using co-extrusion technology. However, it is desirable to use the minimum amount of barrier resins in the multilayer PET containers due to recycling concerns. Barrier resins PA-6 MXD6 and EVOH are suitable candidates for PET multilayer structures, but it is necessary to increase the barrier properties of these resins in order to meet the PET multilayer application requirements.

High barrier resins EVOH [6] and PA-6 MXD [4] have also been successfully converted into nanocomposites via the melt compounding process. EVOH resins are sensitive to processing due to a possible crosslinking reaction of their hydroxyl groups at high temperatures, and mild extrusion conditions are the key parameters to form EVOH nanocomposites. For EVOH, the barrier properties are dependent on the ethylene content. The higher the ethylene content, the lower the barrier properties, whereas process stability of EVOH increases with increasing ethylene content. Therefore it is important to increase the barrier property of the high ethylene content EVOH grades and create better barrier and better processing EVOH grades. This will increase the scope of EVOH resin in multilayer packaging and industrial applications. It is also important to reduce the moisture sensitivity of EVOH, since its barrier performance decreases rapidly when the relative humidity level increases about 70 %.

Table 9.4

Mechanical properties of PA-6 and its nanocomposites (via in-situ polymerization) films

Flow direction				
Testing sample	Weight-% organoclay	Young's modulus (MPa)	Yield strength (MPa)	Strain @ break (%)
PA-6	0	150	12	700
PA-6 nanocomposite	3	270	35	650
Perpendicular direction				
PA-6	0	200	16	600

PA-6 nanocomposite	3	380	28	550

PA-6 MXD6 from Mitsubishi Gas Chemical Co is another high barrier resin. It offers excellent processing characteristics and barrier properties. In particular, the high barrier properties at higher relative humidity levels than EVOH are attractive for many food and beverage packaging applications. A commercial nanocomposite grade of PA-6 MXD6 has been developed under the trade name Imperm 103, and supplied to the marketplace by a commercial alliance between Nanocor and Mitsubishi Gas Chemical Co. [7]. Table 9.5 demonstrates the improvement in barrier properties of Imperm 103. For oxygen and carbon dioxide, the barrier improvement (OTR and CO_2TR) is more than doubled. In addition, the water vapour transmittance rate is halved.

Recently, oxygen scavenger catalysts based on cobalt compounds were incorporated into Imperm 103 with the resulting product branded as Imperm-OS. The oxygen scavenging rate of Imperm-OS offers even more significant oxygen barrier improvements, with the same carbon dioxide barrier properties. The superior combination of barrier properties allows its use in plastic multilayer beer bottles. PA-6 MXD6 nanocomposite is also being evaluated in new military MRE metal-free packaging designs. The "Meal Ready-to-Eat" (MRE) is a self-contained, individual field ration in lightweight packaging, procured by the US military for its personnel to use in combat or other field conditions where organized food facilities are not available.

<table>
<tr><td colspan="6" align="center">Table 9.5
OTR and CO_2TR of PA-6 MXD6, Imperm®-103 and Imperm®-OS as non-oriented monolayer film by Oxygen Permeation Instrument Ox-Tran2/20 from Mocon</td></tr>
<tr><td align="center">Properties</td><td align="center">Units</td><td align="center">PA-6 MXD6</td><td align="center">Imperm 103</td><td align="center">Imperm-OS</td></tr>
<tr><td align="center">OTR
(23 °C, 60 % RH)</td><td align="center">$cm^3 \cdot mm/m^2 \cdot day \cdot atm$</td><td align="center">0.09</td><td align="center">0.02</td><td align="center">0.002</td></tr>
<tr><td align="center">CO_2TR
(23 °C, 60 % RH)</td><td align="center">$cm^3 \cdot mm/m^2 \cdot day \cdot atm$</td><td align="center">0.30</td><td align="center">0.15</td><td align="center">0.15</td></tr>
<tr><td align="center">WVTR
(40 °C, 90 % RH)</td><td align="center">$g \cdot mm/m^2 \cdot day$</td><td align="center">1.36</td><td align="center">0.58</td><td align="center">N/A</td></tr>
</table>

Barrier improvement of plastic with the addition of organoclays also creates new materials for non-food packaging applications. PA-6 and HDPE nanocomposites have been used as barrier materials for fuel tanks and agrochemical storage. In fuel tank applications, significant development work is underway for small engines such as motorcycles, all-terrain vehicles and snowmobiles, to meet tighter emission regulations. Monolayer construction is typical for such small fuel tanks. In addition, organoclays have also been used in new marine tank development. Plastic marine fuel tanks offer excellent corrosion resistance over aluminum tanks, and nanocomposite-based plastic tanks will meet new emission requirements set by the Environmental Protection Agency of the United States.

9.4   Summary

Nanocomposites made from polymers and organoclays are emerging as commercial materials at a rapid speed. Organoclays are effective additives to enhance plastic barrier properties, and this has been demonstrated in variety of plastic resins. With enhanced barrier properties, better shelf-life packages can be designed. More importantly, due to the high barrier properties and enhanced mechanical properties of nanocomposites, down-gauging of plastic packages is feasible. This will contribute to a reduction in plastic waste with reduced impact on the environment.

REFERENCES

1   Clarey, M.; Edwards, J.; Tsipursky, J.; Beall, G.; Eisenhour, D. Method of manufacturing polymer-grade clay for use in nanocomposites. US Patent 6,050,509, 2000.

2   Bagrodia, S.; Germinario, L.; Gilmer, J.; Lan, T. Polymer nanocomposite comprising a matrix polymer and a layered clay material having an improved level of extractable material. US Patent 6,586,500, 2003.

3   Lan, T.; Psihogios, V.; Turner, S.; Connell, G.; Matayabas, J.; Gilmer, J. High barrier amorphous polyamide-clay intercalates, exfoliates, and nanocomposite and a process for preparing same. US Patent 6,376,591, 2002.

4   Lan, T.; Cruz, H. T.; Tomlin, A. Intercalates formed by co-intercalation of onium ion spacing/coupling agents and monomer, oligomer or polymer MXD6 nylon intercalants and nanocomposites prepared with the intercalates. US Patent 6,232,388, 2001.

5   www.jswcompounding-usa.com

6   Lan, T.; Tomlin, T; Tomlin; A.; Psihogios; V. Intercalates formed by co-intercalation of onium ion spacing/coupling agents and monomer, oligomer or polymer ethylene vinyl alcohol (EVOH) intercalants and nanocomposites prepared with the intercalates. US Patent 6,225,394, 2001.

7   www.nanocor.com

Chapter 10

Status of biodegradable nanocomposites for industrial applications

Jo Ann Ratto, Christopher Thellen and Jeanne Lucciarini

U.S. Army Natick Soldier Research, Development and Engineering Center, Natick, USA

10.1 Introduction

This chapter highlights biodegradable nanocomposites and their potential for industrial applications. The chapter addresses processing issues, properties, performance, and commercial potential of these materials. The processing studies of biodegradable polymer/nanoparticle materials that are presented here do not indicate any emerging processing problems due to the nanoparticles. The processing of the nanocomposites has been achieved by solution cast films and melt-processing with little difficulty and with similar processing parameters as the homopolymers. The properties of biodegradable nanocomposites can exceed those of the homopolymer in regards to mechanical, barrier and thermal properties.

Biodegradation testing results for these nanocomposites vary due to the polymer/nanoparticle interaction and morphology of the system. Many of the studies which have been reported do not follow standard test methods from the American Society for Testing and Materials (ASTM) that would confirm acceptable results on biodegradation rates.

Since several biodegradable polymers have low melt and softening temperatures, the end use applications are not as vast as traditional thermoplastic resins. However, nanoparticles can potentially improve their thermal stability and elevate end use temperatures.

There are many biodegradable nanocomposites that have been studied for structure-property relationships with little emphasis on an application. Few publications on biodegradable nanocomposites produce film or barrier properties which are needed for industrial applications. Most studies focus on compression moulded samples through which mechanical properties are obtained. Although functionally successful with valuable contribution to the fundamental research of nanocomposites, this method is not a true representation of what can be done industrially, as it is not a continuous manufacturing process.

Another obstacle that must be considered is that biodegradable polymers do not have attractive barrier properties; therefore, an exfoliated nanocomposite using these polymers may not provide sufficient barrier improvement for a high barrier shelf-stable food packaging applications.

10.2 Biodegradable polymers

Biodegradable polymers are important for the environment and recently there has been a global emphasis on using polymers from renewable resources. Sustainability is currently a dynamic force for industry to pursue the use of green, bio-based, compostable materials. The use of biodegradable polymers that are disposed of through composting can reduce the amount of solid waste generated in most areas of the world. This alone is a significant reason to use biodegradable polymers; however, there needs to be a balance with price, performance, properties and manufacturing methods used to create these polymers. In addition, there is a lack of available composting infrastructure, especially in the United States, to accommodate these polymers. Biodegradable polymers are continuously struggling to perform as well as the more common commodity polymers at a competitive price.

There are several types of biodegradable polymers that are simply classified as natural or synthetic [1]. The natural polymers are based on renewable resources and include polysaccharides, proteins, lipids, cellulose esters to name a few. Synthetic biodegradable polymers are petroleum/fossil fuel based and include: aliphatic polyesters, aromatic polyesters, polyvinyl alcohol, and modified polyolefins. Many of these biodegradable polymers have been combined with nanoparticles to produce nanocomposites that have improved properties over the neat polymer. The most common and commercially available biodegradable polymers that this chapter will focus on are polycaprolactone (PCL), polylactic acid (PLA), and polyhydroxyalkanoates (PHA).

Synthetic polyesters, such as polycaprolactone (PCL) play an important role in the design of biodegradable polymers due to the increasing environmental concern about plastic waste pollution. PCL is a linear polyester with a low melting point (60°C) and poor mechanical properties. PCL is often blended with other polymers to improve thermal stability and mechanical properties, as well as to ease processing [2].

Poly(L-lactic) acid is typically formed through the fermentation of corn and corn-based products. Commercially available PLA materials are copolymers of poly(L-lactic acid) and poly(D-lactic acid) with different ratios affecting the properties. Most studies using PLA are in the crystalline form with D-content less than 6 %. The hydrolysis of this polymer back into its monomer state provides a major advantage over many petroleum-based polymers and polymer resins. Other advantages of PLA include high gloss and clarity, high tensile strength, low coefficient of friction, and high dead-fold and twist retention. These make it a good choice for packaging candy wrappers, beverage cups, bread bags, and short shelf-life bottle applications [2]. Currently, there is increasing interest in using PLA for the preparation of disposable, degradable plastic articles. However, other properties of PLA, such as flexural properties, heat distortion temperature (HDT), gas permeability, impact properties, melt viscosity for further processing, etc. are often not sufficient for various end-use applications [3]. The high glass transition temperature of PLA (55°C) and the high molecular weight of extrusion-grade PLA resins provide the relative stiffness of the material at room temperature. As a result, a plasticizer is frequently required to promote flexibility and control during extrusion processing of PLA.

Polyhydroxyalkanoates (PHA) are a family of naturally occurring polyesters found in many bacterial species. PHA materials have thermoplastic properties that can be tailored to specific applications by varying their chemical structure. These properties are determined by either the length of the pendant groups that extend from the polymer backbones, or from the distance between the ester linkages in the backbone [1]. This chemistry allows PHA to be tailored to provide similar mechanical and thermal properties to traditional thermoplastics such as polyethylene and polypropylene while maintaining biodegradability.

10.3 Nanoparticles

Nanoparticles are defined as having one of more dimensions between 1-100 nanometres. Polymer nanocomposites consist of a nanometre-scale particle phase dispersed in a polymer matrix. Nanocomposites have been shown to yield large improvements in barrier properties, as well as in physical properties such as tensile strength, tensile modulus (values obtained from stress/strain curve) and heat distortion temperature [4, 5, 6, 7].

10.3.1 Structure of montmorillonite layered silicates (MLS)

The nanoparticles typically used are organically modified montmorillonite layered silicates (MLS), mica-type silicates of the 2:1 phyllosilicate family, which consists of sheets arranged in a layered structure. MLS consist of approximately 1 nm thick lamellae with sheets of octahedrally coordinated aluminum atoms sandwiched between two layers of tetrahedrally coordinated silicon atoms. The lamellae are negatively charged with atoms within the layers, such as Aluminum (Al) or Magnesium (Mg). The negative charge is compensated by exchangeable cations most commonly Na^+ that keep the structure together. The interlamellar spaces or galleries can, therefore, be filled with polymer, creating a nanocomposite.

MLS are hydrophilic by nature and, therefore, are not compatible with most polymers. Consequently, the MLS must be chemically modified in order to make its surface more organophilic. The most commonly used surface treatments are ammonium cations, which can be exchanged for existing cations already on the MLS surface. The surface treatments minimize the attractive forces between the agglomerated platelets and increase the interfacial adhesion between the inorganic silicate and the polymer. Chemically modified MLS are commercially available and are typically treated in accordance with the characteristics and polarity of the polymer in which they will be dispersed. Once the surface of the MLS is chemically treated, it can be incorporated into the polymer during polymerization or by melt processing. This chemical modification, in conjunction with shear forces encountered during polymerization and melt mixing, further separates and disperses the MLS platelets in the polymer matrix.

10.3.2 Morphology of polymer/MLS nanocomposites

MLS is commonly used in nanocomposites due to its high cation exchange capacity and its large surface area, approximately 750 m^2/g and large aspect ratio (larger than 50) with a platelet thickness of 1nm [4]. Intercalation occurs when a small amount of polymer moves into the gallery spacing between the clay platelets, causing less than 2-3nm separation between the platelets. This results in a well-ordered multilayer, with alternating polymer/clay layers. Exfoliation occurs when the polymer chains further separate the clay platelets. The separation distance can be from 8-10nm, which results in a well-dispersed nanocomposite with the potential of significantly enhancing the mechanical, thermal and barrier properties. Although typically desirable, complete exfoliation is not required in order to achieve improvements in properties. In many cases, intercalated nanocomposites may provide enough enhancement for the application in which it will be used.

The dramatic reduction in permeability with nanoparticles has been attributed in part to the presence of well-dispersed, large aspect ratio silicate layers, which cause solutes to follow a tortuous path, thus increasing the effective path length for diffusion. In this tortuous path model [8], the concentration of the exfoliated MLS, the length and width of the exfoliated MLS platelets and the orientation of the lamellae influence the permeability in the model. Only a very small loading of nanoparticles (roughly 2-8 weight-%) is required to achieve these property improvements. These low loading levels typically do not cause detriment to processability. A key factor in determining the ultimate improvement in properties is the compatibility of the polymer/nanoparticle and the dispersion of the nanoparticles within the polymer matrix.

10.4 Biodegradable nanocomposites

The recent interest and level of success of polymer nanocomposites in general have stimulated experimentation on nanocomposites that are based on biodegradable polymers as the matrix. Research is ongoing in this field, and this chapter offers a status report, describing some of the most recent and promising results on the topic of biodegradable nanocomposites. Various other review articles and book chapters already cover particular advances in biodegradable nanocomposites [9, 10, 11, 12, 13, 14, 15,16, 17].

Ratto et al. [9] have recently reported a summary on the melt processing of nanocomposites, highlighting biodegradable nanocomposites of PLA, PHA, PCL and poly(butylene succinate)-co-adipate (PBSA). Ratto et al first began studying melt processable biodegradable nanocomposites for the U.S. Navy for improvements in heat deflection temperature for utensil applications as well as for faster rates of biodegradation. Since the nanoparticles have a huge surface area, it was theorized that this would aid in the biodegradation rates if the morphology was exfoliated by increasing the overall surface area for microbial attack. Biodegradation studies with ASTM methods were performed to determine the rates in comparison to the homopolymers. Ultimately, in the many nanocomposite systems studied, there were no significant improvements to biodegradation, due most likely to the lack of exfoliation. This research group then focused on studying the preparation of polymer nanocomposites using melt extrusion. Systematic and comprehensive studies were performed to determine if the melt processing parameters (screw configuration, screw speed, and melt temperature) could influence the polymer and nanoparticle interactions to achieve exfoliated structures [18].

Ray et al. [10, 11] have published several reviews on biodegradable nanocomposites. He stresses how biodegradable nanocomposites are the greening materials for the 21st century for materials science research. He has an extensive review article on biodegradable nanocomposites encompassing the following topics: biodegradable polymers, nanoparticles, mechanical properties, thermal stability, gas barrier properties, biodegradability, rheology and foam processing. In particular, he recently published a review article on PLA nanocomposites emphasizing the preparation with clays and carbon nanotubes, as well as reporting on properties and biodegradability. Cost and low production rates for the biodegradable polymers are cited as reasons why the market and applications have not grown. Ray addresses nanocomposites produced from extrusion, foaming methods, and even electrospinning, but the research is strictly fundamental with the emphasis on the morphology and structure-property relationships for nanocomposite formulations. Ray mentions that PLA nanocomposites are commercially being used, but no further details are given.

Okamoto et al. [12] also have a dedicated book chapter on biodegradable nanocomposites. This chapter covers the preparation of the nanocomposites based on PCL, polyvinyl acetate (PVA), PLA, polybutylene succinate (PBS), PHA, starch, plant oil and chitosan. They examine the structure, preparation methods, mechanical properties, biodegradation, crystallinity, rheology, and processing conditions of these systems. Okamoto believes that there will be future growth in biodegradables since sustainability is such a driving force in industry. In addition, environmental issues with an emphasis on waste disposal also are reasons to consider biodegradable nanocomposites. Okamoto mentions both

NatureWorks™ who produce PLA and Metabolix who produce PHA as leaders in the biodegradable polymer industry seeking niche markets from customers.

Hussain et al.[13] have published a review article with Okamoto that provides a comprehensive report on the structure-property relationships, processability, manufacturability, and applications of all types of nanocomposites. In this review, biodegradable nanocomposites are discussed with some overlap from Okamoto's chapter. Mentioned in the review is the commercialization of PLA/layered silicate nanocomposites by Unitika Ltd, Kyoto, Japan. These PLA nanocomposites were developed with the Toyota Technical Institute for packaging materials for short-term disposable applications. This review also highlights a few studies on the biodegradability of nanocomposites with PCL, PLA and PBS. In addition, this article discusses simulation and modeling of nanocomposites to account for the morphology of the nanoparticles. Hussain emphasizes that the biodegradable nanocomposites will have unique properties and will add new materials to the plastics industry. Both Hussain and Okamoto stress the non-toxicity of biodegradables with microorganisms actively attacking the polymer chain to convert carbon into carbon dioxide, water and biomass as a function of time.

Rhim et al.[14] have an informative review emphasizing natural biopolymers for food packaging applications. Starch, cellulose, chitosan and proteins are used to make nanocomposites with nanoparticles. For these types of polymers, water sensitivity can be an issue, so the nanoparticles may help improve water resistance. Exploration of active packaging that extends shelf life and improves safety while maintaining the quality of the food is the focus. Rhim believes that there is a huge potential for natural polymer based nanocomposites to enhance the quality of the food by increasing barrier properties and using packaging materials with antimicrobials. However, the water resistance and the processability of these materials are always a challenge. Rhim suggests blending bionanocomposites with a water resistant polymer, and even co-extrusion of multilayer products.

Sorrentino et al.[15] investigate biodegradable nanocomposites specifically for food packaging applications. Polyhydroxybutyrate (PHB), PLA, PCL and starch-based systems were discussed, along with edible coating technology with nanoparticles as carriers for antimicrobials. The review emphasizes the positive impact that nanocomposites will make on the food packaging industry, with packaging providing improved mechanical, thermal and barrier properties, as well as potential nano-sensors for tracking shelf life of the food.

Avella et al.[16] have reported work on a melt processed starch/clay nanocomposite to evaluate morphology (intercalation) and mechanical properties for food packaging applications. What is of special interest here is that these films were tested for conformity with actual regulations and European directives for biodegradable materials using migration tests and putting the films in contact with vegetables and stimulants.

10.5 Biodegradability

Biodegradation is the conversion of the carbon on the polymeric chain to carbon dioxide. There are standard test methods established with appropriate conditions (i.e. temperature, moisture, pH, microorganisms) for biodegradation to occur as a function of time. The rate and amount of carbon dioxide evolved correlates with the rate of biodegradation. From the biodegradation studies on biodegradable nanocomposites, there appears to be some misunderstanding on the type of experiment that needs to be performed to specify if indeed biodegradation is occurring. Numerous studies have not observed the biodegradation by standard methods. These studies have simply used weight loss procedures and try to correlate this data to mechanical or molecular weight data. Claims are being made and conclusions drawn from data without substantiation of biodegradation based on existing standard methods of converting the carbon to carbon dioxide.

There are numerous methods from ASTM and International Standards for measuring biodegradability of polymers. These standards describe the degree of biodegradation and the time frame in which it must occur. Examples of these standards are shown in Table 10.1.

Table 10.1		
Examples of standard methods for biodegradation experiments in different environments		
Test	**Environment**	**Property**
ASTM D5209	aerobic sewage sludge	carbon dioxide
ASTM D5210	anaerobic sewage sludge	carbon dioxide/methane
ASTM D5271	activated sewage sludge	oxygen/carbon dioxide
ASTM D5338	controlled composting	carbon dioxide
ASTM D5511	anaerobic digestion	carbon dioxide/methane
ASTM D5526	accelerated landfill	carbon dioxide/methane
ASTM D6691	aerobic marine environment	carbon dioxide
ISO 14855	aerobic under controlled conditions	carbon dioxide
ISO 14852	aerobic in aqueous environment	carbon dioxide
ISO 15985	aerobic in high solids sewage	carbon dioxide

10.5.1 A recent study of PHB nanocomposites

Maita et al. [19] most recently published a valuable research study on PHB nanocomposites with modified montmorillonite or a synthetic fluoro-mica as the nanoparticle. All samples were compounded with a melt extruder and were formed into films with a compression moulder. Typically, the nanocomposites were characterized by differential scanning calorimetry (DSC), thermogravimetric analysis (TGA), dynamic mechanical analysis (DMA), microscopy, and gel permeation chromatography (GPC). Biodegradation of the samples were investigated in compost at a two different temperatures. However, these biodegradation studies were not performed running the standard tests, and therefore that data cannot conclude if the nanocomposites improved biodegradation rates in comparison to the homopolymer. In fact, no conclusions can be made on the rates of biodegradation. The study provided an indication of trends with respect to properties and biodegradation, but primarily by reference to weight loss data. The micrographs of the contrasting surface topology and digestion of the presumably more amorphous component was also instructive. The solution characterization indicating the trend in the molecular weight change of solubles also adds to the study. With all the experimental evidence presented there is still no clear information about biodegradation kinetics which is really only accessible by measurement of CO_2 evolution. Although this work contributes to the field of nanocomposites, there is a need to implement standard methods for biodegradation.

10.6   Processability issues

Most articles do not show that there are any specific processing issues related to the nanocomposites, just the challenges associated with the homopolymers. Many studies over the last decade have looked at the compatibility of nanoparticles and biodegradable polymers. Many studies have compounded the nanoparticles and the polymer using twin screw extrusion, and then compression moulded samples. The following studies are recent investigations focusing on processing of the nanocomposites and their characterization.

10.6.1 A recent study of PCL nanocomposites

There are many research studies on the fabrication and characterization of nanocomposites. For example, Luduena et al. [20] recently published a study of PCL nanocomposites with the montmorillonite Cloisite 30B, addressing the mechanical and morphological properties of nanocomposite films prepared by solvent-casting and melt processing. Cloisite 30B from Southern Clay is a natural montmorillonite organically modified with methyl tallow bis-2hydroxyethyl quaternary ammonium. This is a typical study trying to optimize preparation parameters to achieve more interaction of the polymer and the nanoparticle. The study optimized a solvent (a mixture of dimethylformamide and dichloroethane) and an ultrasonic mixing time. Ultrasonication time actually influenced the Young's modulus values of the nanocomposite. The concentration of the 30B varied from 3.75 weight-% to 10 weight-% for the solution cast films with optimal concentration of 4 weight- % 30B and sonication time of 15 minutes. The solution cast preparation was pursued since the layered silicates disperse in the presence of the solvent. The polymer absorbs onto the delaminated sheets, and as evaporation occurs the sheets reassemble to give the ordered nanocomposite structure. However, scanning electron microscopy (SEM) showed that the nanoparticles contributed to defects in the film during the evaporation process which would minimize the improvement of properties.

For industrial applications, the authors realized that intensive mixing by melt and melt processing is the preferred preparation route. If the nanoparticle is mixed in the melt state and compatibility exists, the polymer can migrate into interlayer spacing with shear and screw speed driving forces, to create an exfoliated / intercalated structure. For this method, nanoparticle content varied from 2.5 weight-% to 7.5 weight-% and 0.4 mm thick compression moulded films were produced. One issue related to melt processing and compression moulding is that once the nanoparticles and the polymer are re-heated to high temperatures, partial agglomeration of nanoparticles can occur. For this study, all nanocomposites had improved Young's modulus in comparison to the pure PCL, while the loading of 5 weight-% 30B showed the most improvement of 70 % in Young's Modulus.

10.7   Attainable properties

Many of the current problems associated with biodegradable and bio-based plastics are that they do not provide the necessary properties for numerous products that are currently produced using cheaper petroleum-based materials. Nanocomposites may provide a way to attain the required material properties currently available from commodity plastics using sustainable materials. As indicated later in Table 10.3 the specific properties of many biodegradable polymers may be improved through the incorporation of nanotechnology to achieve the desired performance for a particular application.

10.7.1 A recent study of PLA/PCL nanocomposites

For example, Cabedo et al. [21] studied amorphous PLA blended with PCL. PCL acted as a plasticizer due to problems with brittleness in PLA. Since the PCL has similar processing temperatures, low tensile strength and high elongation, it is acceptable to process with PLA. This study blends the PLA and PCL at PLA/ PCL concentrations of 80/20, 70/30, 60/40 and 20/80. The nanoparticle was a kaolinite at 4 weight-% loading in all the samples. Characterization of the nanocomposite systems was performed using X-ray, transmission electron microscopy (TEM), dynamic mechanical analysis (DMA), microscopy and oxygen transmission rates. The samples were melt blended in a mixer, milled into powders and formed into sheet by compression moulding. A mixed exfoliated / intercalated system was obtained. Overall, there was an improvement in mechanical properties and thermal stability without significant loss of barrier properties. The optimized system contained a small proportion of PCL, and 4 % kaolinite, and demonstrated better processability and thermal stability than the neat PLA and similar gas properties. For the pure system, the oxygen barrier decreases as a function of the amount of PCL. The nanocomposites showed improved barrier properties and the nanoparticle shows compensation of barrier properties for the blending of the polymers.

10.8   Performance data

Many studies do not include all the performance data that is needed to qualify a film structure for industrial applications. If a nanocomposite film is being considered for industrial application, the film must be produced from cast or blown film extrusion. It should be noted that there is laboratory scale compounding and extrusion equipment available for research and development. The U.S. Army has such an extensive laboratory scale extrusion facility (Polymer Film Center of Excellence, Natick, MA) where there are minicompounders (gram size samples) that produce 2 inch films as well as injection moulded parts, and twin screw and single screw extruders that use kg size samples for both blown and cast film. Their capabilities also include the Collin Teach-Line series, comprising co-extrusion multilayer melt processing equipment which can produce as many as 5 layers in blown film and 9 layers in cast film. This equipment can be used for the researcher who has industrial applications with projects that would eventually transition to scaled up extrusion trials.

Table 10.2 is a list of the characterization tests that should be completed which can relate to the performance of the film and the interaction of the polymer and nanoparticle. These capabilities are crucial for industrial application and the U.S. Army also is well equipped with these characterization methods.

Table 10.2
Methods for characterizing nanocomposite films
Barrier Testing Equipment
ASTM D 3985 Oxygen Transmission Rate
ASTM F 1249 Water Vapour Transmission Rate
Mechanical Properties
ASTM D 882 – Tensile Properties
ASTM F-88 – Seal Strength properties
ASTM D 1709 Impact Resistance of Plastic Film (Dart Drop Impact Tester)
ASTM D1922 Propagation Tear Resistance of Plastic Film (Tear Strength Tester)

ASTM F 392 Flex Durability of Flexible Barrier Materials (Gelbo Flex Tester)
Other Polymer Characterization Techniques
Rheology Equipment – Capillary, Cone and Plate, Melt Flow Index
Thermal Analysis – Differential Scanning Calorimetry, Thermogravimetric Analysis Dynamic Mechanical Analysis
Spectroscopy - Wide and Small Angle X-ray Diffraction Equipment
Microscopy – Transmission Electron Microscopy, Scanning Electron Microscopy, Environmental Scanning Electron Microscopy, Atomic Force Microscopy, Polarizing Optical Microscope

10.9   Commercially viable materials

Commercially viable nanocomposite materials are most likely to be based on biodegradable polyesters because they are all commercially available on a larger scale. Table 10.3 indicates current applications for biodegradable resins and blends along with potential applications which might be achieved using nanocomposites.

Table 10.3

Applications for commercially available biodegradable polymers

Polymer	Current Applications	Nano-improved Properties	Potential Future Applications
Polylactic acid (PLA)	Bottles, films, coated paper, fibres, labels, utensils	Mechanical strength	Durable packaging
Polyhydroxyalkanoate (PHA)	Packaging, disposable items, coatings, housewares, paints	Barrier properties	Food packaging
Polycaprolactone (PCL)	Medical applications, containers, film	Mechanical properties	Stretch wrap, marine use
Thermoplastic starch	Mulch film, bread bags, shopping bags, over-wrap	Moisture sensitivity	High oxygen barrier films
Bio-Polyester / Starch Blend	Agricultural films, pet products, personal care	Mechanical properties	High strength bio-films

10.9.1 A recent study comparing biodegradable nanocomposites to polyethylene terephthalate (PET)

Interesting work conducted by Cava [22] compared biodegradable polymers and PET with respect to barrier properties and sensitivity to the retort process. PET, PCL PLA and polyhydroxybutyrate-co-valerate (PHBV) were used and compared in this study. The retort of pure PHBV (non-nano) was the most promising system for the retort procedure. The article mentions that a new range of tailor-made nano-additives for biopolymers is being tested. In the case of a PLA, 50 % oxygen permeability reduction for nanocomposite in comparison to the homopolymer was reported.

10.10 Applications

Applications for biodegradable materials range from commodity packaging, agricultural products and disposable products to medical, surgical and pharmaceutical items. The biodegradable nature of these polymers make them popular for biomedical applications such as sutures, implants, catheters, bone screws, grafts, scaffolding for tissue engineering, and ligament repair materials. Since packaging is the single largest market for plastics, at 10.88 metric tonnes per year worldwide, with one quarter of total US plastics production being used for packaging, there is bound to be an application for biodegradable nanocomposites for food packaging. Table 10.3 provides a summary of the achievable properties and potential applications.

10.10.1 A recent patent on biodegradable polymeric nanocomposite compositions

Mohanty [23] has a recent patent on blends of PHA, PHB and polybutylene adipate co-terephthalate (PBAT), which is intended specifically for packaging applications. Mohanty compares oxygen and water vapour permeability with thermoplastic resins and controls. He presents data on blown film of 60 weight-% of PLA / 40 weight-% PBAT and 5 weight-% Cloisite 25A. Cloisite 25A is a natural montmorillonite that is organically modified with a quaternary ammonium salt. The organic modifier contains dimethyl dehydrogenated tallow 2-ethylhexyl quaternary ammonium. Water vapour transport properties of a nanocomposite blown film showed 43 % improvement over compression moulded samples. Much of this type of data is presented in the patent; however, it would make more sense to compare cast versus blown film since the processing conditions used to make the films are similar except for the die design and the film forming method. However, this is a valuable study as two biopolymers are blended to create a system with a higher biocontent, which is important today with sustainability having an influential force in industry.

10.11 The future of biodegradable nanocomposites

Global growth rates for nanocomposites have been predicted to be as much as 25% per annum. Biodegradable nanocomposites may offer new materials with some unique high performance and improved properties, and with the added benefit of degrading to carbon dioxide, water and biomass in soil, compost or marine environments. One property, clarity, can be obtained with these nanocomposites with improvement in other properties. A new domain may open up if scientists and engineers can control the nanostructures within the nanocomposites. With the expectation of 500,000 ton/year of polymer nanocomposites in 2009, this is probably a long-term prospect for biodegradable nanocomposites as we are still trying to determine polymer/nanoparticle interactions and need to be concerned with large-scale production and demonstration/validation studies. Nanocomposites may have a strong future once the level of production of the polymers increases and the cost is less significant when exploring a wide range of applications. Overall the improvements in the biodegradable nanocomposites are not enough to replace petroleum based polymers, but with certain applications they are a fit. The global bioplastics market itself reached a value of $1 billion US dollars in 2007 and is expected to grow at a rate of 8-10 % per year to reach approximately 25-30% of the total plastics market by the year 2020. There are currently over 500 bioplastics processing companies available and more than 5000 are anticipated by the year 2020. It is expected that Europe will account for 31 % of this market, while the United States and Asia will account for 28 and 32 % respectively. New applications in the automotive, electronics, and packaging industries will drive the growth of these markets in the years to come. Nanocomposite research utilizing these bioplastics will speed the introduction of commercially available nanocomposites based on bioplastics to the market [24].

10.11.1 Life cycle assessment for biodegradable nanocomposites

Another topic for discussion which has been a major focal point with regard to biodegradable polymers is the materials' life cycle assessment (LCA) and carbon management. Recent publications question the benefit of bio-based and biodegradable plastics on the basis of energy and the environment. Statements such as "biodegradability has a hidden cost in that the biological breakdown of plastics releases carbon dioxide and methane into the atmosphere" question the utility of these materials. Publications by Patel and Narayan review and analyze the LCA of numerous bio-based materials in comparison with the petroleum-based products that they are designed to replace [25, 26]. They summarize that available LCA studies for various materials and environmental assessments strongly support the further development of biodegradable and bio-based polymers. Careful monitoring of environmental impacts continues to be necessary for proper decision-making regarding whether or not to use these materials for any particular application. Some materials show an environmental benefit today, while others show great promise and potential.

10.11.2 Safety of biodegradable nanocomposites

New guidelines for the use of nanocomposites have been initiated by a variety of organizations. For example, the National Institute for Occupational Safety and Health has developed a plan for nanotechnology safety that involves addressing knowledge gaps, developing strategies and providing recommendations. Issues such as the toxicity, epidemiology, risk, exposure and safety are all being reviewed. Education about these new packaging technologies is also an issue. With the development of this new technology new safety risks and hazards will also evolve that will need to be addressed. There are studies investigating human health issues with nanocomposites and many scientists have focused on these studies. This is a consideration for the industrialization of these materials as well as for the safety of the engineers and scientists working with nanoparticles.

10.12 Summary

Table 10.4 summarizes the parameters that can affect the polymer and nanoparticle interactions, the property and performance improvements observed with nanocomposites, and potential applications for these materials.

Overall, this chapter has addressed what needs to be done in the future for biodegradable nanocomposite studies, and in particular, the types of processing and characterization required to fully confirm whether a system has potential industrial application. Many historical and current studies have investigated the properties and performance of nanocomposites, but often without considering their industrial applications. If one keeps the industrial application in mind, then more complete studies could result.

Table 10.4

Summary of processability issues, property and performance improvements and potential applications of biodegradable nanocomposites

Parameters to consider*	Loading of nanoparticle	Temperature of melt processing	Chemistry of nanoparticle and degradation temperature of nanoparticles	Screw speed
Property improvements in nanocomposites	Degree of exfoliation	Barrier (water and oxygen)	Heat deflection temperature	Young's modulus
Performance data improvements	Rate of biodegradation	Improvement for barrier (water and	Increase the thermal stability –	Increase mechanical properties
	by ASTM methods	oxygen) at various conditions	end use temperatures	(Young's modulus)
Applications for biodegradable nanocomposites	Food packaging	Medical applications	Disposable items	Agricultural applications
**** need to produce films or parts by extrusion or injection moulding***				

REFERENCES
1   Clarinval, A. M.; Halleux, J.,In *Biodegradable polymers for industrial applications.* Smith, R. (editor), ISBN 1 85573 934 8, **2005,**Woodhead Publishing Limited, England, pages 3-30.
2   Preeti, D.; Rohindra, D.R.; Khurma, J.R., **2003,***South Pacific Journal of Natural Science,* 21, 47-49.
3   Ogata, N.; Jimenez, J.; Kawai, H; Ogihara, T. J. **1997**, *Polymer Science, Part B: Polymer Physics.* 35, 389-96.
4   Giannelis, E. P. **1996**, *Advanced Materials,*   8, 29-35.
5   Pinnavaia T. J; Beall G. W. *Polymer-Clay Nanocomposites.* **2000,** John Wiley & Sons, USA.
6   LeBaron, P. C.; Wang, Z.;. Pinnavaia, T. J. **1999,***Applied Clay Science,* 15, 11-29.
7   Manias, E.; Touny, A.; Wu, L.; Strawhecker, K.; Lu, B.; Chung, T. C., **2001,***Chemistry of Materials,* 13, 3516-3523.
8   Nielsen, L. E. *Journal of Macromolecular Science, Part A.,* **1967**, 1, 929-942.
9   Ratto, J.; Froio, D.; Thellen, C.; Lucciarini, J. Melt processing of polymer/montmorillonite layered silicates (MLS): nanocomposites films for high barrier food packaging. *Packaging Nanotechnology,* Mohanty, A.; Nalwa, H.S. (editors) American Scientific Publishers (in print).
10   Singh, S.; Ray, S. S., **2007,***Journal of Nanoscience and Nanotechnology,* 7, 2596-2615.
11   Ray, S. S.; Bousmina, M. **2005,***Progress in Materials Science* 50, 962-1079.
12   Okamoto, M. in *Handbook of Biodegradable Polymeric Materials and Their Applications*, Mallapragada, S.; Narasimhan, B. (editors), **2005**, American Scientific Publishers, USA, 1-45.
13   Hussain, F.; Hojjati, M.; Okamoto, M.; Gorga, R. E. **2006,***Journal of Composite Materials*, 40, 1511-1575.
14   Rhim, J. W.; Perry, K. W.; Ng, **2007,***Critical Reviews in Food Science and Nutrition*, 47, 411-433.
15   Sorrentino, A., Gorrasi, G. and Vittoria, V. **2006,***Trends in Food Science and Technology,* 18, 84-95.
16   Avella, M.; De Vlieger, J. J.; Errico, M. E.; Fischer, S.; Vacca, P.; Volpe, M. G.**2005**, *Food Chemistry*, 93, 467-474.
17   Philip, S.; Keshavarz, T.; Roy, I., **2007**, *Journal of Chemical Technology and Biotechnology*, 82, 233-247.
18   Ratto, J.; Lucciarini, J.; Thellen, C.; Froio, D.; D'Souza. N. A. The reduction of solid waste associated with military ration packaging. Technical Report natick/TR-06/023, **2006**, 1-75.
19   Maita, P., Batt, C., Giannellis, E. **2007**, *Biomacromolecules*, 8, 3393-3400.
20   Ludueña, L. N., Alvarez, V. A. and Vazquez, A. **2007,***Materials Science and Engineering: Part A,* 460-461, 121-129.
21   Cabedo, L.; Feijoo, J. L.; Villanueva, M. P.; Lagarón, J. M.; Giménez, E**2006,** Macromolecular Symposia, 233, 191-197.
22   Cava, C.; Gumenez, E.; Gavara, R.; Lagaron, J. M., **2006,***Journal of Plastic Film & Sheeting*, 22, 265-274.

23   Mohanty, A.; Parulekar, Y.; Chidambarakumar, M.; Kostruangchi, N.; U.S. Patent application 2007/0037912.

24 Helmut Kaiser Consultancy, Bioplastics Market Worldwide 2007-2025, hkc22.com/bioplastics.html, accessed 05-06-2008.

25   Patel, M.; Bastioli, C.; Marini, L.; Wurdinger, G.E., in *Biopolymers Volume 10, General Aspects and Special Applications*, Steinbuchel, A., **2003**, Wiley-VCH, Germany, 409-452.

26 Narayan, R, Patel, M. in *Natural Fibers, Biopolymers and Biocomposites*. Mohanty, A.K.; Misra, M.; Drzal, L.T. (editors), **2005**, CRC Press, USA, 833-853.

Chapter 11
Thermal properties of polymers with graphitic nanofibres
E. Hammel, A. Eder, X. Tang
Electrovac AG, Klosterneuburg, Austria

11.1   Introduction

Carbon nanotubes (CNTs) have one dimensional cylindrical structures, formed by rolling up graphene sheets. CNTs with diameters D > 70 nm are usually referred as carbon nanofibres or CNFs. They consist of multilayer wrapped graphene sheets, sometimes in the form of stacked cones (Figure 11.1).

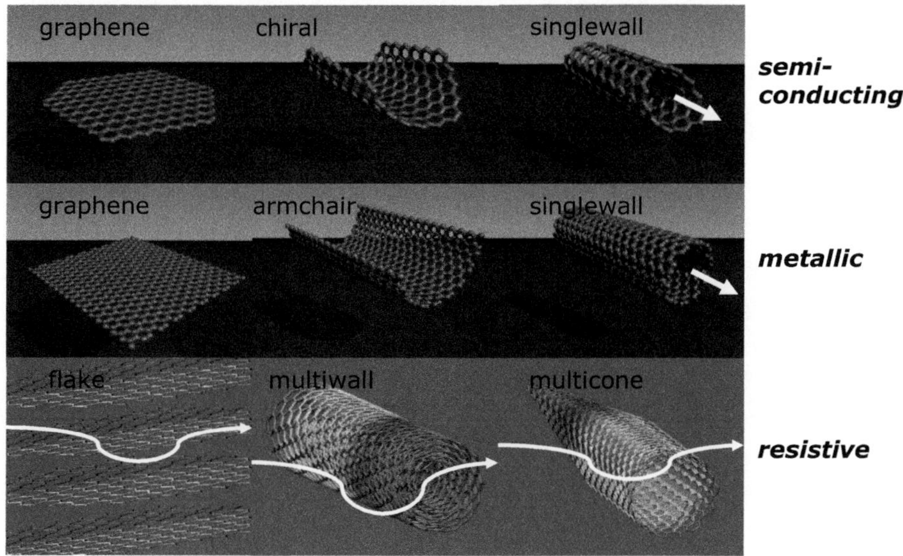

Figure 11.1
Types of planar and tubular graphitic base materials

Single walled CNTs (SWCNTs) are either metallic or semiconducting, depending on how the graphite layer is wrapped into a cylinder. Metallic SWCNTs show ballistic electrical conductivity at room temperature. When electrons are conducted through metals, they usually interact with imperfections in the lattice, scatter and cause electromigration. The ballistic transport allows CNTs to carry very high currents ($>10^9$ A/cm^2) without electromigration failure. Scattering occurs whenever an atom is out of place for any reason. Thermal energy produces scattering by causing atoms to vibrate, and this is the source of the electrical resistance of metals. This usually causes a positive temperature coefficient for the resistivity. Increasing the local current density is referred as the current crowding. Since joule heating is proportional to the square of current density, the current crowding effect leads to a local temperature rise around a void that in turn further accelerates the void growth. The whole process continues till the void is large enough to break the conductor line. The strong bonds between the carbon atoms in the graphene sheets and graphitic tubes simply reduce the void growth rates and permit higher current densities than comparable metal conductors.

Chemical vapour deposition (CVD) is a simple and relatively cheap way to produce non-aligned CNT and highly aligned CNT (ACNT) arrays by using e.g. methane, acetylene or ethylene as carbon sources. Commercial producers use proprietary methods of production, but usually based on CVD. This method is split into fixed and fluidized bed reaction schemes. The reaction processes differ not only due to the particular reactor designs, but also the choice of hydrocarbons, catalyst systems, their supports, the process temperatures and the applied pressures. This has led to a huge variety of proprietary synthetic routes and post-processing conditions from individual suppliers and specific applications for their products. As a rule it cannot be assumed that these materials can simply be substituted for each other in particular applications, when processing variables have so much influence on the final product performance. Similar observations are made with most of the other commercially available polymer additives. At the time of writing (2008) the reader will find a number of references for commercial producers of nanotubes and nanofibres in the Internet, and these are listed in Table 11.1. It is not easy to verify if the claimed capacities correspond to their real volumes being produced. The business is still in an early stage of commercial production and only a few companies have supplied truly commercial products in significant volumes. The authors are certainly aware that companies like Hyperion Catalysis, Bayer Material Science AG, Showa Denko and Pyrograf Products have found applications for their products where the term serial product can be applied. Bayer Technologies is believed to have the largest capacities so far and it will be interesting to see how their business will develop.

Table 11.1
Commercial Suppliers of Nanotubes

Producer	claimed capacity: tons/a
Ahwahnee Technology, Inc.,	-
Applied Sciences/Pyrograf Products	20
Arkema	10
Arry International Group Limited	50
ATMI	-
Bayer Material Science AG	60
Carbolex	1
Carbon Nanotechnologies Inc.	-
Carbon Solution Inc.	-
Carbonnano	45
Catalytic Materials	-
Cheapnanotubes	12
Chengdu Alpha Nano Technology Co., Ltd[Carbon Nano Materials R&D Center]	10
Chengdu Organic Chemicals Co Ltd CAS R&D	8
CNT Technologies	-
Columbian Chemicals	-
Dynamic Enterprise	-
Electrovac	2
Future Carbon	-
G.B. Sciences	-
Hanchow	-
Helix	-
Hyperion Catalysis International	-
Ilgin Nanotech	-
MER	-
Mitsui	40
NanoCarbLab	-
Nanocomp Technologies Inc.	-
NanoCompound AG	-
Nanocyl	40
NanoLab Inc	-
Nanoledge	-
Nanothinx	-
Raymor Industries	3

Polytech & Net Gmbh	3
Rosseter	2
Shenzhen Nanotech Port Co., Ltd.	10
Showa Denko	-
Sinonano Beijing	1
SouthWest NanoTechnologies SWeNT	1
Sumitomo	1
Sunnano	2

Aligned nanofibres and nanotubes are not sold by weight but in the form of applications. The major interests for these materials currently come from the chip industry, but large companies like Intel, IBM and Infineon have their own in-house programmes running to develop ACNTs as high performance transistors with extremely high switching rates or as electrical connectors for metallic connection lines to avoid electromigration and reduce stray capacitances. The idea of using ACNT arrays for thermal management applications is relatively new and a couple of university spin-offs have appeared on the market advertising their services.

For electronic device applications, CVD methods are particularly attractive due to characteristic growth features, such as selective growth, large area deposition and aligned CNT growth. In composites CNTs are usually added by mixing them in random order in low viscosity molten polymers.

We have studied here the effect of adding vapour grown nanofibres with diameters D > 70 nm and length L > 10µm (Figure 11.2).

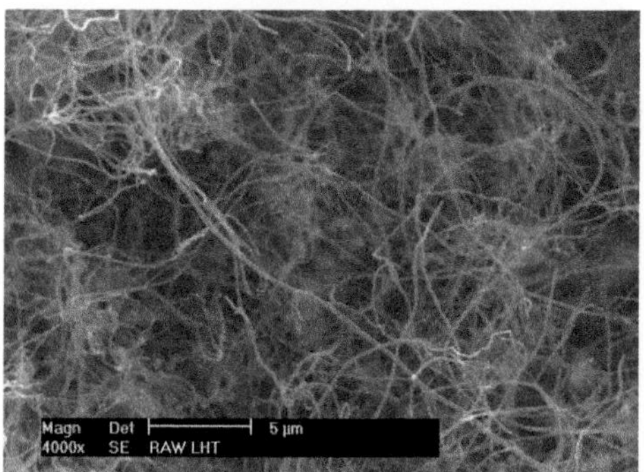

Figure 11.2

Vapour grown nanofibres seen through the scanning electron microscope, SEM (left), and transmission electron microscope, TEM (right)

For electrically conductive composites the percolation concentration limit c_{perc} should be as low as possible, which can be achieved easier by very high aspect ratios at small fibre diameters. To obtain percolation within a CNT network sufficient mass m_{CNT} of CNTs has to be incorporated into the composite mass m_{Total}. Only beyond this limit will we find a continuous increase in the composite's electrical conductivity until we obtain saturation corresponding to the contact resistance between the test body and the sensor (Figure 11.3).

$m_{CNT}/m_{Total} < c_{perc}$

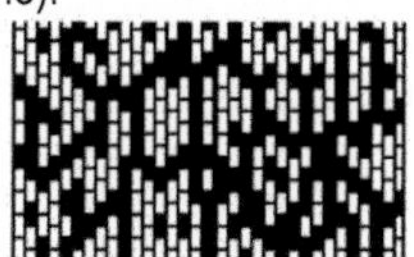

$m_{CNT}/m_{Total} > c_{perc}$

aligned nanotubes (ACNT)

Figure 11.3

Effect of CNT concentrations and orientations on electrical properties of composites

Thermal conductivity in composites is determined by phonon scattering. Phonon wavelengths are considerably larger than inter-fibre gaps even for low fibre concentrations, allowing the phonons to "tunnel" through the composite materials. The thermal conductivity depends strongly on the state of order of the graphite system as a whole, in terms of homogeneity, isotropy and internal graphitic structure.

The thermal conductivity λ is defined by

$$\frac{\Delta Q}{\Delta t} = \lambda \times A \times \frac{\Delta T}{x}$$

where A is the total surface area of conducting surface, ΔT is temperature difference and x is the thickness of conducting surface and $\Delta Q/\Delta t$ is the heat flow. Along the CNT axis x=L, where L is fibre length, and the cross section A=(D/2)$^2 \times \pi$, where D is the fibre diameter.

Berber et al [1] (Figure 11.4) calculated the thermal conductivity of CNTs as a function of the temperature. Phonons propagate easily along nanotubes [3] allowing thermal conductivities of an individual nanotube at room temperature $\lambda_{\parallel}$ >3000 W/m K, which exceeds the conductivity of diamond with theoretically up to 2000 W/m K.

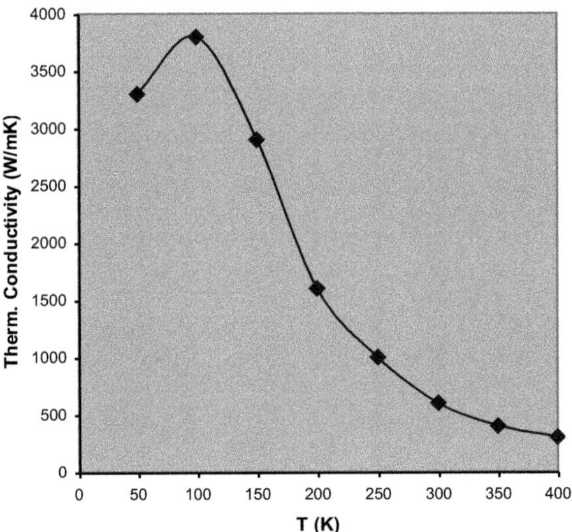

Figure 11.4

Calculated thermal conductivities of CNTs as a function of temperature [1]

For long CNTs we can set A= πLD, where D is the fibre diameter. If we consider only the axial heat flow the thermal resistance of the fibre is given by $R_{\parallel}$ = ΔT / Q = 1/($\pi\lambda_{\parallel}$D). In case of the heat flow perpendicular to the axis we would obtain $R_{\perp}$ = ΔT / Q = 1/($\pi\lambda_{\perp}$L), where $\lambda_{\perp} \approx$ 20 W/m K. These formulae reflect the importance of fibre length and diameter of nanofibres compared to thinner nanotubes for thermal applications. Using thicker CNFs compared to CNTs is equivalent to using better ordered graphite in one single contact. In these experiments we used vapour grown nanofibres in the "as-grown" state (fibres from the CVD process and after low heat treatment (LHT) < 1500°C to remove any residual amorphous carbons) and compared them with fibres after high heat treatment (HHT) (> 2500°C).

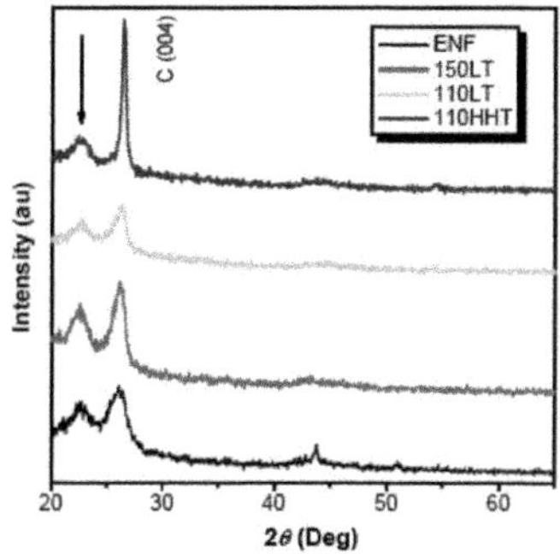

Figure 11.5
XRD spectrum for LHT and HHT fibres

Figure 11.5 shows the results of an x-ray diffraction (XRD) spectrum for LHT and HHT fibres. The graphite peak (004) is clearly enhanced by high temperature treatment of the fibres. Graphite is a very good thermal conductor, and we can assume that this treatment also enhances the thermal properties of the CNFs, and this corresponds well with studies [2] on CVD coated carbon fibres before and after heat treatment (Figure 11.6).

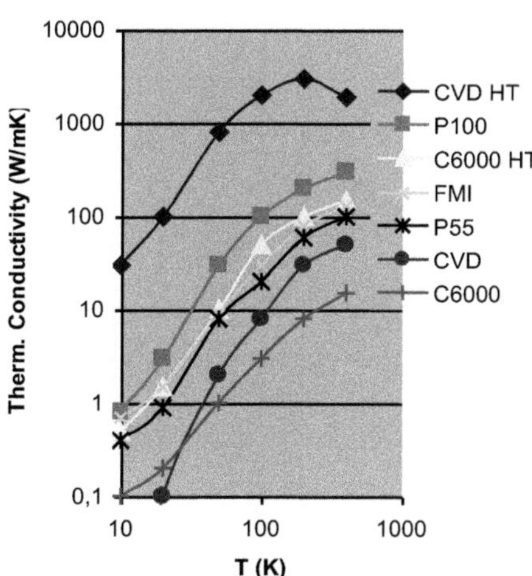

Figure 11.6
Thermal conductivity of different CVD coated carbon fibres

CVD (HT): graphite fibres grown by chemical vapour deposition of methane (graphitized at temperatures up to 3000 °C). (Pyrograf 1 from Pyrograf Products)
P 100: P-100 pitch based graphitized carbon fibres (Amoco)
C 6000 (HT):   PAN-based C6000 (graphitized) carbon fibres
FMI: woven carbon fibres, impregnated with resin or pitch and carbonized (FMI Inc.)
P55: pitch-based P55 carbon fibres (Thornel)

Measurements of the thermal properties of the composite materials were based on the ASTM D 5470 and ASTM E 1461. ASTM E 1461 was used for bulk materials and ASTM D 5470 has been applied for thermal interface materials. We are currently involved into a European framework program under the acronym NANOPACK (www.nanopack.org) where new ways are being developed to accurately measure the thermal properties of nanofiller-enhanced thermal interface materials.

11.2 Thermal interface materials

Thermal interface materials (TIM) like thermal greases are composite materials made with thermally conductive fillers, designed to manage chip cooling by maximising heat transfer. CNTs have been proposed as conductive fillers [4-11] and some results have been obtained for ACNT [12-15]. Alignment of nanotubes can be achieved by various methods within the growth process. The simplest approach useful for thermal management applications is to have a dense catalyst layer on a flat substrate, forcing nanotubes to grow perpendicular to this surface, provided that low temperature and gas flow gradients can be achieved over the substrate area. Fibres will support each other, and highly aligned and dense "CNT grass" can be produced. The catalysts can be applied by chemical and physical deposition methods.

A TIM test-bench shown in Figure 11.7 was used to study samples of bisphenolic base material reinforced by heat treated carbon nanofibres and marketed under the name elNano®. This composite material was placed between two calibrated copper heat flux meters and a heater block below and cold plate above the test column were used to induce a heat flux through the sample. Temperature measurements performed using RTDs (resistance temperature detectors) positioned at uniform intervals in the heat fluxmeters were used to determine the total heat flow rate, Q, and the temperature drop across the joint, ΔT. The thermal joint resistance was calculated by:

$$R = \Delta T / Q$$

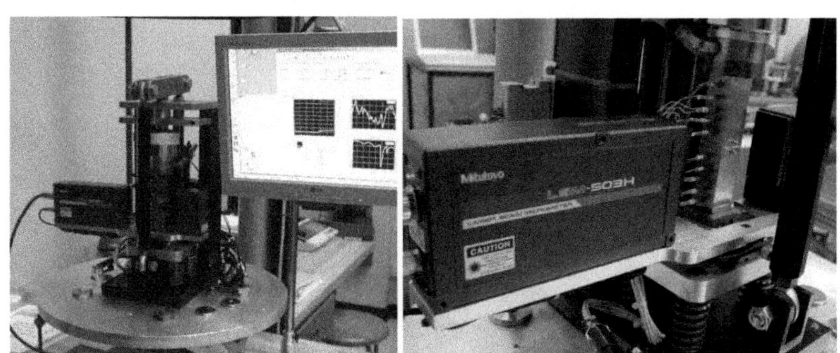

Figure 11.7
TIM test apparatus (Department of Mechanical Engineering, University of Waterloo)

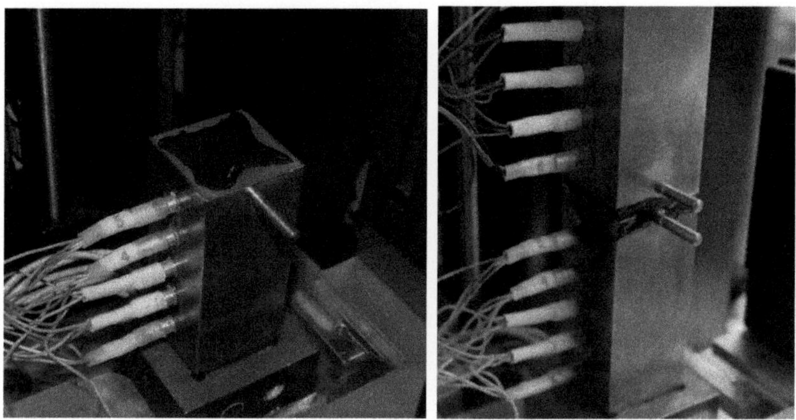

Figure 11.8
Bondline thickness measurement using laser scan micrometer (LSM)

Measurement of bondline thickness is shown in Figure 11.8. Small filler size is important in minimising bondline thickness, and hence thermal resistance. From Figure 11.9 it is apparent that the carbon nanofibre TIM elNano® has significantly lower thermal resistance than the other commercial materials at similar contact pressures. As the load increases, these differences are reduced, to a maximum of 12 % for the Al-particle based material at the largest contact pressure. This result was made possible by squeezing the material down to extreme low bondline thickness (BLT) far below 100 µm. Recently we have determined minimum bondline thicknesses of 10-20 µm. This BLT is in the range of the CNF length applied in the elNano® material (Figure 11.10).

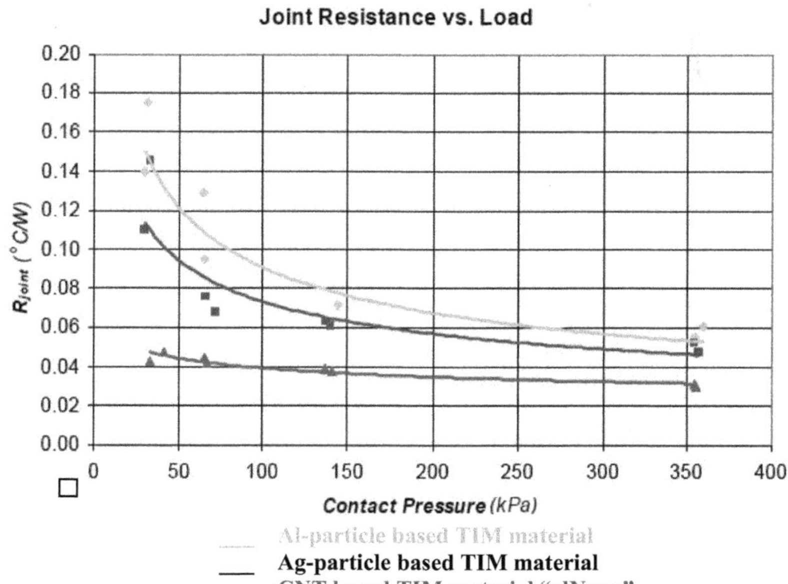

Figure 11.9

211

Thermal joint resistance vs. contact pressure, comparing CNF with commercial materials

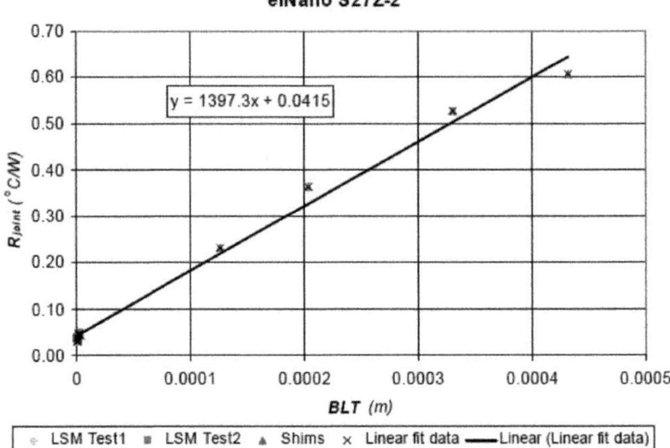

Figure 11.10
Thermal resistance vs. bondline thickness for elNano

The thermal conductivity of elNano[®] is calculated from the slope of a linear fit of the joint resistance versus bondline thickness data in Figure 11.10, using the relationship $R = t/\lambda A$, where t is the material thickness, A is the cross sectional area of the joint (6.54×10^{-4} m^2) and λ is the thermal conductivity determined to be 1.2 W/m K.

Preliminary results of a more detailed evaluation of the material between 20 and 30 µm BLT has resulted in values of λ above 3 W/m K. In this setup the lateral temperature distribution was also carefully evaluated.

Applying pressure up to 355 kPa can force the fibres to orient parallel to the test surfaces (Figure 11.11). Microelectronic applications of such TIM composites show similar results, as evaluations of applications for Pentium V processors in a standard workstation have demonstrated. For non-deformable composite materials we have used other techniques to evaluate the thermal properties of CNF reinforcements.

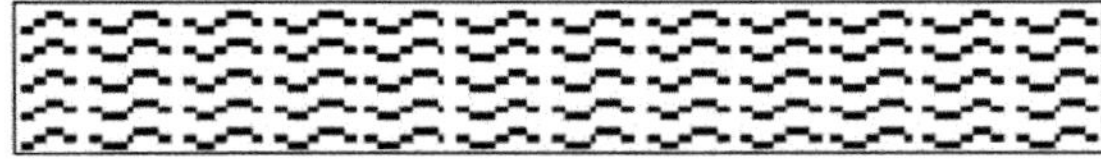

Figure 11.11
Orientation of TIM fibres under contact pressure

11.3   Thermally conductive plastics

Thermally conductive plastics are engineered composites that could replace some metals, ceramics, and conventional plastics in heat-sensitive applications. Whereas conventional plastics have a thermal conductivity of approximately 0.2 W/m K thermal conductive plastics typically conduct 5 to 10 times better (1-2 W/m K) and some conduct more than 20 times as well. In this study we focussed on extrudable and injectable materials only.

One of the most innovative characteristics of these materials is that they can be engineered to work with a wide variety of base polymers including polypropylene (PP), polyamide (PA) and many others. This may allow product designers to add thermal capabilities to their existing base resins, but without making major changes to existing manufacturing processes.

To evaluate the performance of these materials we used the flash technique, one of the most widely used methods for the determination of thermal diffusivity and thermal conductivity of solids. Using this method, the front side of a plane-parallel sample with a well defined thickness is heated by a short light or laser pulse. The resulting temperature rise on the back surface is measured versus time using an infrared detector. Analyzing the measured detector signal with appropriate mathematical models yields information on thermal diffusivity and the specific heat of a material. Together with the density of the material the thermal conductivity can be determined. The flash method is a standardized technique (ASTM E 1461 or DIN EN 821) and uses the equation

$$\lambda(T) = \alpha\,(T)\,c_p(T)\,\rho(T)$$

The thermal diffusivity α can also be measured by flash techniques (Figure 11.12). The front side of a plane-parallel sample is heated by a short light pulse. The resulting temperature rise on the rear surface is measured as a function of time. If the specific heat c_p and density ρ are additionally known, the temperature-dependent thermal conductivity λ of the material can be calculated.

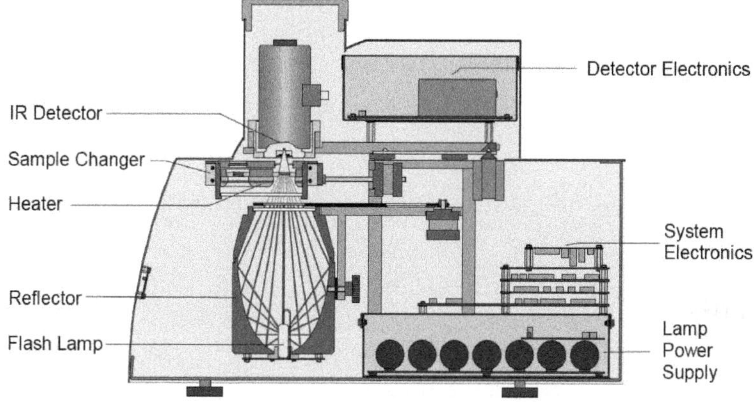

Figure 11.12
Schematic of LFA 447 *NanoFlash®*
(Courtesy of Netzsch-Gerätebau GmbH)

The thermal diffusivity measuring range is 0.01 to 1000 mm²/s, with a reproducibility of approx. +/-3 %. In addition to the thermal diffusivity, by employing a comparative method the specific heat can also be determined using this technique. For the specific heat, a reproducibility of +/-5 % is achieved. If the bulk density is known, a direct determination of the thermal conductivity is possible. The thermal conductivity range is 0.1 to 2000 W/m K.

The method has been applied to the determination of the thermophysical properties of polymer plates between 15°C and 150°C.

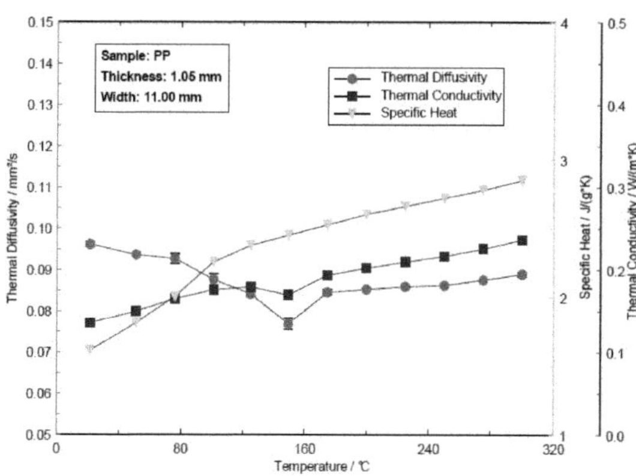

Figure 11.13

NanoFlash® data for a polypropylene bulk sample

Figure 11.13 depicts the thermal diffusivity (red), thermal conductivity (blue) and specific heat (green) of PP as a function of temperature. From room temperature to the onset of melting at 150°C (extrapolated onset of the melting peak from a DSC measurement), the thermal diffusivity decreases significantly from approx. 0.098 to 0.075 mm²/s. After melting, it reaches an almost constant value of 0.085 mm²/s at 250°C. As expected, the specific heat increases in two steps: from 1.5 J/g K at room temperature to 2.2 J/g K at 90°C and, during softening and melting, from 2.3 J/g K to 2.8 J/g K at 250°C. Prior to and after melting, the resulting thermal conductivity shows an increase from 0.14 W/m K to 0.22 W/m K (250°C). Using twin screw extruded compounds of PP with carbon black (CB) as reference and CNF we obtained the results shown in Figure 11.14 for the thermal diffusivity at room temperature.

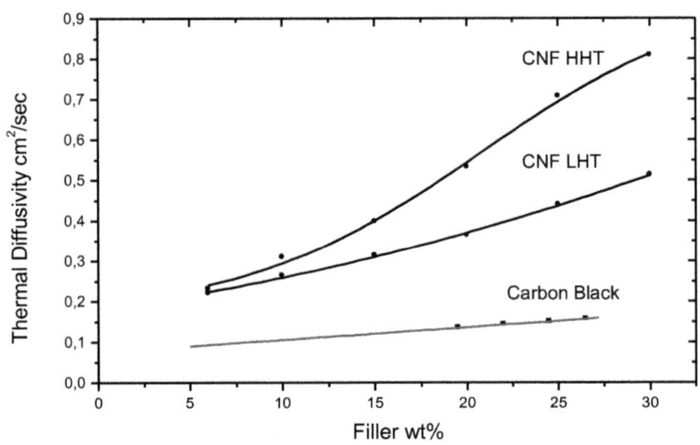

Figure 11.14

Thermal diffusivity versus CNF and CB concentrations

This indicates a very important enhancement of the thermal properties by adding CNFs and further improvement by a high temperature annealing treatment of the fibres. Values this high are usually only obtained by copper fibre reinforcements leading to anisotropic properties (Figure 11.15). In the normal direction CNF HHT reinforced polymers already compete at moderate filling ratios.

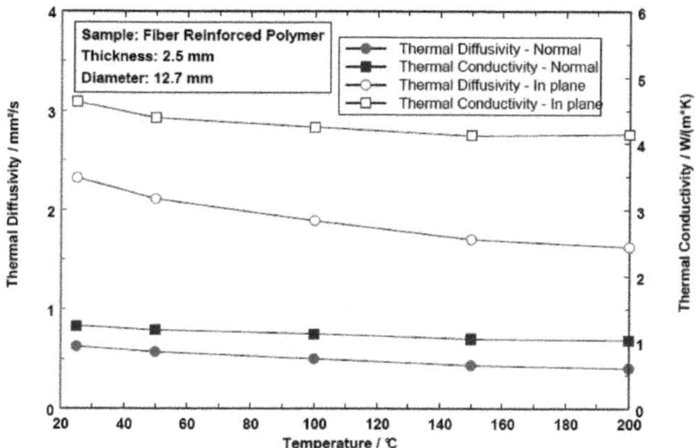

Figure 11.15
Thermal properties of Cu-fibre reinforced polymers

This type of copper fibre reinforced composite material is generally used for fast heat removal in electronic components, where the use of thermally conductive plastics for housings and chassis is becoming increasingly important. Carbon fibre reinforced polypropylene shows much poorer results than carbon nanofibres. Thermal diffusivity and conductivity of CNF reinforced PP are demonstrated in Figure 11.16.

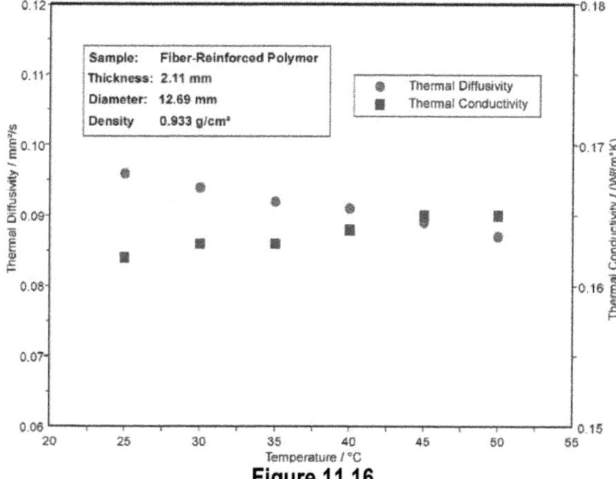

Figure 11.16
Thermal properties of CNF reinforced polypropylene

The CNF-reinforced PP material shows also a decrease of the thermal diffusivity against temperature, while the thermal conductivity slightly increases. We have studied the thermal properties as they develop when reinforcing PP materials with CNF (Figure 11.17 a-d).

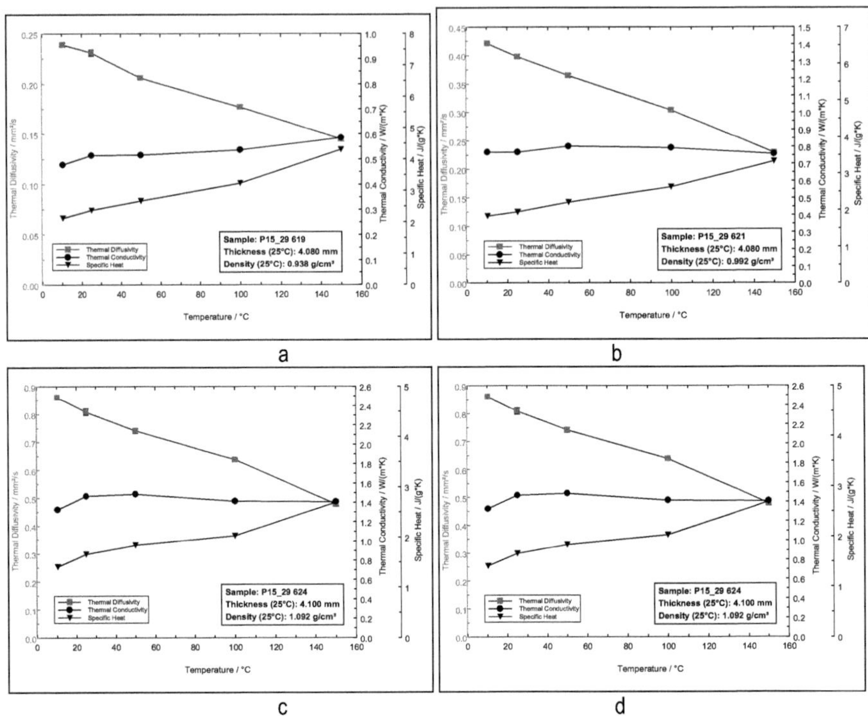

a

b

c

d

Figure 11.17

Thermal properties of 6 weight-% (a), 15 weight-% (b), 20 weight-% (c) and 30 weight-% (d) CNF - HHT reinforced polypropylene

We find that at higher CNF concentrations λ seems to decrease with increasing ambient temperatures. This could be explained by reduced mechanical stability of the system and therefore increased phonon scattering. The positive contributions of the fibres to the total thermal conductivity outweigh mechanical stability losses.

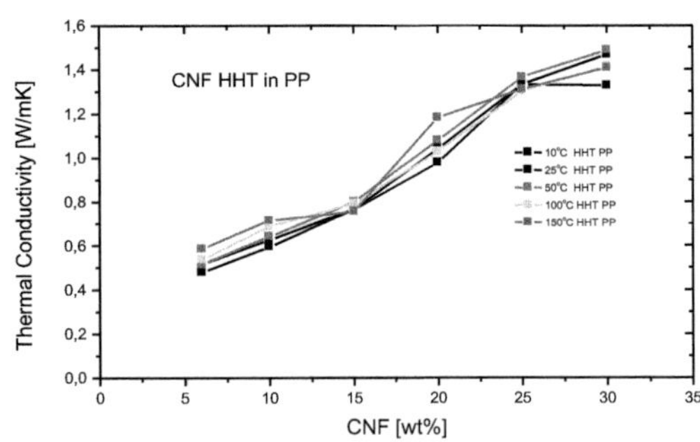

Figure 11.18

Summarized thermal properties of a range of CNF- HHT reinforced PP samples between 10 and 150°C

A maximum thermal conductivity, 1.5 W/m K, was measured for 30 % CNF composite (Figure 11.18). For 20 % CNF composite the value is around 1.2 W/m K, which is quite nicely consistent with the results obtained from TIM elNano® as described in Figure 11.10.
Another system studied was polyamide 6.6, widely used in automotive industry. This system was reinforced with CNFs of the LHT type (Figure 11.19 a-e).

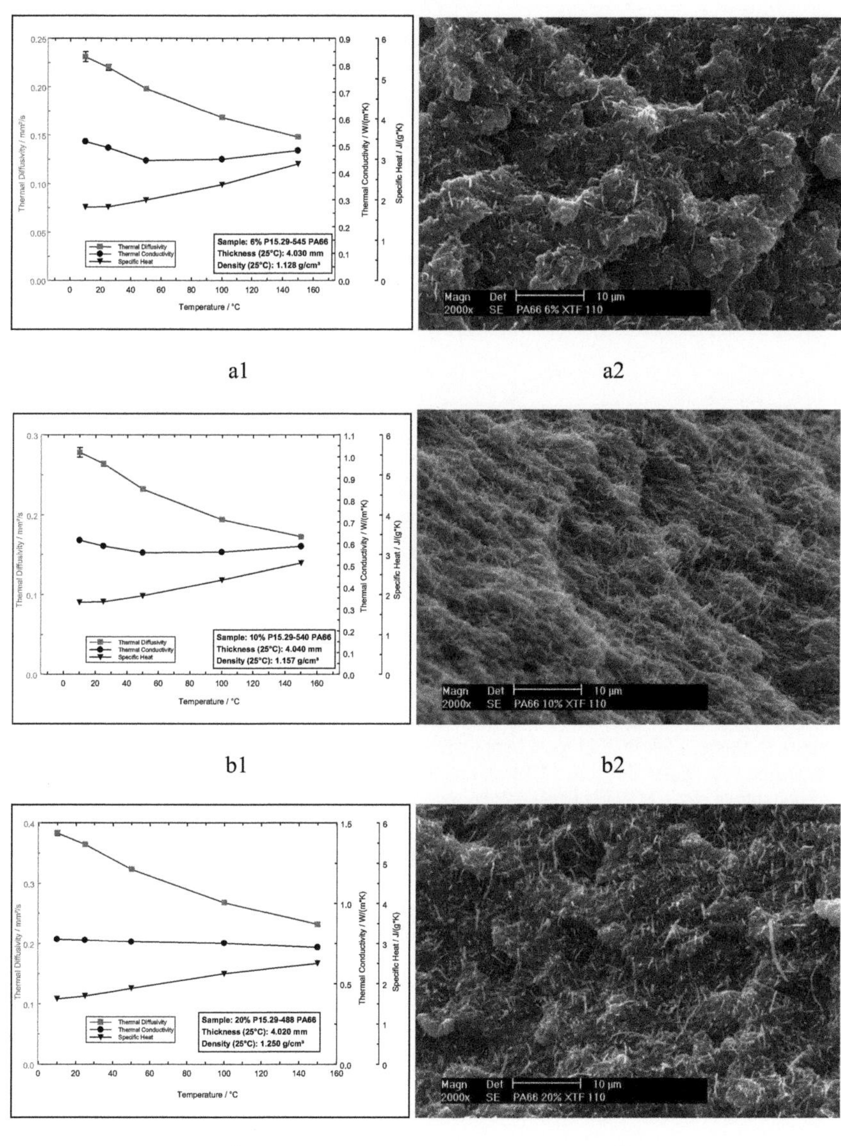

a1

a2

b1

b2

c1

c2

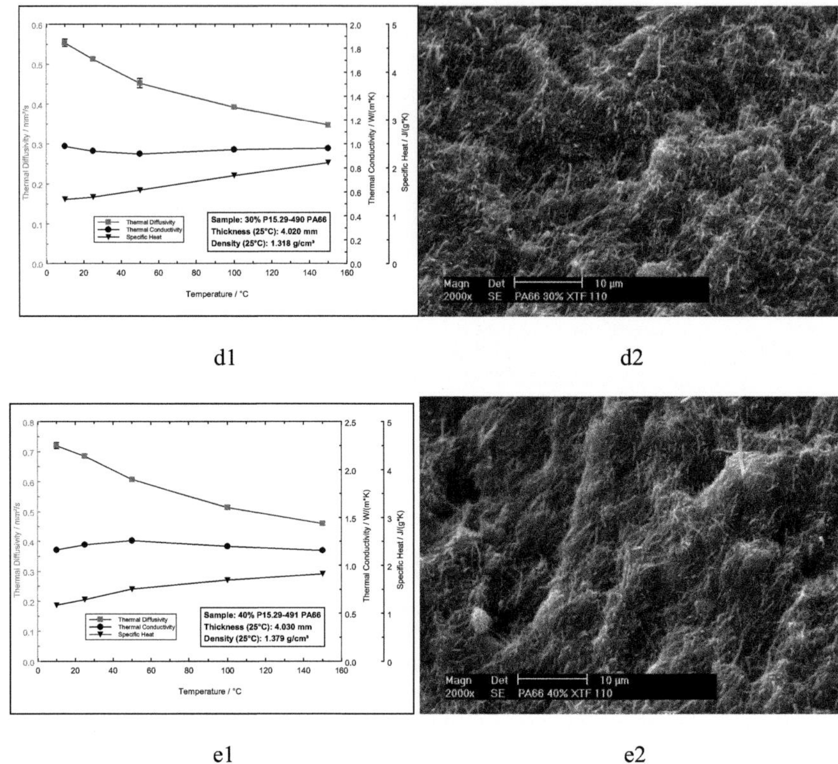

d1   d2

e1   e2

Figure 11.19

Thermal properties of 6 weight-% (a), 10 weight-% (b), 20 weight-% (c), 30 weight-% (d) and 40 weight-% (e) CNF - LHT reinforced polyamide 6.6

Again this system was significantly thermally enhanced by CNF reinforcement. Even the LHT type pushed λ to almost 1.2 W/m K at room temperature and 20 weight-% content.

Summarizing the results of Figure 11.19 we can make similar observations to those for the PP system. High concentration of CNF will slightly deteriorate the mechanical properties of the composite, while the total value of λ is dominated by the increased amount of CNF with its good thermal properties. Figure 11.20 shows the changes of thermal conductivity with the nanofibre content in CNF-polyamide composite at different temperatures.

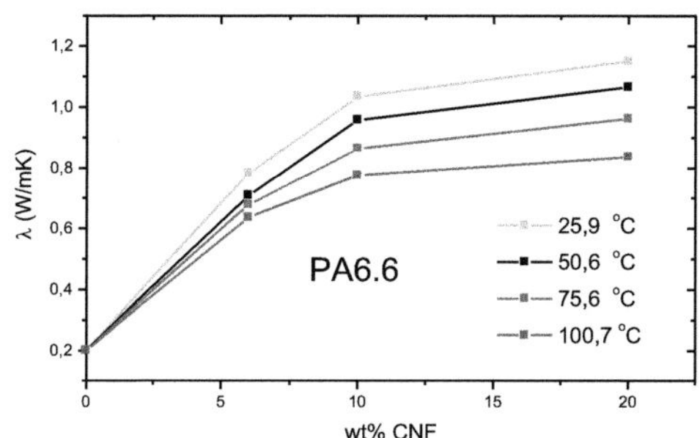

Figure 11.20
Thermal properties of LHT CNF reinforced polyamide 6.6 between 25 and 100 °C

11.4 Conclusion

The thermal conductivity of polymers can be enhanced by using carbon nanofibres. Annealing the fibres at high temperatures increases their effect on composite properties. We have indications that composites compressed to material thicknesses comparable to the nanofibre length show even more thermal conductivity than bulk materials of arbitrary thickness. It has yet to be seen whether further improvements in the λ can be obtained by introducing functional groups and "spacers" for the fibres. Isotropic composites probably will be limited to the thermal conductivities of the tubular graphite perpendicular to their axis. Anisotropic materials could indeed be further enhanced towards the axial values of the perfectly grown nanotubes above the thermal conductivity of diamond.

REFERENCES

1   Berber, S.; Kwon, Y.-K.; Tománek, D. Unusually High Thermal Conductivity of Carbon Nanotubes;*Physical Review Letters,* 2000, **84, 4163 - 4616**

2   Heremans, J., Rahim, I.; Dresselhaus, M. S. *Physical Review B.* Thermal conductivity and Raman spectra of carbon fibers. **1985**, 32, 6742 - 6747

3   Chang, C. W.; Okawa, D.; Garcia, H.; Majumdar, A.; Zettl, A. Nanotube Phonon Waveguide. *Physical Review Letters*, **2007**, 99, 045901

4   Hammel, E.; Tang, X.; Trampert; M.; Schmitt,T.; Mauthner, K.; Eder; A.; Pötschke, P. Carbon nanofibers for composite applications, *Carbon*, **2004**, 42, 1153 –1158 and ICAM/E-MRS 2003 Spring Meeting, 10-13 June 2003, Strasbourg, France

5   Pending Patent, WO04114404, Device comprising at least one heat source formed by a functional element that is to be cooled, at least one heat sink, and at least one intermediate layer located between the heat source and the heat sink and made of a thermally conducting material, and thermally conducting material, especially for use in such a device, **2003**

6   Pending Patent, WO04102659, Composite Material, Electrical Circuit Or Electric Module, **2003**

7   Pending Patent, WO07036805, Method For Treating Nanofiber Material, **2006**

8   Pending Patent, WO07080447, Heat Transport Medium, **2006**

9   Pending Patent, WO02075018, A CCVD Method For Producing Tubular Carbon Nanofibers, **2001**

10   Tang, X.; Hammel, E.; Trampert, M. Study of Carbon Nanofiber Dispersion for Application of Advanced Thermal Interface Materials", Proceedings of 1st Vienna International Conference Micro- and Nano-Technology (Viennano´05), **2005**, March 9-11, Vienna, Austria

11   Hammel, E.; Tang, X.; Trampert, M. Performance of Carbon Nanofiber Based Thermal Grease, IMAPS Advanced Technology Workshop **2005**, Oct.24-26, 2005, Palo Alto, CA, USA

12   Tong, T.; Zhao, Y.; Delzeit, L.; Kashani, A.; Meyyappan, M.; Majumdar, A. Dense vertically aligned multiwalled carbon nanotube arrays as thermal interface materials. IEEE Transactions on Components and Packaging Technologies, **2007**, 30, 92-100

13   Shaikh, S.; Li, L.: Lafdi, K.; Huie, J. Thermal conductivity of an aligned carbon nanotube array. *Carbon*, **2007**, 45, 2608-2613

14   Shaikh, S.; Lafdi, K.; Silverman, E. The effect of a CNT interface on the thermal resistance of contacting surfaces. *Carbon*, **2007**, 45, 695-703

15 Hu, X. J.; Panzer, M. A.; Goodson, K. E. Infrared microscopy thermal characterization of opposing carbon nanotube arrays. *Journal of Heat Transfer*, **2007**, 129, 91-93

Chapter 12
Automotive Industry Applications of Nanocomposites
William Rodgers

Materials and Processes Laboratory, General Motors Corporation Research and Development Center, Warren, USA

12.1   Introduction

Nanotechnology can be defined simply as materials or devices engineered at the molecular level. Within this broad definition, this technology can cover such widely varied topics as bio-inspired materials, nano-sized electronics, metallic alloys engineered at the nano-grain size to control performance, materials engineered to control interfacial phenomena such as friction, bonding, coulomb damping, and the use of nanometre-sized structures in energy storage and electrical devices. The sense in which we will address nanotechnology, however, is none of these. We will be addressing nanotechnology with respect to the subcategory of polymeric nanocomposites.

Nanocomposites are a class of materials where the reinforcing fillers that are used possess at least one dimension on the sub-micrometre scale. The definition has been updated more recently to only include materials with at least one dimension less than 100 nanometres. As a result of this definition, a large variety of materials can be used to manufacture nanocomposite materials. Examples of such can be nanosized calcium carbonate crystals, ultrafine talcs, finely ground micas, halloysites, amorphous silica, polyoctrahedral silsesquioxanes, carbon blacks, carbon nanotubes (single-wall, multi-wall, and nanofibres), synthetic fluoromica, synthetic hydrotalcite, hectorite clays, montmorillonite clays, and many others. While this chapter will mention composites using some of these materials, it will focus on the materials that are based on a class of aluminum silicate clays known as smectites. These clays are described as nanoclays.

The clays used to make reinforcing organoclays are mined in many parts of the world, but the deposits in the USA, Japan, China, and Germany are the ones of the most commercial interest at this time. These deposits were laid down primarily as the deposition of volcanic ash millennia ago. Alteration of these volcanic tuffs by geothermal processes over the course of centuries created the materials that are currently being used [1]. These minerals are mined using conventional mining operations. Each deposit and each stratum within each deposit is analyzed to determine the varieties and quantities of each mineral present. After removing the minerals from the ground, the ore is ground into a fine powder and the minerals are suspended in a water slurry with extensive high energy shear mixing. This allows rocks and impurities to separate from the solution and the montmorillonite to be dispersed into essentially single platelets.

A common representative of the smectite mineral class is montmorillonite. A schematic drawing of the structure of these materials is shown in Figure 12.1. The silicate platelets consist of octahedral sheets of aluminate surrounded on each side with tetrahedral silicate layers. These three layers combine to form a sandwich structure that comprises a single montmorillonite clay sheet. These sheets are separated by a layer of inorganic cations such as Na^+ or Ca^{2+}. Clays that contain Ca^{2+} are often ion exchanged with Na^+ prior to any other substitution. This helps to ensure that the clay sheets can potentially be completely separated or exfoliated when they are processed into composites. The Na^+ ions can also be readily replaced with organic cations such as quaternary alkyl ammonium cations (R_4N^+). Examples of some of the bulky ions used are, di-2-hydroxyethyl, methyl, hydrogenated tallow ammonium bromide, di-methyl, di-hydrogenated rapeseed ammonium chloride, di-methyl, di-dodecylammonium chloride, but the most widely used intercalating agent is di-methyl, di-hydrogenated tallow ammonium chloride. The separation of the clay sheets after ion exchange is based on the structure of the intercalating ammonium salt, but it also varies depending on the concentration of the ammonium species. The charges on the silicate sheets are delocalized and are caused by the isomorphic substitution of an iron or magnesium atom for the aluminum atoms in the octahedral layer. The extent of this substitution determines the charge present on the clay sheets, and is referred to as the charge capacity of the clay, and it is usually expressed as a milliequivalent ratio. This number represents the number of milliequivalents of charge that are neutralized by one gram of clay material. This value is then used to determine how much organic ion is needed to neutralize a given amount of clay. The amount of the intercalating ions can vary based on the differing charges found on clays from different mines. The amount of organic ions present in clay can also be changed by simply adding more material than is necessary to neutralize the charges.

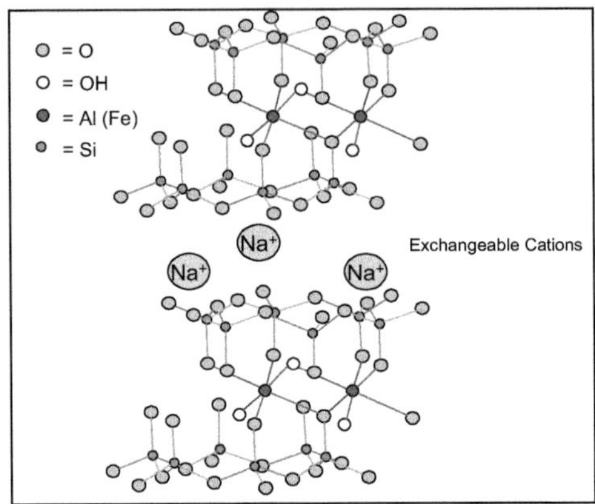

Figure 12.1
Schematic of montmorillonite clay structure

Since the organic ions tend to have the qualities of surfactants, they can form different conformations between the clay platelets. Examples of these conformations are given in Figure 12.2. This replacement increases the spacing between the silicate sheets and improves the compatibility of the filler and the resin system, thereby facilitating exfoliation.

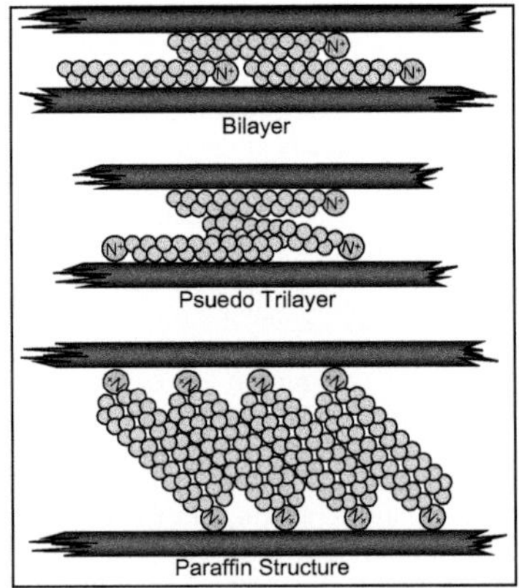

Figure 12.2
Schematics of possible intercalant conformations between clay platelets

When these organically substituted layered silicates are exfoliated properly in an appropriate polymer, they can have size dimensions on the order of 1 nm thick by 100 to 2000 nm long, resulting in aspect ratios on the order of 100 to 2000. This value is extremely high compared to the aspect ratio of conventional fillers such as talc (aspect ratio ~ 1) and glass fibres (aspect ratio ~ 20). Because of this high aspect ratio, comparable properties to conventionally filled materials can be achieved at much lower filler loadings, typically by 2 to 5 %.

Potentially nanoclays can be used in any polymer system; however, some systems are more thermodynamically suitable than others. The work has developed to encompass both thermoplastic and thermoset polymer systems. Examples of thermoplastic systems are nylon 6 [2-4], polystyrene [5,6], polypropylene [7,8], polyethylene [9], thermoplastic polyurethane [10], polycarbonate [11], acrylonitrile-butadiene-styrene (ABS) [12], and poly(ethylene oxide) [13]. Thermosets that have been investigated include epoxy resins [14,15], polybutadiene [16], polyurethane [17], polyisoprene [18,19], polyisobutylene [20], and unsaturated polyester [21]. A wide variety of polymers have been investigated, and many of them have applicability in the auto industry. Many of these investigations were carried out in a search for improved physical properties [22-24], barrier properties [25], thermal properties [26], flame retardant properties [27], or combinations of these. These enhancements are usually a result of the high surface area per unit mass of the fully exfoliated systems. Unfortunately, full exfoliation is never achieved in most of these systems. The only system where the author considers complete exfoliation reliable, at this time, is with nylon. Although there have been reports of complete exfoliation with other polymer systems, few of them hold up under close examination.

The other nanofiller material that seems to be making headway in the automotive industry is carbon nanotubes. These are materials with graphene sheets connected to each other and rolled into a tube. There are several varieties of carbon nanotube available; single wall, multi wall, and what is termed nanofibres. The actual description of the details of how to classify these materials can be found in earlier chapters in this book. Suffice it to say that there are many types of carbon nanotubes. For structural enhancement, the advantage of carbon nanotubes is that they are extremely stiff, with a tensile modulus up to 1 TPa (10^{12} Pascals), and with their relatively low density (~ 2.6 g/cm^3) they also possess very high specific modulus values. The carbon nanotubes have an added feature that the clay based nanofillers are lacking, in that they are electrically conductive. This could be an advantage for charge dissipation when used in areas of the vehicle where static charge build-up can be an issue such as in the fuel system [28]. This could also allow the electrostatic painting of composite parts without requiring a conductive primer or high levels of conductive carbon black. An example of this use will be given later in the chapter.

A number of quite good review articles and books have already been written concerning nanocomposites. They cover all aspects of polymer nanocomposite technology. Alexandre and Dubois [29] wrote one of the earlier review articles on the topic. They have been cited many times by the current author as well as others. Giannelis [30-32] has written a couple of review articles as have Vaia [32-34], Pinnavaia [35], Beall[36], Goettler [37], and Okada [38,39]. These articles tend to focus on reviews of certain polymers or techniques for preparing nanocomposites. The more recent articles have mentioned the use of nanocomposites in various industries. This chapter will focus on the use of nanocomposites in the automotive industry. It will discuss the reasons why nanocomposites are a good choice for use in the automotive arena, and will cite examples where nanocomposite materials are already being used. So sit back, get comfortable, and enjoy what you will be reading.

12.2 Requirements for the automotive industry

A number of commodities and engineered resins are currently used in the automotive industry. These resins have appropriate properties for use in their respective applications. These requirements can include stiffness, reasonably low values for the coefficient of linear thermal expansion, low mass, good surface appearance, and good processibility. These materials work well in the automotive arena, so the question needs to be asked, 'Why nanocomposites?' In this section, we will attempt to answer this question.

12.2.1 Surface appearance

While the general perception is that the automotive industry is an industry that lives by virtue of its engineering capabilities, it is also an appearance industry. Many customers measure the value of their vehicle purchase, at least in part, by the attractiveness of their vehicle. In order to have an outstanding surface appearance when using filled polymer systems, the particle sizes need to be small. 'Small,' in this sense, means small in relationship to what can be seen with our eyes. I will not go into an explanation of the workings of the human eye, but what we usually perceive in a filled polymer system is a lack of distinctness of image, when looking at reflected light, and a haziness or translucency, when looking through a part using transmitted light.

Nanocomposites seem to be tailor-made for this property. By definition, at least one dimension of these particles needs to be less than 1×10^{-7} m or 100 nm. But that is only one dimension. Montmorillonite is known to have a plate like structure, and while the thickness of a well exfoliated plate could well be close to the 0.96 nm as claimed by mineralogists [40], the lateral dimensions may be in the order of 500 nm or somewhat more. Atomic Force Microscopy (AFM) analysis of individual clay platelets has been carried out by Park et al. [41] and although these lateral dimensions are not specifically stated in the paper, it is clear from the figures that they are in the order of a few hundred nanometres, not several thousand nanometres.

Work investigating the dimensional variation of clay platelets under varying shear stress conditions was carried out at the General Motors R & D Center [42]. AFM was employed to examine the topography of individual clay platelets subjected to differing extents of shear stress. Ultrasonication was chosen as the method used to disperse the nanoparticles under low shear stress, and a cyclone mechanical homogenizer was used for the high shear stress samples. Figures 12.3 and 12.4 show the topography and corresponding sectional analysis of exfoliated clay platelets prepared by these methods.

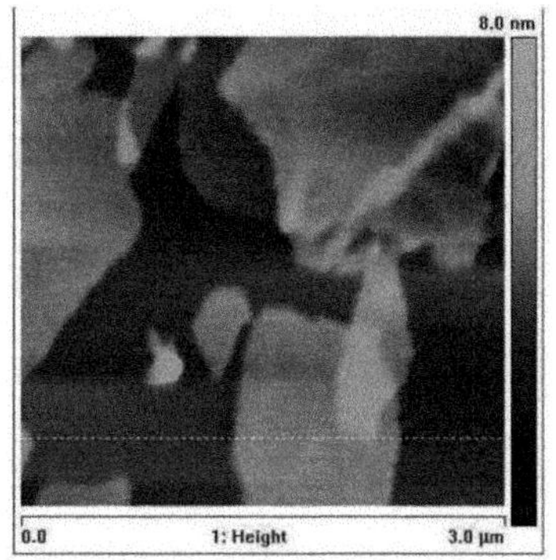

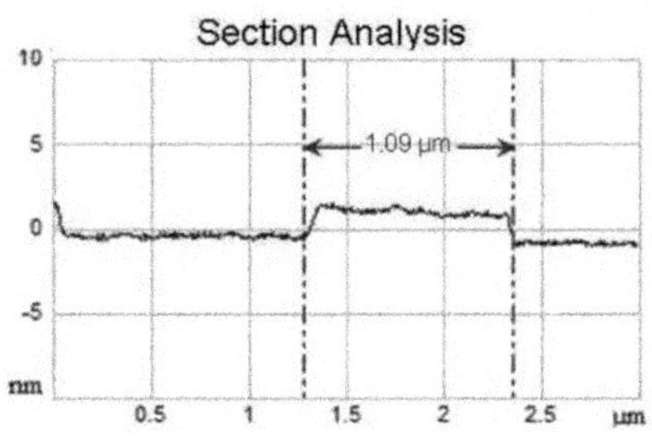

Figure 12.3
AFM height image and section analysis of Cloisite® Na⁺ platelets exfoliated by ultrasonication and deposited on a mica surface. The scan size is 3 µm

Individual and stacked clay platelets are shown on both images. The thickness of individual clay platelets in both images is around 1 nm, which is similar to both the theoretical value and the values reported in the work by Park et al. [41]. It is easily seen that the lateral dimensions of the clay particles decrease considerably after being dispersed using high shear stress. In Figure 12.3, it is difficult to observe an entire platelet in 3 µm × 3 µm scan size. Most of the platelets shown in the image are more than 1.5 µm wide. Even the middle of a small piece has a width of 1.09 µm. In contrast, the platelets that experienced high shear stress, shown imaged in Figure 12.4, show a width less than 0.5 µm. The largest individual platelet observed in the image is approximately 0.57

µm. Clearly the shear stress generated during the processing is a key factor in reducing the platelet size. The size of the platelets after being exposed to reasonably high shear stresses compares favorably with the wavelengths of light; therefore, we would not expect these fillers to interfere with our perception of the material as long as they are reasonably well exfoliated. This is supported by a number of papers in the literature that present figures showing the transparency of montmorillonite filled systems [43-45].

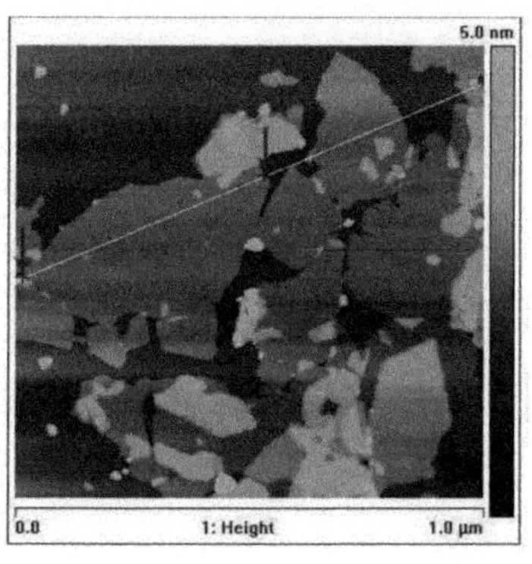
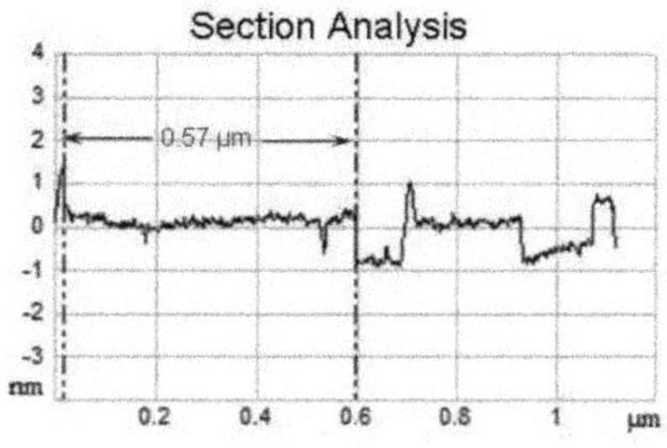

Figure 12.4

AFM height image and section analysis of Cloisite® Na⁺ platelets exfoliated by the use of a cyclone mechanical homogenizer and deposited on a mica surface. The scan size is 1 µm

These advantages also apply to surface distortions. When injection moulded, the filler in filled polymer systems can be carried to the surface due to the flow fields generated during processing. Large particle size fillers can affect the localized shrinkage near the particles causing a waviness or distortion in the surface of the part. This sets up a subtle interference pattern that is seen as a lack of distinctness of image or gloss in the final part. The smaller the particle sizes of the filler material, the smaller the distortions, and the better the distinctness of image and gloss. So, for high gloss or multiple gloss parts, smaller filler sizes are better.

Another advantage of the small particle size is that when a panel in impacted by a stone or other projectile during use, the resulting damage is less visible. An example of this is shown in Figure 12.5. Two thermoplastic olefin (TPO) formulations that had equivalent mechanical properties were subjected to Gravelometer testing [46]. This is essentially an air gun that throws pebbles at the panels at a specified angle. The figure is a photograph of both panels following the testing. The panel on the left was compounded using standard talc filler materials at a level of 28 % by weight; the panel on the right used montmorillonite based organoclay fillers at a level of 2.5 % by weight inorganic content. These levels were chosen so that the materials would have similar physical properties. The differences in performance are easily seen. Both panels have dents in the surface from the impacts, but they are much less visible on the nanocomposite panel.

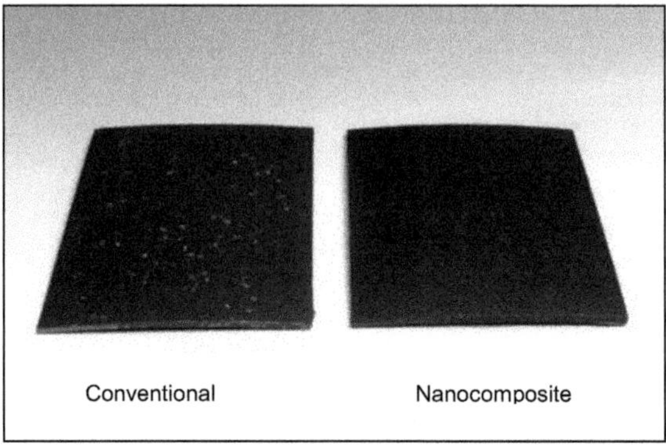

Figure 12.5
Photograph of conventional and nanocomposite filled TPO panels after gravelometer testing
It is difficult to determine whether the difference is more related to the different concentration of filler materials or more related to the size of the filler, but we believe that particle fracture can be a large part of the answer. Fracture of talc fillers has been cited in the literature as the cause of this type of whitening post-impact [47].

12.2.2 Measurement techniques
We have already given an example of the measurement of the size of the organoclay fillers in the previous section where atomic force microscopy was used to determine the size of individual clay platelets. While this method is good for measuring individual particles when special techniques are used to process them, efforts to use AFM on the type of samples normally used have been able to measure agglomerated clay particles, but have had difficulty measuring dispersed organoclay fillers [48]. This field of microscopy is advancing and may well develop into a more useful technique in the future. Until that happens, transmission electron microscopy (TEM) is the method of choice for determining the sizes of dispersed nanoclay particles. The majority of articles published in this field have at least a few micrographs embedded in the articles. Usually, these are used to give an example of the quality of the dispersion of the filler materials, and are almost exclusively used as qualitative support for this assertion.

Work in Prof. Paul's group at the University of Texas at Austin provides much of the quantitative information that has been developed from TEM images [49,50]. The technique that is used is to convert the TEM image into a binary image through the use of image analysis software. The length of each of the particles is measured and divided by the thickness of the particles to arrive at an aspect ratio for the particles. The early work in this system was done on nylon 6 nanocomposites, but more recently this technique has been used to determine the aspect ratio of the nanofillers in polypropylene and thermoplastic olefin matrices [51-54]. This work has provided insight into the exfoliation that has been achieved in different polymers. The work with the nylon systems shows, at worst, a well exfoliated system with many single platelets present in the final material. The results for the olefin materials are not so positive. These values show that the nanofillers are intercalated and partially exfoliated, but it is difficult to put a value on the extent of the exfoliation. It is clear that the aspect ratio achieved in most of the polyolefin based systems is in the order of 30 to 50 as opposed to the desired 80 to 100. These values are interesting, but how close is this to full exfoliation?

More recent work in the India Science Laboratory of the General Motors R&D Center is trying to answer this question [55]. This analysis relies on the use of stereology to provide a more statistically satisfying answer to the extent of exfoliation in these materials. This group starts at the same place as the Paul group, TEM images taken at high resolution at a magnification where single platelets can be observed as well as tactoid structures. The image is then turned into a binary image and the particle length, circumference, and area are measured and summed. This method provides a parameter, ξ, which is a measure of the extent of the exfoliation of the clay.

Both of these methods are important and they can provide a good indication of the quality of exfoliation of nanoclay filler materials when used together. Unfortunately, both methods are also time consuming, and will need further development before they can be used for quality control.

12.2.3 Aspect ratio

To show reasonable property enhancement it is necessary for the reinforcing materials to possess high aspect ratios. Theoretical frameworks that relate the aspect ratio of filler materials to performance, typically modulus enhancement, have been described in the literature for many years [49,56-59]. These theories are based on the idea that high modulus materials can be dispersed in a polymer matrix and that the resulting composite will develop a combination of the properties of both materials. The general thinking is that as the aspect ratio of the filler materials increases, the modulus enhancement increases. The Halpin-Tsai [56] equations are based on the micromechanics developed by Hermans and Hill [57,58]. Halpin and Tsai further developed these theories so that they are applicable to a variety of discontinuous fillers of differing geometries. The other predominant theory in the literature is the Mori-Tanaka average stress theory [59]. This theory is based on the use of Eshelby's inclusion model for the prediction of stress fields around ellipsoidal inclusions. Both of these theories are based on the assumptions that 1) the matrix is firmly bonded to the filler material, 2) the filler particles are perfectly aligned, and 3) particle – particle interactions are negligible. The two theories also have a few differences. The Halpin-Tsai theory treats fibres as fibres and disks as rectangular platelets, while the Mori-Tanaka theory treats all filler particles as ellipsoidal particles. The Halpin-Tsai equations also do not consider the effects of the Poisson's ratio of the matrix or the filler, while these are taken into consideration by the Mori-Tanaka theory. A paper has been published by Fornes and Paul [49] that describes these theories in more detail than will be discussed here.

In this chapter, we are more interested in the predictions resulting from these theories. Figure 12.6 shows the increase in the tensile modulus that can be expected in the flow direction of a reinforced composite plotted against the aspect ratio of the reinforcing filler material. The modulus is expressed as the ratio of the composite modulus to the modulus of the unreinforced matrix material. The parameters used to prepare these predictions were based on a composite with a polypropylene matrix filled with nominally 5 % by weight (2.9 % by weight inorganic content) organoclay material. Other pertinent values for the generation of the data used to produce this chart are a modulus of the polypropylene of 1.37 GPa, a density of the polypropylene of 0.89 g/cm^3, a Poisson's ratio of the polypropylene of 0.35, a modulus of the filler of 178 GPa, a density of the filler of 2.2 g/cm^3, and a Poisson's ratio of the filler of 0.20. Looking at Figure 12.6, note that as the aspect ratio of the filler approaches unity, both theories approach the value that would be calculated by the inverse rule of mixtures, and as the aspect ratio approaches infinity, both theories approach the value calculated by the rule of mixtures. Both theories show a rapid increase in the modulus with increasing aspect ratio and are within 20 % of each other despite the differences in the calculations.

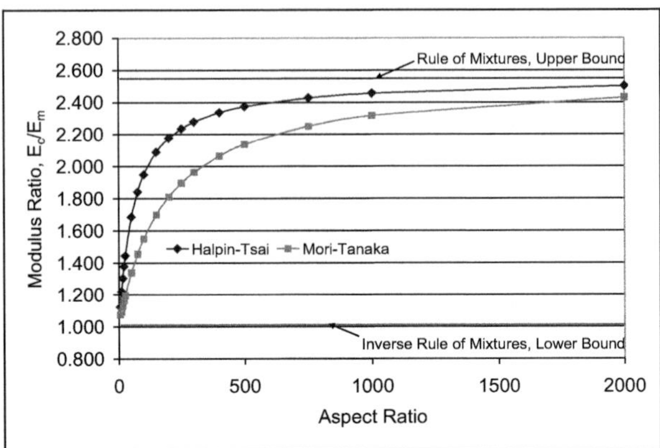

Figure 12.6
Predictions of modulus increase by micromechanical modeling theories

From this it is clear that the higher the aspect ratio of the filler materials, the higher the stiffness of the resulting composite can be expected to be. This forms the basis of the industry-wide interest in achieving high extents of exfoliation in nanocomposite materials.

The effects of increasing aspect ratio on the coefficient of linear thermal expansion can be seen in Figure 12.7. There are few models in the literature that can be used to predict thermal expansion; however one that can be used is a model developed by Chow [60,61]. We will not go into a description of the model here; it is described well in the original papers by Chow as well as in a paper by Yoon, Fornes, and Paul [62]. As the aspect ratio of the filler increases, the coefficient of thermal expansion decreases in the direction in which the filler is aligned.

So it is clear that the stiffness of a composite increases and the coefficient of linear thermal expansion decreases as the aspect ratio of the filler material increases. How does this come into play in the automotive industry? An increase in the stiffness of a composite material, assuming that other important properties are not compromised, enables a couple of applications. Since the stiffness is higher, the polymer may be able to be used in applications where it could not be used previously, for example in semi-structural seat applications or as exterior skins for vertical body panels. It may also permit the displacement of higher cost polymer materials from certain applications. Alternatively, the enhanced stiffness may allow the production of parts using a thinner wall thickness. This would save on the amount of material that is used in an application, thereby lowering cost. The reduction of the coefficient of linear thermal expansion (CLTE) means that the material will not expand or shrink as much with changes in temperature. Therefore, exterior panels made of these materials would be able to be designed with smaller gaps between adjacent panels. This improves the overall appearance of the vehicle resulting in cleaner lines and less chance of interference between panels. All of these are quality improvements that would be well received by customers.

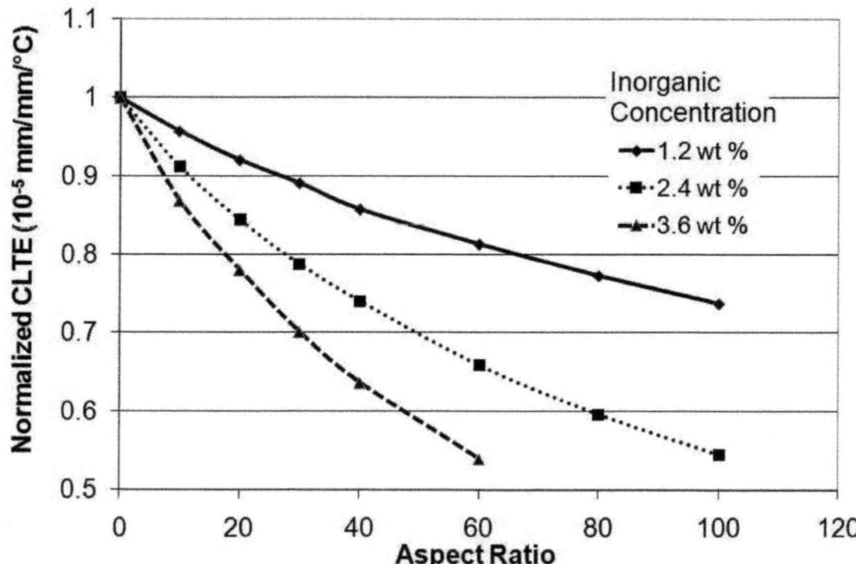

Figure 12.7
Predictions of normalized coefficient of linear thermal expansion with increasing aspect ratios from the Chow theory

12.2.4 Minimization of mass

In order to minimize the mass of a reinforced part, low contents of high modulus fillers are required. In this section, we will show that nanocomposites meet this requirement. In order to do this we again rely on the Halpin-Tsai equations described in the previous section. In the previous section we described how the increase in modulus depends on the aspect ratio of the filler materials. The next three figures (Figures 12.8, 12.9, and 12.10) show the increase in the tensile modulus that can be expected in the flow direction of a reinforced composite plotted against various reinforcing filler properties. An additional assumption to those included in the structure of the Halpin-Tsai equations was used in the preparation of these figures, and that assumption is that the fillers are assumed to be completely distributed and exfoliated.

Figure 12.8 shows the increases predicted by the Halpin-Tsai equations when the filler modulus is increased while holding the aspect ratio of the filler constant. Clearly as the filler modulus increases, the line representing the predictions moves to the lower filler contents to achieve a particular modulus increase. Also note that as the modulus increases, further increases become less effective in lowering the amount of filler required to double the modulus of the system.

Figure 12.9 shows the increases predicted when the aspect ratio of the filler is increased while holding the filler modulus constant. The shapes of the curves are similar to those shown in figure 12.8. As the aspect ratio increases, the filler content needed to achieve a fixed increase in the modulus of the composite is predicted to decrease, and, as was seen with the modulus, these increases become less effective as the aspect ratio gets higher.

Figure 12.10 shows the predicted increases in modulus for various commercially available filler materials. The filler parameters used for generating this figure are considered to be typical of these filler materials given the assumptions stated above and are given in Table 12.1. There are significant differences between the various fillers because for each different filler material, both the modulus and the aspect ratio increase. Therefore, we find that nanotubes are much more efficient reinforcing agents than organoclay fillers which are more efficient than short glass fibres which are more efficient than talc. Increasing the aspect ratio of any particular filler will make it more efficient, but in order to make large changes, the mineral, or synthesized filler particles need to have a different fundamental structure so that their inherent modulus can increase as well. We need to point out that this only holds for the direction in which the fillers are aligned. The calculations can be done in the direction orthogonal to this. In this direction, the aspect ratio of the fibre fillers (short glass and carbon nanotubes) would drop to 1, since these are unidirectional filler materials. The aspect ratios for the organoclay fillers and the talcs would depend on the particular shape of the particles that were used, but would tend to be greater than 1. In fact, since montmorillonite is considered to be a platy filler, it is likely that it would retain a significant portion of its reinforcing capability.

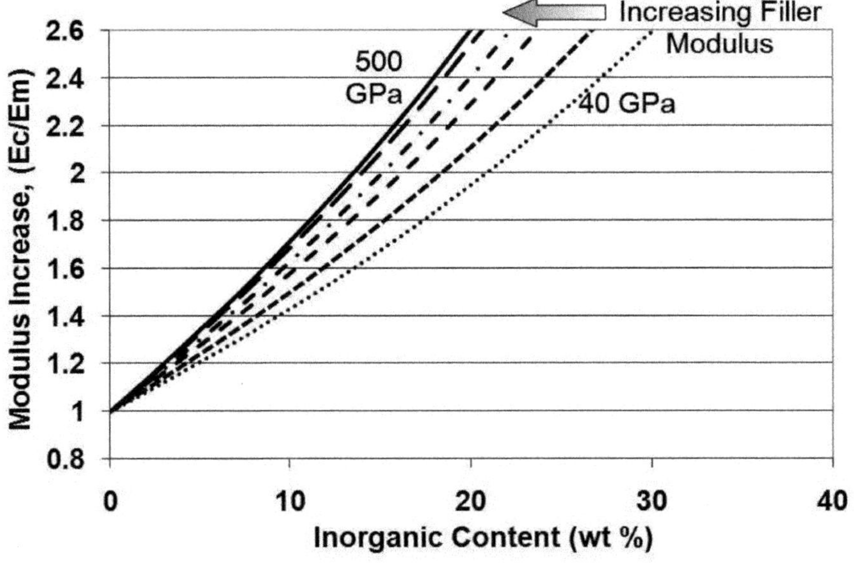

Figure 12.8
Predictions of modulus increase with increasing filler modulus at constant filler aspect ratio. For these calculations the filler aspect ratio was held constant at 10. The different curves represent filler modulus values of 40 GPa, 60 GPa, 100 GPa, 150GPa, 300 GPa, and 500GPa

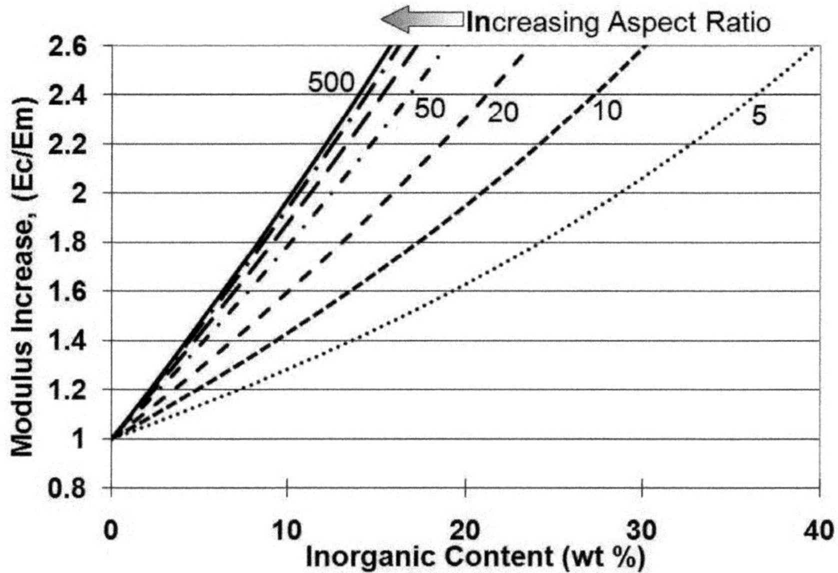

Figure 12.9
Predictions of modulus increase with increasing filler aspect ratio at constant filler modulus. For these calculations the filler modulus was held constant at 40 GPa. The different curves represent aspect ratios of 5, 10, 20, 50, 100, 200, and 500

Table 12.1
Property values used to predict the modulus increases with various commercial filler materials (Figure 12.10)

	Resin	Talc	Milled glass	Nanoclay	Carbon nanotube
Modulus (GPa)	1.37	41	72.4	178	500
Density (g/cm^3)	0.89	2.75	2.54	2.83	2.62
Aspect ratio	---	7	20	50	500

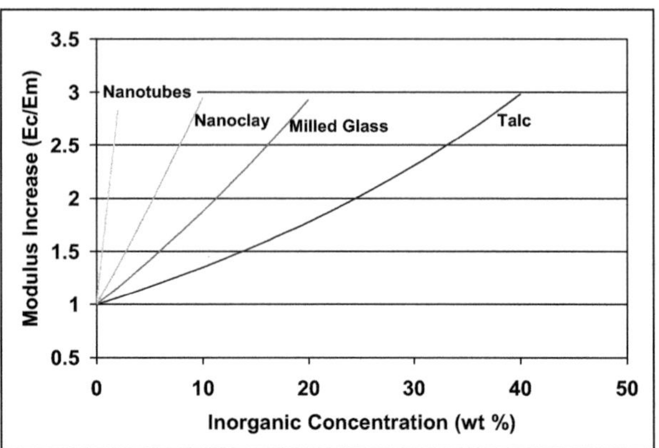

Figure 12.10
Predictions of modulus increase with various commercial filler materials

So what are the implications of this information for filled polymer systems? Simply, it means that the ideal filler will have the highest possible modulus and the highest possible aspect ratio, while staying small enough to not interfere with reflected or transmitted light in a significant way. This ideal filler will also associate into particles that are easy to handle in standard manufacturing equipment while also easily delaminating into primary particles when mixed in conventional mixing equipment. Talc has this latter quality, but its modulus is relatively low when compared with other filler materials and its aspect ratio is relatively low. Carbon nanotubes clearly have very high modulus and high aspect ratios, but they are one dimensional fibres and they tend to be difficult to disperse into primary particles.

Several synthetic nanofillers are being pursued by different organizations. Ube Materials Ltd and Mitsui have partnered to develop MOS Hige filler materials. These are nanofibres based on magnesium oxysulfate. They suffer from the same issues as the other uni-directional fillers. Their cost also leads to their use primarily in conjunction with conventional fillers such as talc. Recent experiments in our laboratory have shown that this is a promising combination of materials that may be useful for automotive interior parts. CBC Co., Japan, markets several semi-synthetic and fully synthetic platy fillers. The semi-synthetic material is based on extending naturally occurring talc with sodium fluorosilicate to make a fluoromica that is marketed under the trade name Somasif[TM]. They also make the fully synthetic hectorite that is marketed under the trade name Lucentite[TM]. Somasif[TM] has been reported in the literature to give good properties [7]. Topy Co. has developed a fully synthetic fluoromica using a high temperature process that directly converts the metal salts into the synthetic mineral, Topy-4C. Kunimine Ind. manufactures a fully synthetic saponite that is marketed under the trade name Sumecton®-SA. The goal of these companies seems to be the production of nanofillers without the impurities that are found with the naturally occurring materials while providing superior aspect ratio materials. These materials may provide significant advantages, particularly with regard to retention of low temperature ductility, but they seem to be limited in supply and are therefore difficult to procure for side-by-side evaluation. Akzo Nobel Inc. has also entered the arena with their Perkalite[TM] materials [63], which are modified layered double hydroxides. They have gone away from the fluoromica/montmorillonite types of mineral to a hydrotalcite based material. The interesting point about these materials is that instead of the platelets carrying a partial negative charge, in the hydrotalcites, the platelets carry a partial positive charge. This means that the counter ion needs to be negatively charged. This opens the door for the use of alternative compatibilizing materials, and perhaps even using the polymer resin system as the compatibilizing agent. These materials are just entering the market at present, and it remains to be seen whether they will provide advantages over the more established nanofillers. An interesting attempt at the preparation of synthetic nanofiller materials has been going on in the laboratories of Andreas Stein and Christopher Macosko at the University of Minnesota [64-66]. Their work is to prepare synthetic materials which incorporate compatibilizing groups in the structure of the minerals. They have successfully incorporated isobutyl groups into layered silicate materials and although it remains to be seen if these materials will prove useful, it is an interesting attempt at using an in-situ compatibilizer instead of adding one during manufacture of the nanocomposites.

We have now discussed some of the advantages that can be gained by using nanocomposite materials. Other advantages such as acting as a flame retardant, or improving recyclability over conventionally filled materials will be discussed in other chapters in this book. We will now move on to discuss how nanocomposite materials used in the automotive industry are manufactured.

12.3 Manufacture of nanocomposite systems

As with any polymer system, there are a variety of ways that nanocomposites can be made. With these materials, however, two methods predominate. These two methods are in-situ polymerization and melt mixing. While it is not the intention of this article to describe these methods in depth, a brief discussion of these preparation methods will provide the reader with an understanding of the processing that is necessary to provide useable materials.

12.3.1 In-situ polymerization

In-situ polymerization was one of the earliest methods used to prepare nanocomposite materials [23,67,68]. This method, in short, incorporates the exfoliation of the organoclay filler material in the polymerization scheme of the polymer. Therefore, no additional steps are necessary in order to prepare the nanocomposite material. In this method, an organoclay capable of swelling with polar materials is suspended in a monomer solution. The monomer is allowed to penetrate into the gallery spacing between the clay sheets and then is polymerized. The polymerization of the monomer provides the force necessary to exfoliate the organoclay into its single sheets. In practice, the monomer is usually ε-caprolactam, which when polymerized results in the commercially important material, nylon-6. Usuki et al. [69] showed that the extent of swelling of the montmorillonite was dependent on the material used as the initial intercalant for the organoclay. They also found that clays with higher ion exchange capacities exfoliated more efficiently during polymerization. Therefore, montmorillonite was considered better than saponite or hectorite for this production technique. Commercial materials such as Ube Nylon 5034 are prepared in this way.

Recently this type of production has been adapted to olefin based materials. Initially, the work was carried out by Dow Chemical [70,71] and it is currently being investigated with some success by Simon's group at the University of Waterloo [72]. This is an adaptation of the nylon methods to handle high pressure gas phase reactions. These methods have not yet produced more than laboratory quantities of materials, but the method is showing promise. The main drawback so far has been the reduction of the activity of the polymerization catalyst systems. If this could be overcome, then large quantities of olefin based nanomaterials could be prepared at reasonable costs.

12.3.2 Melt processing

A large number of investigations have been carried out using melt mixing to prepare nanocomposite systems [5, 73-76]. Typically twin screw extruders were used to do the processing although other types of processing equipment have been used [77,78]. The advantage of using melt mixing is that any polymer or blend of polymers can be used for the production of nanomaterials, as opposed to in-situ polymerization, which is normally optimized for the polymerization of a single monomer. The effectiveness of melt processing for the production of nanocomposites has been the subject of investigations with nylon 6 [3], nylon 66 [79], polyethylene [9], polypropylene [7], and polystyrene [8]. Fornes et al. [80] published an excellent overview of how the chemistry of the specific organic cation can affect the final properties of nylon-6 nanocomposites. This paper provides data that suggests that for the nylon system used, a single long alkyl chain substituted onto the quaternary ammonium salt is more effective at exfoliation during melt processing than intercalants with multiple long alkyl chain substituents. Data is also given that indicates that organoclays intercalated with quaternary ammonium salts at their exchange capacity have improved performance over organoclays that have had an excess of intercalant used. In general, this paper provides good guidance for the choice of intercalating species used in the preparation of organoclays for use in nylon-6 systems.

In a similar study carried out in a polypropylene based system, Reichert et al.[8] concluded that the alkyl chain length of the intercalant must exceed 8 carbon atoms in order to be effective at promoting exfoliation. This paper also studied the effectiveness of different maleated polypropylenes as compatibilizers for the organoclays. They concluded that increased amounts of anhydride functionality were beneficial in the formation of nanocomposite materials. Neither of these papers, however, examined the processing conditions used with the various materials.

This information is provided for nylon based materials by Dennis et al. [81]. In this study, four different extruders with two or three different screw designs per extruder were used to investigate the effectiveness of different processing equipment on the melt blending of organoclays. Their conclusion was that screw designs that provided medium shear were the best at promoting exfoliation. This seems to support a position that organoclays, unlike conventional talc fillers, do not benefit from extremely high shear extruder configurations, at least in nylon based systems. They also propose that whatever screw design is used, time in a low shear environment is needed in order to peel the clay sheets apart into an exfoliated structure.

An example of what can happen if the processing conditions are incorrect is given in a paper by Fasulo et al. [82]. Figure 12.11 shows a transmission electron micrograph of an organoclay agglomerate that was formed during extrusion processing of a nanocomposite resin using improper processing conditions. The dark dots in the figure are particles of TiO_2 that were fed in the extruder during the processing of the nanocomposite material. As can be seen from the figure, the TiO_2 particles have been incorporated into the organoclay agglomerate. Due to the thickness of the section needed for transmission electron microscopy, it is unlikely that the TiO_2 particles are either above or below the agglomerate. The only way that the TiO_2 particles could have been incorporated into the agglomerate is if the agglomerate was formed during processing in the extruder. The article goes on to investigate extrusion conditions that can be used to avoid this type of agglomeration. It is concluded that improvements in dispersion can be related to processing conditions that result in higher specific energies during mixing.

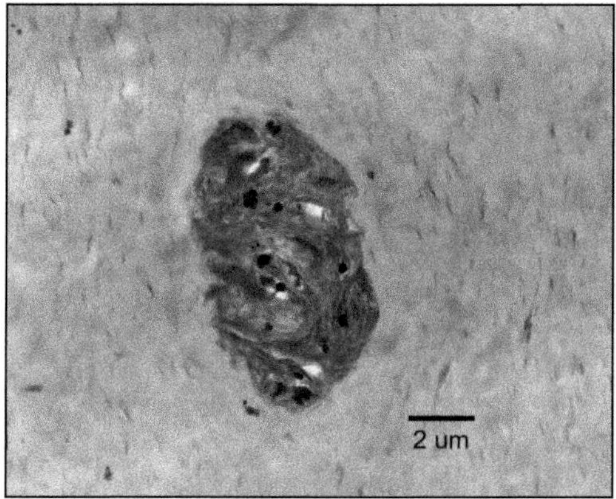

Figure 12.11
Transmission electron micrograph of an organoclay agglomerate in a thermoplastic olefin matrix
This was clearly not an exhaustive analysis of the ways in which organoclay nanocomposites can be prepared. For more detailed investigations into other preparation methods the reader is directed to the other chapters of this book and to the review articles mentioned in the introduction of this chapter.

12.3.3 Injection moulding

While there has been much investigation into the ways that nanocomposite resins can be prepared, little has been done with respect to the ways in which these materials can be turned into useable articles. Extrusion is the method that is primarily used in the film industry; however, in the automotive industry, injection moulding is the primary method in which resins are turned into useful articles. While there are many papers that measure properties of injection moulded samples, there are only a few papers in the literature that investigate the effect of the injection moulding parameters on the properties of the final article. Some of these papers investigate the effects of processing by the microcellular injection moulding process [83,84]; however, the best paper to describe the effect of injection moulding process parameters on the properties of nanocomposite materials is by Okonski [85]. This paper compares the processing of a conventionally filled thermoplastic olefin (TPO) to that of an organoclay filled TPO. Some of the results are shown in Figure 12.12. The first thing to notice is that the processing window for the nanocomposite material is significantly larger than that for the conventionally filled material. This results in more flexibility in choosing the moulding conditions used to fabricate an article and increases the ability of the moulder to vary the processing of an article in order to provide a defect free part. Superimposed on the chart in Figure 12.12 are two important processing areas. The oval area on the left is the area in which warpage of parts can be reduced or eliminated. Warpage of moulded parts is a serious issue since it can render parts unusable. The oval area on the right is the area in which tiger striping of the parts can be reduced or eliminated. Tiger striping is a surface effect that degrades the appearance of a moulded part. Clearly the larger processing window for the nanocomposite provides a greater likelihood of being able to process out part issues.

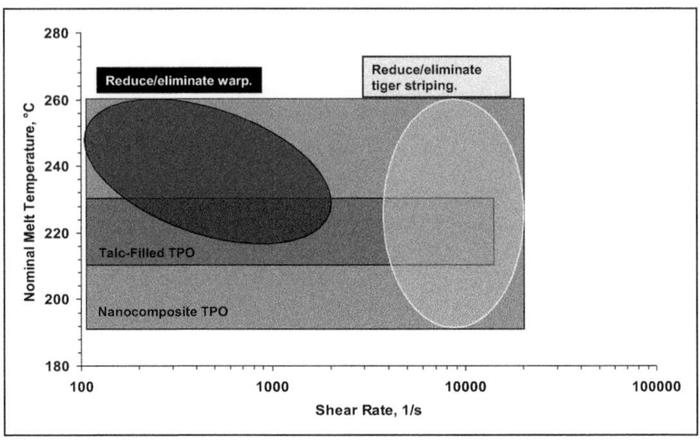

Figure 12.12

Processing window comparison of organoclay and conventionally filled TPO formulations also showing key processing zones for minimization of injection moulding issues

The next section gives examples of nanocomposite materials that have been and are being used in the automotive industry.

12.4 Applications of nanocomposites in the automotive industry

Nanocomposites have been used in the automotive industry almost since its inception. Certainly the use of carbon black as a reinforcing filler in tires and hoses meets the definitions for being nanocomposites. Automotive coatings have been using silica to control thixotropy for almost as long. One can even argue that the wooden bodies of the early cars were not only nanocomposites of cellulose whiskers but were also bio-renewable. More recently PPG Industries has commercialized a nanocomposite coating that is reported to be more durable than other coatings[86]. Research is also continuing in the coating arena to prepare coatings with even more enhanced performance by using organoclay nanofillers [87,88]. While these are legitimate applications of nanotechnology, these are not the applications that we will be examining in this section. In this section we will discuss applications of nanocomposite materials that are replacing conventionally filled materials. Some applications of carbon nanotubes will be discussed as well as organoclay applications.

12.4.1 Applications using carbon nanotubes

There are two main applications currently in the automotive industry that use carbon nanotubes. Both applications use multi-wall nanotubes and the electrical conductivity of the nanotubes is a key feature of both applications. The first application is in automotive fuel lines [28]. The nanotubes used in this application are produced by Hyperion Catalysis. It is claimed that the high electrical conductivity helps to dissipate any electric charge that may build up due to fuel flowing through the hose. Almost certainly the low percolation threshold for these materials is also a consideration. It was claimed that in 2003 [28] that 60 % of the cars on American roads have fuel lines that contain carbon nanotubes. The percentage is likely to be higher today as the same method may be used to increase the conductivity of fuel line connectors and housings.

Sabic Innovative Plastics (formerly GE Plastics) reportedly commercialized a conductive version of its Noryl GTX line of poly(phenylene oxide) – nylon blend that used carbon nanotubes from

Hyperion Catalysis [89]. This material was developed to enhance the ability of the plastic to dissipate electrical charge, thereby enhancing its ability to be painted using electrostatic spray painting without the need for the application of a conductive coating [90]. Another article [28] reports that these materials were used on the Renault Clio and Megane models. In 2002, it was reported that the Nissan X-Trail was produced using nanotubes in the fender [91]. This was reported to increase the strength of the polymer while retaining flexibility and its ability to recover from deformation. It is also likely that the low concentration of nanotubes required to form a percolating network lowered the amount of material needed to make the substrate electrically dissipative. This allowed the manufacture of a part that could be electrostatically painted without the need for the application of conductive coatings. If correct, this would result in a simplified production procedure with an improvement in the overall economics.

Economic forces will determine if the use of carbon nanotubes will become widespread in the automotive industry. The nanotubes themselves are expensive, and the challenge will be to find ways to offset the impact of this cost. If this challenge can be met, it is likely that carbon nanotubes will be used in an increasing number of applications.

12.4.2 Applications of organoclay nanocomposites

12.4.2.1　　　　Underhood applications

It is generally recognized that the first commercial application for nylon 6 nanocomposites was the production of a timing belt cover by Toyota in collaboration with Ube in 1991 [22,92-94]. The nanocomposite improved the mechanical performance of the nylon material. Significant improvements were observed in tensile strength (55 %) and tensile modulus (91 %) [22]. Additionally, and perhaps most significantly, the heat distortion temperature for the nanocomposite nylon 6 increased from 65°C to 145°C. This allowed the nanocomposite to be used in an application where the nylon had traditionally been highly glass filled resulting in reduced mass for the part as well. The values for the improvements in these systems have been reported differently, but the increase in the heat distortion temperature has always been of this same level.

Shortly after the Toyota/Ube announcement, Unitika Co. introduced its nylon 6 nanocomposite. The Unitika material used a synthetic silicate material where the Ube material used naturally occurring montmorillonite. The Unitika material was reported to be used on Mitsubishi Motors GDI engines as an engine cover. This is another application where the high heat distortion temperature and low mass would be advantageous. It was reported [93,95] that neither of these applications was continued due to the economics surrounding the nanocomposite materials.

Despite their reported demise, both Ube and Unitika are still manufacturing nanocomposite nylon. Ube's nanocomposite (Ube Nylon 5034) is used as an inner layer in their ECOBESTA material which is used for fuel lines. These materials are multilayer composites designed to minimize the permeation of fuel through the walls of the fuel line [96]. Unitika continues to manufacture nanocomposite nylon materials as well. The production of their original nanocomposite, M2350, is continuing, and this material has been joined by at least 3 other offerings [97]. In addition, on their applications page [98], five nanocomposite nylon 6 applications are shown. Interestingly, these applications appear to be on Toyota/Lexus vehicles as opposed to Mitsubishi. Therefore, while it seems that there may not be broad acceptance of the nanocomposite nylon materials for under hood applications, there are applications in the marketplace, and one can only assume that they are economically viable.

12.4.2.2　　　　Exterior applications

After the announcement of the commercialization of nanocomposites based on nylon materials for under hood applications, researchers investigated the possibility of making nanocomposites from materials based on polypropylene. Polypropylene is widely used in the automotive industry for exterior applications and it was hoped that a nanocomposite would improve the mechanical and thermal properties of this commodity polymer in the same way as it did for the nylon materials. Work had been going on in this area for many years [99-102], but it was not until the announcement by General Motors (GM) in 2001 that a commercial product had been developed [93,95,103]. This initial application was a step assist cover manufactured in thermoplastic olefin (TPO) for the 2002 GMC Safari and Chevrolet Astro vans. In general, TPOs are blends of polypropylene with an elastomer that acts as a toughening agent. This development was made possible by the collaborative work between General Motors, Basell Polyolefins, Southern Clay Products, and Blackhawk Automotive Plastics. The application saved over 10 % of the total mass of the part while at the same time maintaining low temperature ductility and improving the appearance of the part. The material was also designed so that the same tooling that had been used for the conventionally filled material it replaced could still be used to produce the part; thereby, saving on the capital costs of the switchover. This application also showed that polypropylene based nanomaterials could be moulded at standard injection moulding houses with similar production rates to conventional materials. Unfortunately, this product is no longer in the marketplace as GM decided not to compete in the small van segment of automotive market.

GM's next application assured that the nanocomposite material could be handled at production rates. In early 2004, GM announced its next nanocomposite application, the body side mouldings for the 2004 Chevrolet Impala [104]. As a result of the work by the same team that made the initial application possible, this application allowed the usage of polypropylene based nanocomposites to reach the half million pound level on GM's highest volume vehicle. While this application resulted in lower mass parts, the reason nanocomposite materials were used was that they reduced the warpage of the parts resulting in more consistent, more dimensionally stable parts.

This introduction was followed later that year by the introduction of nanocomposite material on the 2005 HUMMER H^2 SUT [105]. This was the first GM vehicle where the intent was to use nanocomposite materials from the start, and while the primary team remained the same, the moulding house was Sport Rack Automotive. The nanocomposite material was the same material used for the Impala application and was used on a total of five parts; the centre-bridge, two sail panels, and two bed rail protectors with a total of just over 3 kg of material. This application also resulted in lower mass for the parts, but the reduction of the coefficient of linear thermal expansion for these materials meant that the parts would not change shape as much with changes in temperature [106]. A result of this would be that the gaps engineered in between the parts would be tighter, giving a better appearance to the bed surround.

Although there has been research in the use of nanoclay fillers with thermoset materials almost as long as the work for thermoplastics [14-20], there are few references for commercial applications. One such reference, however, did appear in 2004 [107], where it was reported that International Truck and Engine Corporation was using a nanocomposite SMC material from Core Molding Technologies to make a front end for a truck. The material enabled the production of large parts due to its rheology and, due to its reduced density, was reported to reduce the mass of this moulded part by approximately 13.5 kg.

12.4.2.3   Interior applications

Interior applications face many of the same performance challenges as the exterior applications, so it seems natural that nanocomposites would also be good materials for interior applications. This was confirmed with the announcement in 2004 of an interior application for nanocomposite polypropylene from Noble Polymers [105,108]. The material that was used was from Noble's Forte™ line of polypropylenes that were commercialized late in 2003. The application was the seat backs in the 2004 Honda Acura TL, where the nanocomposite replaced a glass filled polypropylene composite. By targeting glass filled materials, improvements in appearance, warpage, mass, and cost could potentially be realized, and these improvements were reported by representatives of Noble Polymers [108].

Earlier in that year it was reported that another polypropylene nanocomposite was being used for an interior door handle by DaimlerChrysler. The material was one of the Scancomp line of nanocomposites offered by Polykemi AB (Sweden) [109]. This article reports that the nanocomposite met the part requirements where traditional materials did not. A quote from the article states, *"Immediately on testing of the nanocomposite it was seen that previous problems with sink-marks and poor surface had disappeared.*" Clearly this is a reference to the improved processing offered by the nanocomposite materials.

The most recent application from GM is also on the interior of the vehicle where a Forte™ resin from Noble Polymers is used to make the centre console for the 2006 Chevrolet HHR [110]. In this application, the nanocomposite also replaced a glass filled polypropylene resin. The improvements seen with this application are much like those cited in the DaimlerChrysler application above, improved aesthetics including improved scratch and mar performance, and as a bonus, the minimization of warpage from the moulding of the part.

This section has gone into some detail describing applications of nanocomposite materials. This was not intended to be an exhaustive list, but rather a representative list of applications that have been documented in the industry. Announcements of new applications of organoclay nanocomposites are not now happening as frequently as in the 1999 to 2004 timeframe. It is the author's opinion that this is due to the fact that the benefits of the materials have been proven in applications, and the materials have reached a point in their development where they are being used where they are appropriate from a business and performance perspective without press announcements being made. Of course, polymer materials are seldom perfect, and new developments are always taking place to improve materials.

12.5 The future of nanoclay composites

The use of nanocomposites in automotive applications appears to have a good foothold in the industry. The benefits have been shown to be real; however, organoclay nanocomposites face a number of challenges to their continued use.

12.5.1 Alternative conventional filler materials

Conventional composite fillers materials are improving their performance. The producers of talc based fillers learned from the nanofiller industry about the benefits of small particle sizes and are providing fillers with higher surface areas. HTPultra5c from IMI-Fabi is an example of these sub-micron sized talcs. This is highly milled talc that is reported to have 99 % of the material with a particle size less than 5 μm and 75 % of the material with a particle size less than 1 μm and a median particle diameter of 500 nm. With this reduction in particle size it is clear that the surface area to volume ratio of these materials will increase which should lead to improved properties with a lower volume fraction of filler. It is likely that these materials are being used in Basell Polyolefin's high performance TPO materials such as Pro-fax TRN779. In other cases, Rio Tinto Minerals has made the effort to produce micron sized talcs that possess high aspect ratios. These materials make up Rio Tinto's line of aptly named HAR (High Aspect Ratio) talcs. With the HAR talcs, Rio Tinto is not trying to create very small particle size materials, but rather talc particles with high levels of lamellarity. These materials can provide some of the benefits of organoclay nanocomposites while using a filler material familiar to processors. The use of these materials is also able to lower the density of the resulting conventional composite materials, making them competitive with organoclay nanocomposite densities. Resin suppliers are turning to these materials as increasing levels of performance are being required of polymeric parts.

12.5.2 Exfoliation issues with olefinic resins

Another issue facing the continued use of nanocomposite materials is the apparent inability to completely exfoliate the materials in olefinic resins. This is especially evident when higher concentrations of nanofiller are desired. Figure 12.13 shows a plot of the organoclay particle aspect ratio and length versus the clay concentration. It is clear from this plot that reasonably good exfoliation is observed at the lowest organoclay concentration. This changes as the concentration of the organoclay increases, to the point where, at the highest concentration, the aspect ratio was lower by a factor of two. This indicates that the efficiency of the organoclay fillers decreases with increasing concentration, clearly a difficulty for increasing performance of the nanocomposites. This is an issue that requires further research to overcome. There are reports of highly exfoliated organoclay materials in olefinic materials [111]. These reports show very well exfoliated materials; however, the resin systems that are used are often impractical for commercial use. But these studies are the starting points for the research that will eventually lead to our understanding of the exfoliation mechanism well enough to utilize the full potential of organoclay nanocomposites.

12.5.3 New nanomaterials

In addition to the competition from improved conventional filler materials, there are a number of new nanofillers that in the coming years may present alternatives to organoclay nanocomposites. Perhaps the most obvious is the challenge posed by carbon nanotubes. More producers of good quality carbon nanotubes (or nanofibres) are entering the market and the prices for these materials are dropping. The availability of predispersed concentrates of the nanotubes eliminates the difficulties of processing using loose nanotubes. The electrical conductivity of the nanofibres and the low concentration needed to form a percolating network could eliminate the need for conductive primers in painted applications. This could provide a mechanism to offset some of the remaining price issues limiting their use in the automotive industry. It remains to be seen if this will develop into a thriving industry.

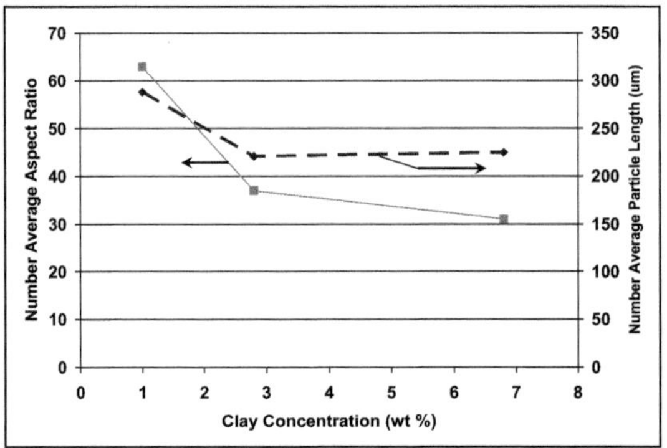

Figure 12.13

Plot of particle aspect ratio and particle length versus clay concentration in an olefin nanocomposite (data from reference 50)

Other nanomaterials that are being investigated include exfoliated graphite available from XG Sciences (www.xgsciences.com). This material is reported to maintain the electrical properties of the carbon nanotubes while being in a platelet form. This could help to control the anisotropy associated with one-dimensional fillers. Nano-calcium carbonate available from Specialty Minerals under the trade name EMforce™ is also being investigated. While EMforce™ is reported to be less effective at enhancing stiffness than other materials, they appear to be better at retaining low temperature ductility. Halloysites are available from Natural Nano and Nanodynamics. Halloysites are essentially silicate analogs to carbon nanotubes. These are naturally occurring minerals that have been used for centuries in the production of fine porcelain. While there are few articles in the literature that use these materials at this time, they provide another nanofiller capable of enhancing physical properties. Whether any of these materials will be able to develop into a commercial success will only be determined after significant work by the scientific community to prove their capabilities.

12.6 Concluding remarks

In conclusion, we have tried to give some background concerning organoclay nanocomposites and their use in the automotive industry. We hope that we have been able to present an unbiased narrative of the reasons why nanocomposites, in general, and organoclay nanocomposites, in particular, are being used in the automotive industry. We have also presented some challenges that remain concerning the continued use of organoclay nanocomposites, the most important of which is, in the author's opinion, the inability to achieve complete exfoliation in a number of the polymeric resins that are used in large volumes in this industry.

Acknowledgements

The author would like to acknowledge the invaluable assistance of Ms. Ting Cao and Dr. Charles H. Olk for their work on the Atomic Force Microscopy of dispersed nanoparticles, Ms. Paula D. Fasulo, Mr. David A. Okonski, and Dr. Howard W. Cox for their discussions and assistance in the collation of the materials used to prepare this chapter.

REFERENCES

1 Cuadros, ·J., Caballero, E., Huertas, F. J., deCisneros, C. J., Huertas, F., Linares, J. Experimental alteration of volcanic tuff: Smectite formation and effect on ^{18}O isotope composition, *Clays and Clay Minerals*, **1999**, 47, 769-776

2 Liu, L., Qi, Z., Zhu, X. Studies on Nylon 6/clay nanocomposites by melt-intercalation process, *Journal of Applied Polymer Science,* **1999**, 71 1133-1138

3 Cho, J. W., Paul, D. R. Nylon 6 nanocomposites by melt compounding, *Polymer,* **2001**, 42 1083-1094

4 Wang, H., Zeng, C., Elkovitch, M., Lee, L. J., Koelling, K. W., Processing and properties of polymeric nano-composites, *Polymer Engineering & Science,* **2001**, 41, 2036-2046

5 Vaia, R. A., Ishii, H., Giannelis, E. P. Synthesis and properties of two-dimensional nanostructures by direct intercalation of polymer melts in layered silicates, *Chemistry of Materials,* **1993**, 5, 1694-1696

6 Hoffmann, B., Dietrich, C., Thomann, R., Friedrich, C., Mulhaupt, R. Morphology and rheology of polystyrene nanocomposites based upon Organoclay, *Macromolecular Rapid Communications,* **2000**, 21, 57-61

7 Kawasumi, M., Hasegawa, N., Kato, M., Usuki, A., Okada, A. Preparation and mechanical properties of polypropylene-clay hybrids, *Macromolecules,* **1997**, 30, 6333-6338

8 Reichert, P., Nitz, H., Klinke, S., Brandsch, R., Thomann, R., Mulhaupt, R. Poly(propylene)/organoclay nanocomposite formation: Influence of compatibilizer functionality and organoclay modification, *Macromolecular Materials and Engineering,* **2000**, 275, 8-17

9 Gopakumar, T. G., Lee, J. A., Kontopoulou, M., Parent, J. S. Influence of clay exfoliation on the physical properties of montmorillonite/polyethylene composites, *Polymer,* **2002**, 43, 5483-8491

10 Liff, S. M., Kumar, N., McKinely, G. H. High-performance elastomeric nanocomposites via solvent-exchange processing, *Nature Materials,* **2007**, 6, 76-83

11 Landry, C. J. T., Coltrain, B. K. Organic-inorganic composites prepared via the sol-gel method: Some insights into morphology control and physical properties, *Polymer Preprints,* **1990**, 32, 514-515

12 Akelah, A., Salahuddin, N., Hiltner, A., Baer, E., Moet, A. Preparation and characterization of butadieneacrylonitrile/montmorillonite nanocomposite, *Polymer Preprints*, **1994**, 35, 739-740

13 Vaia, R. A., Vasudevan, S., Krawiec, W., Scanlon, L. G., Giannelis, E. P. New polymer electrolyte nanocomposites: Melt intercalation of poly(ethylene oxide) in mica-type silicates, *Advanced Materials*, **1995**, 7, 154-156

14 Messersmith, P. B., Giannelis, E. P. Synthesis and characterization of layered silicate-epoxy nanocomposites, *Chemistry of Materials*, **1994**, 6, 1719-1725.

15 Wang, M. S., Pinnavaia, T. J. Clay-polymer nanocomposites formed from acidic derivatives of montmorillonite and an epoxy resin, *Chemistry of Materials*, **1994**, 6, 468-474

16 Surivet, F., Lam, T. M., Pascault, J. P., Mai, C. Organic-inorganic hybrid materials. 2. Compared structure of polydimethylsiloxane and hydrogenated polybutadiene based creamers, *Macromolecules*, **1992**, 25, 5742-5751

17 Wang, Z., Pinnavaia, T. J. Nanolayer reinforcement of elastomeric polyurethane, *Chemistry of Materials*, **1998**, 10, 3769-3771

18 Jeon, H. S., Rameshwaram, J. K., Kim, G., Weinkauf, D. H. Characterization of polyisoprene-clay nanocomposites prepared by solution blending, *Polymer*, **2003**, 44, 5749-5758

19 Jeon, H. S., Rameshwaram, J. K., Kim, G. Structure-property relationships in exfoliated polyisoprene/clay nanocomposites, *Journal of Polymer Science Part B: Polymer Physics*, **2004**, 42, 1000-1009

20 Liang, Y., Wang, Y., Wu, Y., Lu, Y., Zhang, H., Zhang, L. Preparation and properties of isobutylene-isoprene rubber (IIR)/clay nanocomposites, *Polymer Testing*, **2005**, 24, 12-17

21 Kornmann, X., Berglund, L. A., Sterte, J., Giannelis, E. P. Nanocomposites based on montmorillonite and unsaturated polyester, *Polymer Engineering & Science*, **1998**, 38, 1351-1358

22 Kurauchi, T., Okada, A., Nomura, T., Nishio, T., Saegusa, S., Deguchi, R. Nylon 6-clay hybrid – Synthesis, properties and application to automotive timing belt cover, *SAE Technical Paper*, #910584, SAE International, http://www.sae.org/, **1991**

23 Kojima, Y., Usuki, A., Kawasumi, M., Okada, A., Fukushima, Y., Kurauchi, T., Kamigaito, O. Mechanical properties of nylon 6-clay hybrid, *Journal of Materials Research*, **1993**, 8, 1185-1189

24 Shi, H., Lan, T., Pinnavaia, T. Interfacial effects on the reinforcement properties of polymer-organoclay nanocomposites, *Chemistry of Materials*, **1996**, 8, 1584-1587

25 Gorrasi, G., Tortora, M., Vittoria, V., Pollet, E., Lepoittevin, B., Alexandre, M., Dubois, P. Vapor barrier properties of polycaprolactone montmorillonite nanocomposites: effect of clay dispersion, *Polymer*, **2003**, 44, 2271-2279

26 Laus, M., Camerani, M., Lelli, M., Sparnacci, K., Sandrolini, F., Francescangeli, O. Hybrid nanocomposites based on polystyrene and a reactive organophilic clay, *Journal of Materials Science*, **1998**, 33, 2883-2888

27 Gilman, J. Flammability and thermal stability studies of polymer layered-silicate (clay) nanocomposites. *Applied Clay Science*, **1999**, 15, 31-49

28 Jamieson, V. Open Secret, *New Scientist*, **2003**, 177 (2386), March 15 issue

29 Alexandre, M. and Dubois, P. Polymer-layered silicate nanocomposites: preparation, properties and uses of a new class of materials. *Materials Science and Engineering R-Reports*, **2000**, 28, 1-63

30 Giannelis, E. P. Polymer layered silicate nanocomposites, *Advanced Materials*, **1996**, 8, 29-35

31 Giannelis, E. P. Polymer-layered silicate nanocomposites: Synthesis, properties and applications, *Applied Organometallic Chemistry*, **1998**, 12, 675-680

32 Vaia, R. A., Giannelis, E. P. Polymer nanocomposites: Status and Opportunities, *MRS Bulletin*, **2001**, 26, 394-401

33 Winey, K. I. and Vaia, R. A. Polymer nanocomposites, *MRS Bulletin*, **2007**, 32, 314-319

34 Krishnamoorti, R., Vaia, R. A. Polymer nanocomposites, *Journal of Polymer Science Part B: Polymer Physics*, **2007**, 45, 3252-3256

35 LeBaron, P. C., Wang, Z., Pinnavaia, T. J. Polymer-layered silicate nanocomposites: an overview, *Applied Clay Sciences*, **1999**, 15, 11-29

36 Powell, C. E., Beall, G. W. Physical properties of polymer/clay nanocomposites, *Current Opinion in Solid State and Materials Science*, **2006**, 10, 73-80

37 Goettler, L. A., Lee, K. Y., Thakker, H. Layered silicate reinforced polymer nanocomposites: Development and Applications, *Polymer Reviews*, **2007**, 47, 291-317

38 Okada, A., Usuki, A. Twenty years of polymer-clay nanocomposites, *Macromolecular Materials and Engineering*, **2006**, 291, 1449-1476

39 Okada, A., Usuki, A. Twenty-year review of polymer-clay nanocomposites at Toyota Central R&D Labs., Inc., *SAE Technical Paper*, #2007-01-1017, SAE International, http://www.sae.org/, **2007**

40 Piner, R. D., Xu, T. T., Fisher, F. T., Qiao, Y., Ruoff, R. S. Atomic force microscopy study of clay Nan platelets and their impurities, *Langmuir*, **2003**, 19, 7995-8001

41 Park, H-M., Liang, X., Mohanty, A. K., Misra, M., Drzal, L. T. Effect of compatibilizer on nanostructure of the biodegradable cellulose acetate/organoclay nanocomposites, *Macromolecules*, **2004**, 37, 9076-9082

42 Cao, T., Rodgers, W. R., Fasulo, P. D., Olk, C. H., data to be published.

43 Vasiliu, E., Wang, C.-S., Vaia, R. A. Preparation of optically transparent films of poly(methyl methacrylate) (PMMA) and montmorillonite, *Materials Research Society Symposium Proceedings,* **Fall 2001**, 703, Symposium V, Nanophase and Nanocomposite Materials IV, 6.4

44 Kashiwagi, T., Morgan, A. B., Antonucci, J. M., VanLandingham, M. R., Harris, R. H., Awad, W. H., Shields, J. R. Thermal and flammability properties of a silica-poly(methylmethacrylate) nanocomposite, *Journal of Applied Polymer Science,* **2003**, 89, 2072-2078

45 Beecroft, L. L. Ober, C. K. Nanocomposite materials for optical applications, *Chemistry of Materials,* **1997**, 9, 1302-1317

46 ASTM Standard D3170-03 (reapproved 2007), "Standard Test Method for Chipping Resistance of Coatings," ASTM International, West Conshohocken, PA, www.astm.org

47 Chu, J., Xiang, C., Sue, H.-J., Hollis, R. D. Scratch resistance of mineral-filled polypropylene materials, *Polymer Engineering and Science,* **2000**, 40, 944-955

48 Sadhu, S., Bhowmick, A. K. Morphology study of rubber based nanocomposites by transmission electron microscopy and atomic force microscopy, *Journal of Materials Science,* **2005**, 40, 1633-1642

49 Fornes, T. D., Paul, D. R. Modeling properties of nylon 6/clay nanocomposites using composite theories. *Polymer,* **2003**, 44, 4993-5013

50 Fornes, T. D., Hunter, D. L., Paul, D. R. Effect of sodium montmorillonite source on nylon 6/clay nanocomposites. *Polymer,* **2004**, 45, 2321-2331

51 Lee, H.-S., Fasulo, P. D., Rodgers, W. R., Paul, D. R. TPO based nanocomposites. Part 1. Morphology and mechanical properties. *Polymer,* **2005**, 46, 11673-11689

52 Lee, H.-S., Fasulo, P. D., Rodgers, W. R., Paul, D. R. TPO based nanocomposites. Part 2. Thermal expansion behavior, *Polymer,* **2006**, 47, 3528-3529

53 Kim, D. H., Fasulo, P. D., Rodgers, W. R., Paul, D. R. Structure and properties of polypropylene based nanocomposites: Effect of PP-*g*-MA to organoclay ratio, *Polymer,* **2007**, 48, 5308-5323

54 Kim, D. H., Fasulo, P. D., Rodgers, W. R., Paul, D. R. Effect of the ratio of maleated polypropylene to organoclay on the structure and properties of TPO-based nanocomposites. Part I. Morphology and mechanical properties, *Polymer,* **2007**, 48, 5960-5978

55 Basu, S. K., Tewari, A., Fasulo, P. D., Rodgers, W. R. Transmission electron microscopy based direct mathematical quantifiers for dispersion in nanocomposites. *Applied Physics Letters,* **2007**, 91, article 053105

56 Halpin, J. C. Kardos, J. L. The Halpin-Tsai Equations: A review, *Polymer Engineering and Science,* **1976**, 16, 344-352

57 Hill, R. Elastic properties of reinforced solids: Some theoretical principles, *Journal of the Mechanics and Physics of Solids*, **1963**, 11, 357-372

58 Hill, R. Theory of mechanical properties of fibre-strengthened materials I. Elastic behavior, *Journal of the Mechanics and Physics of Solids*, **1964**, 12, 199-212

59 Mori, T., Tanaka, K. Average stress in matrix and average elastic energy of materials with misfitting inclusions, *Acta Metalllurgica*, **1973**, 21, 571-574

60 Chow, T. S. Effect of particle shape at finite concentration on the elastic moduli of filled polymers, *Journal of Polymer Science Part B: Polymer Physics*, **1978**, 16, 959-965

61 Chow, T. S. Effect of particle shape at finite concentration of thermal expansion of filled polymers, *Journal of Polymer Science Part B: Polymer Physics*, **1978**, 16, 967-970

62 Yoon, P. J., Fornes, T. D., Paul, D. R. Thermal expansion behavior of nylon 6 nanocomposites, *Polymer*, **2002**, 43, 6727-6741

63 Akzo Nobel Polymer Chemicals, http://www.akzonobel-polymerchemicals.com/ProductGroups/Perkalite+-+Properties+and+Applications.htm site visited January 2008

64 Chastek, T. T., Que, E. L., Shore, J. S., Lowy, R. J., Macosko, C., Stein, A. Hexadecyl-functionalized lamellar mesostructured silicates and aluminosilicates designed for polymer-clay nanocomposites. Part I. Clay synthesis and structure, *Polymer*, **2005**, 46, 4421-4430

65 Chastek, T. T., Stein, A., Macosko, C. Hexadecyl-functionalized lamellar mesostructured silicates and aluminosilicates designed for polymer-clay nanocomposites. Part II. Dispersion in organic solvents and in polystyrene, *Polymer*, **2005**, 46, 4431-4439

66 Chastek, T. T., Que, E. L., Jarzombeck, P., Macosko, C. W., Stein, A. Synthesis of lamellar isobutyl silicates and dispersion in polypropylene melts, *Journal of Applied Polymer Science*, **2007**, 105, 1456-1465

67 Usuki, A., Kojima, Y., Kawasumi, M., Okada, A., Fukushima, Y., Kurauchi, T., Kamigaito, O. Synthesis of nylon 6-clay hybrid, *Journal of Materials Research*, **1993**, 8, 1179-1184

68 Yano, K., Usuki, A., Okada, A., Kurauchi, T., Kamigaito, O. Synthesis and properties of polyimide-clay hybrid, *Journal of Polymer Science Part A: Polymer Chemistry*, **1993**, 31, 2493-2498

69 Usuki, A., Kawasumi, M., Kojima, Y., Okada, A., Kurauchi, T., Kamigaito, O. Swelling behavior of montmorillonite cation exchanged for ω-amino acids by ε-caprolactam, *Journal of Materials Research*, **1993**, 8, 1174-1178

70 Alexandre, M., Dubois, P., Sun, T., Garces, J. M., Jerome, R. Polyethylene-layered silicate nanocomposites prepared by the polymerization-filling technique: synthesis and mechanical properties, *Polymer*, **2002**, 43, 2123-2132

71 Sun, T., Garces, J. M. High-performance polypropylene-clay nanocomposites by in-situ polymerization with metallocene/clay catalysts, *Advanced Materials*, **2002**, 14, 128-130

72 Shin, S.-Y. A., Simon, L. C., Soares, J. B. P., Sholz, G. Polyethylene-clay hybrid nanocomposites: in situ polymerization using bifunctional organic modifiers, *Polymer*, **2003**, 44, 5317-5321

73 Andersen, P. G. Twin Screw Extrusion Guidelines for Compounding Nanocomposites. *SPE ANTEC 2002*, San Francisco, May **2002**, 48, 219-223

74 Vaia, R. A., Jandt, K. D., Kramer, E. J., Giannelis, E. P. Kinetics of polymer melt intercalation, *Macromolecules*, **1995**, 28, 8080-8085

75 Vaia, R. A., Giannelis, E. P. Lattice model of polymer melt intercalation in organically-modified layered silicates, *Macromolecules*, **1997**, 30, 7990-7999

76 Lertwimolnun, W., Verges, B. Influence of screw profile and extrusion conditions on the microstructure of polypropylene/organoclay nanocomposites, *Polymer Engineering and Science*, **2007**, 47, 2100-2109

77 Merinska, D., Vaculik, J., Kalendova, A., Kristkova, M., Simonik, J. Properties of polypropylene nanocomposite prepared by different way of compounding, *SPE ANTEC, 2003*, Nashville, May **2003**, 3, 2744-2750

78 Andersen, P. G. Processing nanocomposites on a kneader reciprocating single screw compounding system, *SPE ANTEC 2003*, Nashville, May **2003**, 1, 283-289

79 Liu, X., Wu, Q. Polyamide 66/clay nanocomposites via melt intercalation, *Macromolecular Materials and Engineering*, **2002**, 287, 180-186.

80 Fornes, T. D., Yoon, P. J., Hunter, D. L., Keskkula, H., Paul, D. R. Effect of organoclay structure on nylon 6 nanocomposite morphology and properties, *Polymer*, **2002**, 43, 5915-5933

81 Dennis, H. R., Hunter, D. L., Chang, D., Kim, S., White, J. L., Cho, J. W., Paul, D. R. Effect of melt processing conditions on the extent of exfoliation in organoclay-based nanocomposites, *Polymer,* **2001**, 42, 9513-9522

82 Fasulo, P. D., Rodgers, W. R., Ottaviani, R. A., Hunter, D. L. Extrusion processing of TPO nanocomposites, *Polymer Engineering and Science*, **2004**, 44, 1036-1045

83 Yuan, M., Turng, L-S., Spindler, R., Caulfield, D., Hunt, C. Microcellular nanocomposite injection molding process, *SPE ANTEC, 2003*, Nashville, May **2003**, 1, 691-695

84 Wee, D., Seong, D. G., Youn, J. R. Processing of microcellular nanocomposite foams by using a supercritical fluid, *Fibers and Polymers*, **2004**, 5, 160-169

85 Okonski, D. A. Injection molding a polyolefin-based nanocomposite versus a talc-filled TPO, *Materials Research Society Symposium Proceedings, Fall* **2001**, 702, Symposium U, Advanced Fibers, Plastics, Laminates and Composites, 7.5

86 Wilson, R. T., PPG Protects: Paint and coatings supplier develops new clearcoats to improve paint protection, *Automotive Industries*, **2003**, 183, 14-15

87 Nobel, M. L., Mendes, E., Picken, S. J. Acrylic-based nanocomposite resins for coating applications, *Journal of Applied Polymer Science*, **2007**, 104, 2146-2156

88 Nobel, M. L., Picken, S. J., Mendes, E. Waterborne nanocomposite resins for automotive coating applications, *Progress in Organic Coatings*, **2007**, 58, 96-104

89 Conductive nanotubes show promise in new applications, *Modern Plastics International*, **2000**, 30(2)

90 Corbett, B. Carbon nanotubes target fuel systems, bumpers, *Ward's Auto World*, **2002**, 38, 30-31

91 Kunii, I., Japan: A tiny leap forward. Nanotech may revive its industry, *Buiness Week Online*, **2003**, April 14 issue, 3828, p58

92 Sherman, L. Nanocomposites. *Plastics Technology*, **1999**, 45, 52-57

93 Edser, C. Auto applications drive commercialization of nanocomposites, *Plastics Additives & Compounding*, **2002**, January issue, 30-33

94 Stewart, R. Toyota gets early credit, *Plastics Engineering*, **2004**, 60(5), 24

95 Leaversuch, R. Nanocomposites broaden roles in automotive, barrier packaging, *Plastics Technology*, **2001**, 47, 64-69

96 Ube Industries, http://northamerica.ube.com, follow links for Ube Nylon and Ubesta (Nylon 12), site visited January 2008

97 Unitika Plastics, http://www.unitika.co.jp/plastics/E/nylon/nano-nylon6/05.html, site visited January 2008

98 Unitika Plastics, http://www.unitika.co.jp/plastics/E/applications/app01_car/car_01.html, site visited January 2008

99 Usuki, A., Kato, M., Okada, A., Kurauchi, T. Synthesis of polypropylene-clay hybrid, *Journal of Applied Polymer Science*, **1997**, 63, 137-139

100   Kurokawa, Y., Yasuda, H., Oya, A. Preparation of a nanocomposite of polypropylene and smectites, *Journal of Materials Science Letters*, **1996**, 15, 1481-1483

101   Kato, M., Usuki, A., Okada, A. Synthesis of polypropylene oligomer-clay intercalation compounds, *Journal of Applied Polymer Science*, **1997**, 66, 1781-1785

102   Hasegawa, N., Kawasumi, M., Kato, M., Usuki, A., Okada, A. Preparation and mechanical properties of polypropylene-clay hybrids using a maleic anhydride-modified propylene oligomer, *Journal of Applied Polymer Science*, **1998**, 67, 87-92

103   Miel, R. Nanocomposites applications stepping up, *Plastics News*, **2001**, 13, 4

104   Wilson, R. Let's get small, *Automotive Industries*, **2004**, 184, 29

105   Sherman, L. Chasing nanocomposites, *Plastics Technology*, **2004**, 50, 56-61

106   Advanstar Communications Inc., Press Release posted on http://www.off-road.com/hummer/news/2004_06/h2sut.html, site visited January 2008

107   Stewart, R. Automotive Plastics: New materials and processes help OEMs build better vehicles, *Plastics Engineering*, **2004**, 60, 24-32

108   Maniscalco, M., March 2004, "Next up for Nanocomposites," *Injection Molding Magazine*, available from http://www.immnet.com, site visited January 2008

109   Promise of compounds containing nanoclays becoming reality? *Modern Plastics*, **2004**, 81, 25

110   Visnic, W. Size not only matters, it's everything, *Ward's Auto World*, **2006**, 42, 46-50

111   Hasegawa, N., Okamoto, H., Usuki, A. Preparation and properties of ethylene propylene rubber (EPR)-clay nanocomposites based on maleic anhydride-modified EPR and organophilic clay, *Journal of Applied Polymer Science*, **2004**, 93, 758-764

Chapter 13
Nanocomposites in aerospace applications

Michael Meador

Polymers Branch, Structures and Materials Division, NASA Glenn Research Center, Cleveland, USA

13.1   Background

Weight, durability, and functionality are critical concerns for any aerospace system. Reductions in aircraft weight can enable increased payload capacity, improved manoeuvrability and decreased fuel consumption and emissions. Decreases in spacecraft weight lead to reduced launch costs and enable increased payload capacity. At the present time, it costs between $10,000 and $20,000 per pound to launch an item on the Space Shuttle. Durability is an important criterion not only because it affects vehicle safety, but also because it can impact maintenance costs and schedule.   Unexpected delays in the launch of the Space Shuttle can costs millions of dollars in operations costs. For long duration space missions, failure in a critical component can jeopardize the success of the mission.

Performance and functionality are also key drivers for aerospace systems. Vehicle performance impacts fuel consumption and emissions as well as survivability for military aircraft. The use of multifunctional materials and components is becoming increasingly desirable, combining load-bearing ability with other functions such as lightning strike protection or damage sensing, because this can lead to enhanced performance as well as weight reductions. Adaptive materials (shape memory alloys and polymers, piezoelectric materials) are also receiving considerable attention because they can enable radical new designs for aircraft and spacecraft that incorporate "morphing" structures, such as aircraft wings that can bend or change their shape and aerodynamics, thereby eliminating the need for ailerons. Use of adaptive structures can enable enhanced vehicle performance and fuel efficiency, and lead to reduced maintenance and improved durability. These materials could find use in advanced aircraft designs, such as the Silent Aircraft concept being developed by a team from the Massachusetts Institute of Technology [1].

While improvements in vehicle weight, durability, performance and functionality have been achieved with conventional materials, the use of nanostructured materials offers another approach to satisfying these needs. Significant improvements in materials' properties have been achieved with the addition of very small amounts (less than 5 weight %) of nanoscale fillers. In addition, nanotechnology offers the possibility of discretely tuning a specific property of a material without adversely affecting its other desirable properties. This review will cover recent advances in the development of polymer-based nanocomposites and their applications in aerospace systems. Definitions, formulae and chemical structures of all the monomers and polymers mentioned are provided at the end of the chapter.

13.2 Clays

13.2.1 Background

Clays have been used as additives to enhance the properties of polymeric materials for several decades [2]. However, it wasn't until the groundbreaking research by scientists at Toyota [3] who demonstrated the benefits of adding organically modified clays to nylon-6 that the development of polymer-clay nanocomposites became an area of intense interest. In that study, the authors reported that addition of 4.6 weight % of dodecylammonium salt-modified montmorillonite clay to nylon-6 resulted in significant increases in tensile strength (from 68.6 to 97.2 MPa), tensile modulus (from 1.11 to 1.87 GPa), flexural strength (from 89.3 to 143 MPa) and flexural modulus (from 1.94 to 4.34 MPa). While conventional approaches to enhancing strength usually result in decreased toughness, these nanocomposites had a higher Izod impact strength than the base resin (20.6 vs. 18.1 J/s). In addition, the heat distortion temperature of the nylon nanocomposite was more than twice that of nylon-6 (152° compared to 65°C).

Since this groundbreaking report, organically modified clays have been added to a variety of polymers to enhance their mechanical properties [4]. For example, Balakrishnan and co-workers [5] have reported that addition of an octadecyl ammonium exchanged Na-montmorrillonite clay to a rubber toughened epoxy led to increased tensile strength and modulus as well as enhanced ductility. Addition of Cloisite 30 B montmorillonite organoclay to an allylphenol-modified bismaleimide resulted in a two-fold increase in its impact strength [6].

Other polymer properties can be enhanced through the addition of small amounts of organically modified clays. Yano and co-workers [7] reported that addition of a dodecylammonium-modified montmorillonite to Kapton polyimide resulted in a six-fold reduction in oxygen permeability. Yasmin and co-workers [8] report that addition of 3 weight % octadecyl trimethylammonium-modified montmorillonite to a DGEBA-based epoxy reduced its coefficient of thermal expansion (CTE) by 16 %.

The remarkable effect of clays on polymer properties is due in large part ot the structure of the clay additive [9]. Naturally occurring clays are comprised of layers of negatively charged silicates with metal cations, typically sodium or potassium, sandwiched in between each silicate layer. The individual clay layers or galleries are about 1nm in thickness, hundreds of nanometres in length, and have aspect ratios ranging from 10-1000. The high aspect ratio and high surface area of the clays allow them to interact strongly with the polymer matrix and give rise to significant changes in the structure and properties of the polymer.

However, the key to obtaining these property changes lies in how well the individual clay galleries are dispersed within the polymer [10]. Replacement of the metal ions in between the clay galleries with the salt of an organic cation, typically an alkylammonium derivative, by a simple ion exchange process leads to a swelling of the clay. This clay is frequently called an organoclay. It is then possible to start to insert polymers chains between the individual clay galleries, forming a nanocomposite. Typically clays disperse in a polymer matrix to give rise to two types of geometries – intercalated and exfoliated. In the intercalated geometry, the clay galleries are separated by a few layers of polymer chains but the long range stacking order between the clay galleries is retained. In the exfoliated geometry, the individual galleries are completely separated and are randomly distributed throughout the matrix. In general, the exfoliated geometry is the one that gives rise to the most changes in polymer properties with a minimal amount of clay additive.

Because of the property enhancements that can be realized with a small amount of added clay and because of the ready availability of these clays, polymer-clay nanocomposites are attractive materials for a variety of aerospace applications. A few of these applications will be discussed in the following sections.

13.2.2 Cryotanks

Cryogenic propellant tanks account for more than half of the dry-weight of any launch vehicle. Use of composites in place of metals in these tanks would lead to vehicle weight reductions of as much as 20-30% [11]. While the utilization of composites in cryotanks has been the subject of intense effort, these activities have met with limited success. This is due primarily to the inherent permeability of polymers and composites to low molecular weight gases such as hydrogen, and the propensity of composites to microcrack during thermal cycling due to the large differences in the coefficients of thermal expansion of the polymer matrix and the fibre reinforcement. Formation of microcracks in the inner walls of a composite cryotank can lead to loss of propellant and gives rise to an effect known as cryopumping. Cryopumping is the undesirable transfer of propellant from inside the tank to the honeycomb or foam core of the composite tank structure and can lead to delamination and failure of the cryotank. In order to address the permeability and microcracking problems with composites, a thin metal foil liner is often used on the inner wall of the tank. However, such a liner adds significant weight to the tank (as much as 50%), and adds complexity to the manufacturing process which can lead to higher costs. It is also a potential site for defects, since the metal liner can delaminate from the tank wall due to a coefficient of thermal expansion mismatch between the metal and composite substrate. Use of polymer-clay nanocomposites as the matrix resin for cryotank structures has the potential for overcoming both permeability and microcracking problems and thereby eliminating the need for a metal liner [12].

13.2.2.1   Permeability

Significant reductions in the permeability of gases through polymer films have been achieved by the addition of clays. This has been attributed to the clay platelets organizing to create a more tortuous path for diffusion of a gas through the polymer film [13]. Miller and co-workers have found that addition of 2 weight % Cloisite 30B montmorillonite organoclay to Epon 862 epoxy resin reduced its helium permeability at room temperature by 25% [12]. Carbon fibre-reinforced composite tanks were prepared from this nanocomposite and leak tested (Figure 13.1). The nanocomposite tank had one-fifth the helium leak rate than a comparable tank prepared with the epoxy without any added clay. Significantly lower permeabilities can be obtained by ordering the clay platelets perpendicular to the diffusion path of the gas. For example, Pinnavaia has reported the development of epoxy-clay fabrics that have a 2-3 fold lower oxygen permeability than the pristine epoxy film [14].

Figure 13.1

Graphite-fibre-reinforced epoxy/clay nanocomposite pressure vessels manufactured by filament winding

13.2.2.2   Toughness

Reducing the susceptibility of composites to microcracking can be achieved by reducing the coefficient of thermal expansion of the resin and enhancing its toughness. Miller has reported a 25% reduction in the coefficient of thermal expansion of Epon 862 by addition of 5 weight % Cloisite 30B [12]. This same nanocomposite had twice the notched Izod impact strength over that of Epon 862. Pang and co-workers have recently reported that addition of 1 weight % of an organically modified montmorillonite to a DGEBA-based epoxy suppressed the formation of micocracks due to thermal cycling at cryogenic temperatures [15]. At higher weight percentages, the clays tended to agglomerate and the ductility of the resin was diminished. Zhang and co-workers have also reported that addition of 1-3 weight % of cetyl trimethylammoninum-modified montmorillonite to a PMDA-ODA polyimide enhanced its ductility at cryogenic (liquid nitrogen) temperatures and reduced microcrack formation [16].

13.2.2 Other structures

As discussed above, addition of clays can significantly enhance the mechanical properties and improve the ductility of a variety of resins used in aerospace applications. However, clay additives can also improve other resin and composite properties including thermal oxidative stability, moisture and solvent resistance, and flammability.

High temperature polymers, in particular polyimides, and fibre-reinforced composites are finding increased usage in propulsion components and hot airframe structures. The primary driving force for these applications is weight reduction. For aircraft, a one pound reduction in engine weight can lead to between ten and twenty-five pound reduction in the overall weight of the vehicle. This is a cascading effect – reducing the engine weight enables reduction in pylon and wing weight, which, in turn, enable reductions in fuselage and landing gear weight. Engine components that are well suited for use of polymeric materials, include fan blades, cases and ducts – structures which are typically some of the largest and heaviest within the engine.

The two major barriers to the use of polymeric materials at higher service temperatures are (1) the inherent susceptibility of organic-based materials to oxidative degradation at high temperatures and (2) the glass transition temperature of the material. Typically, the intended use temperature of these materials should be at least 25°C below the material's glass transition temperature. Efforts to improve high temperature oxidative stability and increase the glass transition temperature have been confounded by the fact that those chemical structures which impart good high temperature stability and performance often render the polymer difficult to process. In addition, these approaches often employ fluorinated monomers which are costly to produce, leading to resins which are too expensive for many applications.

Thermal oxidative degradation of polyimides at high temperatures has been shown to occur primarily on the surface of the resin or composite and is limited by the rate of diffusion of oxygen into the polymer [17]. Oxidation resistant coatings, typically metals or ceramics, have been employed in the past as a way of limiting access of oxygen to the surface of the composites. While these coatings work well for a few hundred hours at high temperatures, they eventually crack or spall off due to mismatches between the coefficients of thermal expansion of the coating and composite substrate. Addition of clays to high temperature composites would make the matrix resin its own oxidation barrier thereby eliminating the need for any coating.

Campbell has shown that addition of 5 weight % of a montmorillonite modified by a cation mixture derived from octadecylamine and methylene dianiline to PMR-15, a high temperature polyimide, reduced its oxidative weight loss after 1000 hours at 288°C by 25% [18]. Addition of 1 weight % of the same clay to a T650-35 8-HS carbon fabric-reinforced PMR-15 composite produced the same reduction in oxidative weight loss. In addition, the carbon fabric-reinforced nanocomposite had a 30% higher flexural strength and modulus both at room and high temperature (288°C) than the corresponding composite prepared without clay. The effect of the addition of clay on the formation of surface oxidation layers in graphite fibre-reinforced polyimides is shown in the SEM images depicted in Figure 13.2. These samples were aged in air for 500 hours at 288°C. While the sample without clay showed the formation of a pronounced oxidation layer as well as oxidation induced microcracks, the sample with clay had a much thinner oxidation layer and no apparent microcrack formation.

Figure 13.2

SEM photomicrographs of aged graphite-reinforced polyimide composites with (right) and without (left) organoclay

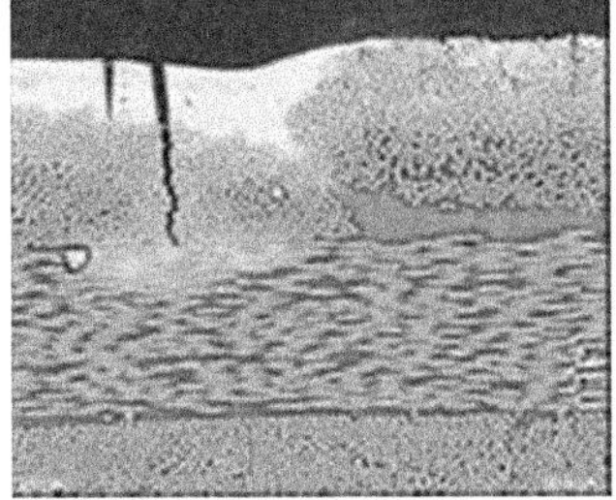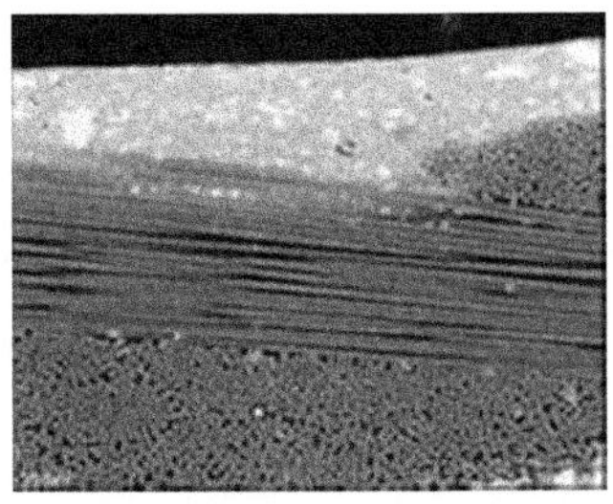

The use of the proper organic modifier is crucial for these applications, since high temperature polyimides require processing temperatures as high as 370°C. This temperature is well above that at which typical alkylammonium salts begin to degrade. To circumvent this, more stable modifiers such as tetraphenylphosphonium bromide have been explored for use in high temperature nanocomposites [19]. Kurose and co-workers reported the use of 4,4'-bis(4-aminophenoxy)diphenylsulfone, BAPS, as a more stable modifier for use in polyimides [20]. Montmorillonite clays modified with BAPS survived processing temperatures of above 300°C. Cations derived from other aromatic diamines have been successfully used as high temperature resistant modifiers.

Montmorillonite modified with the hydrochloride salt of 1,3-bis(4-aminophenoxy)benzene reduced the oxidative weight loss of 6-FDA derived poly(ether imide)s after 300 h at 300°C by as much as 58 %. The coefficient of thermal expansion in films prepared from these nanocomposites was reduced by 23 % over that of the neat poly(ether imide) films, and an increase in tensile modulus by up to 38 % was observed upon addition of organoclay. However, addition of these clays did make the polyimide films more brittle – elongation at break was reduced from 12.6 to 1.6 % with the addition of 5 weight % clay. Tyan and co-workers examined the use of an aryl triamine as a modifier for montmorillonite, and found that addition of 5 weight % of this clay to a BTDA-ODA polyimide increased the onset of decomposition by 24°C (from 583 to 607°C), reduced CTE by 30 %, and increased tensile strength and modulus [21].

Reduced charge clays, prepared by substituting some of the sodium ions in montmorillonite with lithium ions, can also give rise to more thermally stable organoclays. Delozier and co-workers have reported that a reduced charge clay modified with the dihydrochloride salt of 1,3-di(4-aminophenoxybenzene), APB, was successfully incorporated into APB-BPDA polyimide films and survived processing at 300°C [22]. Campbell has also reported the use of reduced charge organoclays in a thermosetting polyimide, DMBz-15 [23].

Organoclays have also been used to enhance the flame retardancy of polymers which could be used in the interiors of aircraft and spacecraft [9]. Gilman and co-workers have reported that addition of less than 3 weight % of organically modified montmorillonite or fluorohectorite clays to nylon, polystyrene and polypropylene-*graft*-maleic anhydride significantly reduced (as much as 70%) the peak heat release rate of these polymers [24].

13.3   Carbon-based nanostructured additives

13.3.1 Carbon nanotubes

Since the reported discovery by Ijima [25] in 1991 of single-wall carbon nanotubes, carbon nanotubes, CNTs, have attracted considerable attention because of their remarkable mechanical, and electrical and thermal properties [26]. Use of carbon nanotubes in nanocomposites has the potential to provide materials with significantly enhanced mechanical properties, electrical and thermal conductivities. Application of carbon nanotube-based nanocomposites in aerospace components has the potential to significantly reduce vehicle weight. A comparison of the specific strength and modulus of an aerospace grade aluminum alloy, a conventional carbon fibre-reinforced epoxy composite, and the projected specific strength and stiffness values for a CNT-reinforced polymer composite and single crystal CNT fibres is given in Figure 13.3. A study performed by NASA in the late 1990s estimated, based upon these projected property values, that the use of carbon nanotube-reinforced composites could enable an 82% reduction in the weight of a Space Shuttle-like reusable launch vehicle, and as much as a 45% reduction in the empty weight of a blended wing body subsonic commercial aircraft [27].

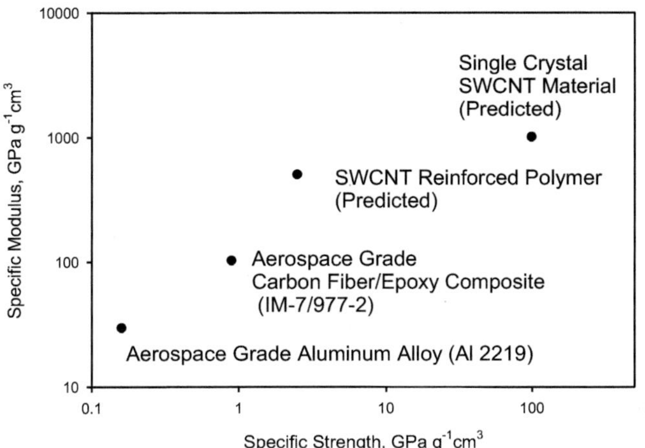

Figure 13.3

Comparison of specific strength and modulus of various aerospace

materials

Carbon nanotubes have been produced in a number of forms – single wall

(SWCNT), double wall (DWCNT) and multi-wall (MWCNT). The properties of

CNTs, in particular SWCNTs, are determined by their length, diameter and

chirality. Assembly of an SWCNT can be envisioned as taking a sheet of carbon

atoms and rolling it into a cylinder along two vectors, a and b, called lattice

vectors (Figure 13.4). The chirality and diameter of an SWCNT is determined by

the roll-up vector, r, where

r = m**a** + n**b.**

Two special chiralities for SWCNTs are armchair SWCNTs where m=n, and zig-zag SWCNTs where n=0, other types of SNTS are designated as "chiral". Optical, thermal and electrical properties of SWCNTs are dictated by their chirality. For example, metallic electrical conductivity is achieved in SWCNTs when

$$2n + m = 3q \quad \text{where q is an integer.}$$

This means that all armchair SWCNTs have metallic conductivity, and about 1/3 of all zig-zag SWCNTs are metallic [28].

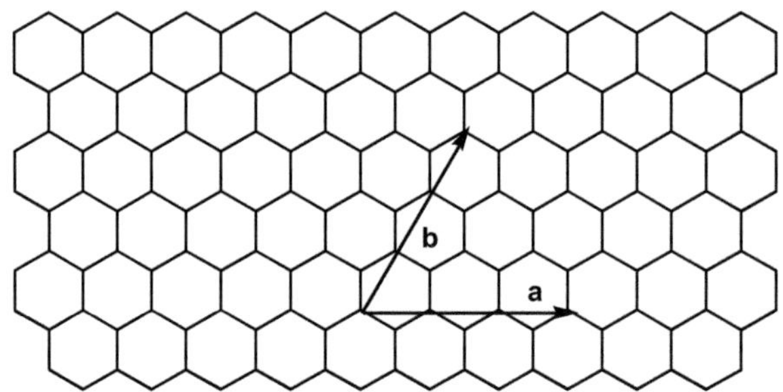

Figure 13.4

Chirality of carbon nanotubes is determined by cutting and folding a sheet of carbon atoms along the two lattice vectors, a and b

A comparison of the mechanical properties of SWCNTs and MWCNTs with a conventional intermediate modulus carbon fibre (IM-7) is given in Table 13.1. A considerable number of theoretical and experimental studies have focused on measuring and predicting the mechanical properties of carbon nanotubes, and also on understanding the effects of nanotube chirality and size on mechanical behaviour [29,30]. Troiani and co-workers examined the tensile failure of a SWCNT in a field emission transmission electron microscope (TEM) and found that tensile loading leads to necking in the tubes until a single chain of carbon atoms forms between the tube fragments [31]. This chain eventually breaks. Tensile failure in MWCNT occurs via a "sword and sheath" mechanism in which the outer nanotube slides away from the core nanotubes [32].

Table 13.1

A comparison of the mechanical properties of SWCNTs and MWCNTs with those of a conventional carbon fibre

MATERIAL	YOUNG'S MODULUS, GPA	TENSILE STRENGTH, GPA	STRAIN AT FAILURE, %	TOUGHNESS, J/G
SWCNT	320-1470[b]	10-52[b] 17.64[c]	5.3	770
MWCNT	270-950[a]	11-63[a] GPa	12%	1240
IM-7 Carbon Fibre	300[d]	5.3[d]	1.8[d]	--
References: [a] Ref. 32 [b] Ref. 33, [c] Ref. 34, [d] Ref. 26				

13.3.1.1 Synthesis methods

Carbon nanotubes have been prepared by a number of techniques [35]. Ijima was the first to report the presence of MWCNTs in the soot formed from arc-discharge evaporation of carbon from graphite electrodes [25]. The arc-discharge method has been scaled up by Ebbesen and Ajayan [36] to produce

MWCNTs in 75% yield, relative to the amount of graphite starting materials. Arc-discharge synthesis of SWCNTs was reported simultaneously by Ijima [37] and Bethune [38] and was later scaled up by Journet [39]. SWCNTs have also been produced by laser ablation of a graphite target containing small amounts of Ni and Co [40]. Vander Wal has reported the combustion-based synthesis of a mixture of SWCNTs and MWCNTs [41]. In this method, Fe or Ni catalyst particles, generated by a thermal evaporation technique, are entrained into a flame produced in a CO/H_2 mixture to produce SWCNTs or ethane/H_2 to produce MWCNTs.

Another method for the production of CNTs involves the catalytic decomposition of a feedstock gas at high temperatures, typically 1000°C. Smalley and co-workers developed the HiPCO (high pressure carbon monoxide disproportionation) method to produce SWCNTs, based upon the Boudard decomposition of carbon monoxide to carbon in the presence of a $Fe(CO)_5$ or $Ni(CO)_4$ catalyst [42]. Similarly, Resosko and co-workers have reported the production of SWCNTs by Co and Mo catalyzed decomposition of CO at 700-950°C in a fluidized bed reactor [43]. By adjusting the relative amounts of Co and Mo catalysts and using a silica support, the CoMoCat process can provide nanotubes with narrower chirality distributions than those produced by the HiPCO process [44].

13.3.1.2   Purification

Successful use of carbon nanotubes in nanocomposites has required the development of methods to purify and functionalize them. As prepared nanotubes can have a significant amount of impurities in them, including non-nanotube forms of carbon and spent catalyst. These impurities can adversely affect both the properties and the high temperature stability of the nanotubes. Over the years, a variety of different methods have been developed for the purification of both SWCNTs and MWCNTs. Removal of the catalyst is particularly challenging because these particles often become encased in carbon during the nanotube synthesis. The challenge is to develop purification approaches that remove these impurities without damaging the nanotubes.

The most common methods involve some type of oxidation treatment to remove the non-nanotube carbon as well as the carbon coating on the catalyst, followed by acid treatment to remove the metal catalyst. Hauge and co-workers reported such an approach which involved three sequential oxidation treatments at 225°, 325°, and 425°C, respectively [45]. In between each of these steps the nanotubes were either refluxed in concentrated HCl or treated in concentrated HCl in a water sonicating bath. The authors reported that iron content, as estimated by thermal gravimetric analysis, was reduced from 5.6 atom % (22.2 weight %) in the crude HiPCO nanotubes to 0.03 atom % (0.14 weight %). The final step in the purification process was an annealing treatment in argon at 800°C. This step was intended to "heal" any defects introduced into the nanotubes during earlier steps in the purification process.

Attempts to scale up this method from the reported 100mg to gram-scale quantities of crude HiPCO tubes led to decomposition of the tubes after the annealing step [46]. TEM images taken after each step in the purification process revealed that as the purity of the nanotubes increased, they tended to rope up forming tight bundles that trapped some of the catalyst particles. As the SWCNTs were heated to 800°C, the iron catalyst particles catalyzed their degradation leaving behind iron oxide. A modification of Hauge's method [47] utilized high intensity ultrasonication with a titanium horn to break up the nanotube bundles and replaced the acid reflux steps with Soxhlet extraction with concentrated HCl. The resulting process reduced the iron content in HiPCO tubes from 22.3 weight % to 0.03 weight % with no visible damage to the nanotubes. In addition, this method was capable of being scaled-up to purify 2.0 gram batches of HiPCO tubes.

In another approach, Hauge used a combination of oxidation at successively higher temperatures to remove the carbon coating on the catalyst particles followed by treatment of the exposed iron oxide with a mixture of SF_6 and $C_2H_2F_4$ to convert it to FeF_3.[48] Formation of FeF_3 renders it inactive in the oxidation of SWCNTs. The FeF_3 is then removed in a final step by Soxhlet extraction with concentrated HCl. Using this method, the overall iron content can be reduced from 30 to 0.3 – 1.3 weight % by treatment with a mixture of 20% O_2 and 5% SF_6 with an overall yield of purified CNTs of 62 % (based on the weight of crude nanotubes corrected for the weight of the catalyst). Other approaches to nanotube purification include magnetic filtration [49], sulfidation [50], and a combination of acid treatment and hydrogenation [51].

In addition to removing impurities, there have been some efforts invested in the development of techniques to purify nanotubes by size or properties, e.g., metallic vs. semiconducting. Miyata and co-workers have reported a method to selectively oxidize semiconducting nanotubes with hydrogen peroxide [52]. This aproach yields a mixture of SWCNTs that is more than 80% metallic. Ramesh and co-workers demonstrated that binary mixtures of chlorosulfonic acid and methane sulfonic acid in varying concentrations could be used to selectively solvate SWCNTs of certain diameters [53]. Wrapping SWCNTs with DNA [54], conjugated polymers [55] or surfactant mixtures [56] can enable extraction of SWCNTs of a specific chirality or produce SWCNT mixtures with a reduced distribution of chiralities.

13.3.1.3   Functionalization

Functionalization of carbon nanotubes is essential for their use in nanocomposites. Carbon nanotubes have limited or no solubility in common organic solvents and water, and are only soluble in strong acids such as fuming sulfuric acid (oleum). In addition, carbon nanotubes, in particular SWCNTs, tend to aggregrate or "rope" due to strong van der Waal's attractive forces, making them difficult to disperse into organic solvents and polymers. Interfacial bonding between the CNTs and the matrix, an essential requirement for good mechanical properties, can also be problematic due to the low surface energy of the nanotubes. A variety of different approaches have been developed for the functionalization of CNTs [57]. These fall into two categories – covalent functionalization, attachment of functional groups to the ends or side walls of the nanotubes, and non-covalent functionalization, adsorption of large molecules or polymers onto the side walls of the nanotubes. A few examples of each method will be discussed in the next paragraphs.

Perhaps the most common approach to the functionalization of SWCNTs involves oxidation of their ends. Treatment of SWCNTs by prolonged sonication in a mixture of nitric and sulfuric acids "cuts" the nanotubes into smaller fragments leaving carboxylic acid groups at the cut ends which can be readily converted to amides [58] or carboxylic esters [59]. Nanotubes have been grafted onto polymers through these carboxylic acid groups. For example, carboxylic acid functionalized SWCNTs were copolymerized with caprolactam to produce a nanotube grafted nylon. A three-fold increase in Young's modulus was achieved by copolymerization of just 1.5 weight % of carboxylate functionalized SWCNTs into the nylon [60]. Carboxylic acid functionalized SWCNTs have also been co-reacted into a polyimide matrix [61].

Carbon nanotubes are susceptible to covalent side-wall functionalization via a number of different chemistries. Substituted aromatic diazonium compounds have been used extensively in SWCNT side-wall functionalization. These compounds can be electrochemically reduced onto a SWCNT "Bucky-paper" electrode to give aryl radicals which functionalize up to 20 carbons on the SWCNT side-walls [62]. Alternatively, aryl diazonium intermediates have been generated *in situ* and used to functionalize SWCNTs in the absence of any solvent [63].

Other reactive intermediates such as carbenes [64] and nitrenes [65], quinodimethanes [66], and alkyl radicals [67] have also been used in SWCNT functionalization.

Margrave and co-workers have demonstrated that SWCNTs can be directly fluorinated with gaseous fluorine [68]. These fluorinated SWCNTs have then been alklyated with methyl, *n*-butyl or *n*-hexyllithium to produce SWCNTs that are soluble in common organic solvents such as tetrahydrofuran (THF) [69]. Displacement of the fluorides with other chemical species is quite facile. Reaction of fluoro-SWCNTs with diols pretreated with LiOH, NaOH, or KOH or with ω-aminoalcohols in the presence of a pyridine derivative has been shown to lead to the formation of ω-hydroxyalkyl functionalized SWCNTs [70]. Similarly, *N*-alkylidene amino functionalized SWCNTs can be prepared by treatment of fluoro-SWCNTs with the corresponding diamine in the presence of pyridine [71]. Side-wall functionalized SWCNTs prepared by these routes have been incorporated into a variety of polymers, including epoxies and nylon [72].

Reductive alkylations have been used for direct side-wall functionalization of SWCNTs. Treatment of SWCNTs with lithium and an alkyl halide in liquid ammonia has been shown to provide alkylated SWCNTs with a high degree of coverage of alkyl groups [73]. The resulting nanotubes are soluble in common organic solvents. In addition, the authors claim that this approach is quite effective in debundling of SWCNTs. Recently, SWCNT lithium salts have been used as anionic initiators for the polymerization of methyl methacrylate to produce graft SWCNT/poly(methyl methacrylate) (PMMA) copolymers [74]. Atomic force microscopy images show that the nanotubes in these copolymers are well exfoliated.

Non-covalent functionalization methods have also been intensively studied since these methods do not alter the chemical bonding within the nanotubes. These non-covalent methods take advantage of intermolecular forces, such as π-π interactions, to complex the nanotubes with the functionalizing agent. In this way, the surface properties of the nanotubes can be tuned to increase solubility in both aqueous and organic media, to mitigate bundling and roping of the nanotubes and to promote enhanced interations with a given polymer matrix. A variety of different complexing agents have been employed ranging from polycyclic aromatic hydrocarbon derivatives [75], to synthetic polymers, to DNA [56]. A few examples of these will be discussed in the following paragraphs.

Polycyclic aromatic hydrocarbons, such as perylenes or pyrenes, can form stable complexes with SWCNTs via π-π interactions [74]. Chan and co-workers demonstrated the successful formation of π-complexes between SWCNTs and the succinimidyl ester of 1-pyrenebutanoic acid for potential use in biosensors [76]. This approach was modified by Lebron-Colon and Meador to make SWCNT/pyrene complexes which were then incorporated into high temperature polyimides [47]. Polymers substituted with pyrenes as part of the backbone or as pendant groups have also been used to functionalize carbon nanotubes. Lou and co-workers have reported the non-covalent functionalization of MWCNTs with a random block copolymer of methyl methacrylate and pyrenyl methacrylate [77]. Wang and co-workers have described the synthesis of a pyrene substituted "polysoap" that forms water soluble complexes with SWCNTs [78]. Complexation of nanotubes has also been reported with anthracenes [79], porphyrins [80] and mixed systems [81].

Non-covalent functionalization of CNTs by "wrapping" with conjugated polymers has also received a considerable amount of attention. Wrapping of SWCNTs by conjugated polymers, including poly(*m*-phenylene vinylene) [82], poly(phenylacetylene) [83], and poly(9,9-dialkylfluorenes) [54] have been shown to render them soluble in organic solvents. Stoddart has also reported the non-covalent functionalization of SWCNTs by trapping them in cavities generated within dendrimers [84] and hyperbranched polymers [85].

Wrapping of nanotubes with more polar polymers has been shown to lead to enhanced solubility both in water and polar organic solvents. The classic example is wrapping of SWCNTs with poly(styrene sulfonate) and poly(vinyl pyrrolidone) [86]. The resulting nanotubes have solubility in water as high as 4 g/litre. Delozier and co-workers have described the wrapping of SWCNTs with a poly(tetraphenyl pyridinium co-9,9-dioctylfluorene) ionomer and their subsequent use in polyimide nanocomposites [87]. Both single and multiwall carbon nanotubes have been wrapped with polycation salts of poly(vinyl pyridine)s to render them soluble in aqueous media [88].

13.3.2 Carbon nanotube-based nanocomposites

Functionalized single, double and multiwall nanotubes have been dispersed in a variety of polymers to enhance mechanical properties, as well as thermal and electrical conductivity. As discussed in Section 3.1, substitution of a conventional fibre-reinforced polymer matrix composite with a SWCNT-reinforced polymer in an advanced aircraft or launch vehicle can lead to significant vehicle weight reductions. These payoffs assume that the nanotubes are utilized in a high strength fibre. However, significant improvements in mechanical properties can be obtained by simple addition of nanotubes to a polymer matrix. In addition, the possibility of achieving enhanced electrical or thermal conductivity in conjunction with improved mechanical properties enables the development of multifunctional structures which could also lead to significant component and vehicle weight reductions.

13.3.2.1   Electrical and thermal conductivity

As discussed previously, increased use of composites in spacecraft and aircraft is desirable due to the significant reductions in weight that can be achieved with these materials. For example, more than half of the airframe for Boeing's 787 "Dreamliner" will be made from polymeric composites.

The Joint Strike Fighter, currently under development, will also have heavy utilization of composites in its airframe. NASA is currently exploring the possibility of using composites in the inner stage, core stage, earth departure stage and payload shroud for the Ares V launch vehicle in an attempt to reduce launch vehicle weight and enable an increase in payload capacity (Figure 13.5).

Figure 13.5

Artist's rendering of the Ares V heavy launch vehicle

Increased use of composites in each of these applications carries with it special concerns about the management of thermal and electrical energy in these structures, since conventional aerospace resins and composites have poor electrical and thermal conductivity. Inherently conductive polymers, such as polyaniline, do not have the necessary environmental stability and mechanical properties suitable for structural applications. Consequently other engineering approaches have been employed to enhance electrical and thermal conductivity in these components. For example, composite airframe structures are susceptible to catastrophic damage from lightning strike, since the composites cannot effectively conduct electrical charge away from the strike area. To mitigate against damage from lightning strikes, airframe manufacturers apply a thin veil of a copper mesh to the surface of the airframe. While this approach provides suitable protection for the airframe, it adds significant weight to the vehicle and also leads to increased manufacturing costs.

Addition of carbon nanotubes, both SWCNTs and MWCNTs, to engineering polymers and composites have been shown to lead to increases in electrical and thermal conductivity without adversely affecting mechanical properties or processability. Thostenson and Chou reported a decrease in the DC volume resistivity of EPON 862 epoxy from 10^{16} to 100 Ωcm with the addition of 5 weight % MWCNT [89]. The percolation threshold of the nanocomposite was achieved at MWCNT loading levels below 0.01 weight %. The authors also reported a 50% increase in the thermal conductivity of the epoxy with 5 weight % added MWCNT. Moisala and co-workers compared the effect of addition of chemically functionalized or ball-milled SWCNTs and MWCNTs on the thermal and electrical conductivity of a bisphenol A-derived epoxy [90]. The electrical percolation threshold of the nanocomposites was achieved at 0.05 weight % for the functionalized SWCNTs and at 0.25 weight % for ball-milled SWCNTs, presumably due to the smaller lengths of ball-milled nanotubes. By contrast, electrical percolation was observed at 0.0025 weight % in the MWCNT nanocomposites. While the thermal conductivity of the epoxy resin increased with addition of MWCNTs, it decreased with addition of both functionalized and ball-milled SWCNTs. The authors suggest that poor interfacial interactions between the SWCNTs and epoxy matrix led to inefficient phonon coupling and poor thermal conductivity.

Wang and co-workers prepared SWCNT-epoxy nanocomposite films by solution casting from ethanol and aligned the nanotubes by stretching the films while they were still wet with solvent [91]. The DC volume conductivity was about 1000 times higher parallel to the alignment direction than perpendicular to alignment. Ghose and co-workers reported that addition of MWCNT to an ethylene-vinyl acetate copolymer led to enhanced thermal conductivity [92]. The

largest increase was observed in samples in which the nanotubes were aligned

by extruding the nanocomposite into ribbons. The authors also demonstrated

that these materials could be extruded into flexible tubing which had a high

radius of curvature. Use of this tubing in the thermal control garment for

spacesuits would enable enhanced cooling efficiency and could reduce weight in

the spacesuit.

Addition of CNTs to polyimides has also been shown to enhance electrical

and thermal conductivity. Yu and co-workers reported that addition of 1 weight %

of carboxylate functionalized SWCNTs to a 6FDA/1,3-APB polyimide reduced its

volume resistivity from 10^{16} to 10^{7} Ωcm [93]. The same nanocomposite had a 9 %

higher tensile strength and a 90 % higher Young's modulus than the base

polyimide resin. Addition of SWCNTs to this polyimide reduced the elongation-

to-break of the polyimide from 4.5 to 2.6 %, indicating a loss of toughness.

Lebron-Colon and Meador studied the effect of addition of unfunctionalized and

non-covalently functionalized SWCNTs on the electrical and mechanical

properties of a BTDA/BAPP polyimide.[47] While the surface resistivity of the

polyimide film dropped from > 10^{14} to 10^{10} Ωcm with addition of 5 weight %

unfunctionalized SWCNTs, addition of the same amount of non-covalently

functionalized SWCNTs reduced surface resistivity to 10^{5} Ωcm. Addition of

SWCNTs also led to enhanced tensile strength. At a loading level of 0.4 weight

%, the tensile strength of the functionalized SWCNT nanocomposite was twice

that of the neat polyimide, while addition of the same amount of unfunctionalized

SWCNTs only resulted in a 12 % increase in tensile strength. The authors attributed the higher improvements in electrical and mechanical properties of the functionalized SWCNT nanocomposites to better nanotube dispersion. Ounaies and co-workers reported that addition of 0.1 weight % of SWCNTs prepared by laser ablation to LaRC CP-2 polymide increased DC conductivity from 10^{-18} to 10^{-7} Scm^{-1} [94]. Watson and co-workers demonstrated that spray coating of SWCNT suspensions in tetrahydrofuran (THF) or dimethyl acetamide (DMAc) onto a series of polyimide films enhanced their electrical conductivity without adversely affecting optical clarity [95]. The resulting films were fairly robust – no loss in conductivity was observed upon bending, folding or crumpling of the films. The authors suggest that these materials could be useful for coatings to mitigate electrostatic charge build-up in spacecraft.

Carbon nanotubes have also been used to develop "smart" nanocomposites with the capability for self-sensing and self-actuation [96]. Park and co-workers [97] compared the use of MWCNTs and carbon nanofibres as additives to impart load and damage sensing capability to an epoxy matrix, by monitoring changes in the electrical resistivity of the nanocomposite. The authors found that the MWCNT nanocomposite (0.5 volume % MWCNT) had better sensitivity to stress and strain, measured by an electro-pull out test, and was more sensitive to micro-damage than the corresponding nanocomposite made with carbon nanofibre. Boger and co-workers [98] reported the use of DWCNTs and MWCNTs to monitor interlaminar shear stress and tensile stress and strain in a glass fibre-reinforced epoxy composite as well as residual strain after cyclic loading. The electrical resistivity of these samples increased with an applied mechanical load due to disruption of the nanotube percolation networks within the composites. This disruption is an irreversible process. Increases in resistivity were observed under cyclic tensile loading after the fifth cycle.

Yun and co-workers [99] demonstrated SWCNT-epoxy nanocomposite actuators that could be activated electrochemically in an aqueous sodium chloride solution. Several groups have utilized Nafion (sulphonated tetrafluoroethylene copolymer) as a solid electrolyte for nanotube-based actuators [100]. Lee and co-workers [101] described an actuator constructed of a MWCNT-Nafion nanocomposite coated with platinum. The blocking force of the nanocomposite actuator was evaluated as a function of MWCNT loading. The highest blocking force and elastic modulus was observed at a loading level of 1 weight %. Electroactive shape memory effect was recently reported in a MWCNT-polyurethane film that had been coated on either side with polypyrrole [102]. A nanocomposite film with 2.5 weight % MWCNT showed 90-95 % recovery in 20 s under an applied voltage of 25 V.

13.3.2.2   Mechanical properties

Addition of carbon nanotubes to polymers has also been found to lead to improvements in mechanical properties [103]. While some mention of increases in tensile strength and Young's modulus has already been made in this chapter, a few more examples will be given in this section.

Brinson and co-workers have studied the effects of addition of unfunctionalized and functionalized MWCNT on the mechanical properties of a polycarbonate matrix [104]. Mechanical properties of nanocomposites prepared with the functionalized nanotubes were superior to those made with unfunctionalized MWCNT. Addition of 5 weight % functionalized MWCNT led to almost a doubling of the Young's modulus (from 2.0 to 3.8 MPa) and a 30 % increase in tensile yield strength (from 59 to 78 GPa). The ductility of the nanocomposite was poorer than that of the neat resin – strain to failure was reduced from >100 to about 10 %.

Kim and co-workers demonstrated that the crack growth resistance of a carbon fibre-reinforced epoxy composite could be enhanced by the addition of MWCNTs [105]. Addition of 0.7 weight % MWCNTs to the epoxy laminates reduced the crack density after thermomechanical cycling from RT to -150°C by 28 %, and increased the Mode I interlaminar fracture toughness at RT by 6.4 % and by 30.8 % at -150°C. These improved materials could find applications in linerless cryogenic propellant storage tanks.

Improvements in mechanical properties by addition of nanotubes have also been observed with polyimides. One technique that has worked well is to disperse the nanotubes into a suitable organic solvent using ultrasonication, introduce the appropriate monomers, and polymerize the mixture to create the polyimide nanocomposite. Using this *in situ* method, Hu and co-workers demonstrated that addition of 0.5 weight % MWCNTs to an ODA/BPDA polyimide led to a 10% increase in room temperature tensile strength and modulus with no decrease in toughness [106]. Park and co-workers [107] demonstrated that addition of 1 weight % SWCNTs to LaRC CP2 polyimide led to a 40% increase in storage modulus. Lebron-Colon and Meador found that addition of 0.4 weight % of non-covalently functionalized SWCNTs to a BAPP/BPADA polyimide film led to a doubling of its tensile strength [47].

Zhu and co-workers [108] reported that addition of SWCNTs to a glass fibre-reinforced vinyl ester led to significant improvements in Z-axis properties, such as shear. Inclusion of 0.015 weight % SWCNTs into the middle ply of a 10-ply glass fabric-reinforced composite resulted in a 45 % increase in shear strength over the laminate without nanotubes.

Addition of carbon nanotubes has also been shown to improve the friction and wear properties of polymers. Liu and co-workers have reported that addition of 7 weight % of carboxylate functionalized MWCNTs to a bismaleimide reduced its friction coefficient from 0.8 to 0.65 and its wear rate by 45 % [109]. In addition the micro-hardness of this nanocomposite was 30% higher than that of the neat resin.

13.3.3 Carbon nanotube-based fibres

A considerable amount of research has focused on the development of carbon nanotube-based fibres and on carbon nanofibres. Fibres derived from the continuous growth of single wall carbon nanotubes are highly desirable, since it was projected that they would have mechanical properties (strength and modulus) approaching that of individual single wall carbon nanotubes. While a significant amount of progress has been made in producing these materials, the properties of the resulting fibres have fallen short of those of SWCNTs. For example, Zhu and co-workers [110] reported the direct synthesis of SWCNT fibres with lengths ranging from 10-20 cm. They employed a catalytic CVD process (chemical vapour deposition) using a floating catalyst method in a vertical furnace with hexane as the carbon source and a mixture of thiophene and ferrocene as the catalyst. Fibres prepared by this method had a tensile strength of about 1.0 GPa and Young's modulus ranging from 49-77 GPa.

Dry spinning of fibres from vertically aligned arrays of nanotubes has been reported by several groups. Zhang and co-workers [111] described the fabrication of MWCNT fibre yarns by dry spinning of arrays of vertically aligned MWCNT (MWCNT forests). These fibres had tensile strengths as high as 460 MPa. The dry-spinning of fibres from a SWCNT aerogel prepared by CVD has been reported by Motta [112]. Mechanical properties of these fibres were dependent upon both the concentration of iron catalyst and the nature of the hydrocarbon feedstock (hexane, ethylene glycol, or ethanol). The best conditions (0.008 atom % Fe and hexane feedstock) gave fibres with a tensile strength of 1.46 GPa and a tensile modulus of 30 GPa. Peng and co-workers have recently described the preparation of vertically aligned pearl-like carbon nanotube arrays and their dry-spinning into fibres [113]. The authors maintain that the shape of the fibres facilitates the dry spinning process. The maximum tensile strength of fibres obtained from these arrays was 0.35 GPa and could be increased to 1.24 GPa by wetting the fibres with polydiacetylenes.

Fibres have also been obtained by wet spinning of carbon nanotubes in various media. Dalton and co-workers [114] have reported the coagulation spinning of 100 m long yarns of SWCNT/poly(vinyl alcohol) with tensile strengths as high as 1.8 GPa and impact energy absorption ten times that of Kevlar. Fibres of blends of HiPCO SWCNTs and poly(phenylene benzobisoxazole) (PBO) prepared by dry-jet wet spinning from polyphosphoric acid had tensile strengths nearly twice that of fibres made from neat PBO as well as improved tensile modulus and compressive strength [115]. Recently, Zhou and co-workers have reported the *in situ* preparation of a MWCNT-PBO copolymer and its spinning into fibres [116]. The tensile strength and modulus of this fibre were only slightly higher than those of a neat PBO fibre prepared under the same conditions (1.18 vs. 1.48 GPa for strength and 65.6 vs. 70.2 GPa for modulus). The tensile modulus of the fibres increased upon heat treatment in nitrogen at 480°C for 1 minute under a 20 MPa tensile load.

Carbon nanotubes have also been grafted onto the surface of conventional fibres in an attempt to improve interfacial bonding between the fibres and polymer matrixes. Qian and co-workers [117] deposited iron catalyst particles onto the surface of oxidized carbon fibres and then grafted carbon nanotubes onto the fibre surface using a CVD process. Fibre pull-out and push-out tests were conducted to measure the effect of nanotube grafting on the interfacial shear strength with an epoxy matrix. While a significant improvement in pull-out shear strength was observed with the nanotube grafted fibre (118.3 compared to 75.2 MPa for the pristine fibre), fibre push-out shear strengths were nearly identical. The authors indicate the apparent dichotomy in behaviours is due to a different type of failure mechanism for fibre pull-out versus fibre push-out. In the pull-out test, failure occurs at the fibre-matrix interface, while failure in the push-out test occurs via cohesive failure within the fibre itself. Examination of SEM images of failure surfaces after these tests reinforces this explanation.

Mathur [118] has deposited MWCNTs by CVD onto carbon fibres, fabrics and felts and used these to reinforce phenolic composites. Unlike the previously described study, the iron catalyst particles are not coated directly onto the fibre but are injected into the reactor as ferrocene which then thermally decomposes to iron. In these materials, MWCNTs are initially attached to the fibre surface but, as the growth process continued, they became wrapped around the fibre substrate. Flexural strength and modulus are improved in composites made with these reinforcements. Unidirectional composites made with MWCNT coated T-300 carbon fibres (9.1 weight %) had a 20 % higher flexural strength and 28 % higher flexural modulus than composites made with bare carbon fibre. Laminates made with MWCNT coated T-300 carbon plain weave fabric showed a 75 % higher flexural strength and 54% increase in flexural modulus over those made with bare carbon fabric.

Garcia and co-workers [119] have described the *in situ* growth of aligned CNT coatings onto woven alumina cloth. Epoxy composites made with these laminates had 69% higher interlaminar shear strength than those made with bare alumina fabric. Electrical conductivies of the CNT coated fabric laminates were also higher – a 10^6 times increase in in-plane and 10^8-fold improvement in through the thickness electrical conductivity compared to that of the pristine fabric laminates.

13.3.4 Other nanoscale carbon additives

Nanoparticles derived from graphite have also attracted interest in the nanocomposites community because they are plentiful, have high aspect ratios and a platelet geometry similar to that of clays. In addition to enhancing mechanical and gas barrier properties, they also offer the potential to improve electrical and thermal conductivity of materials due to their high degree of electronic conjugation. Naturally occurring graphite (flake graphite) is difficult to exfoliate and disperse into polymers. However, chemical treatment of the graphite can lead to exfoliation and, thereby improve its ability to disperse in polymeric materials. This section will discuss three graphite derivatives – expanded graphite, graphite oxide and functionalized graphene sheets.

13.3.4.1   Expanded graphite and nanocomposites

Expanded graphite is formed by intercalating flake graphite with a mixture of sulfuric or nitric acid and then rapidly heating the intercalated material (called expandable graphite) either in a furnace or a microwave oven to temperatures at or above 900°C to vapourize the acid and, thereby, expand the distance between the graphite sheets. The resulting worm-like material can be broken up into smaller aggregates by ultrasonic treatment in a suitable solvent. The size of these aggregates can be further reduced by ball-milling for extended periods of time [120].

Expanded graphite (EG) has been used as an additive in a variety of polymer nanocomposites. Kim and Macosko have reported the effects of expanded graphite addition on the mechanical, thermal and electrical properties of poly(ethylene-2,6-naphthalate), PEN [121]. In this study, the authors prepared mixtures by cryo-milling up to 20 weight % (12.7 volume %) EG in PEN. These

mixtures were then melt processed in a twin-screw extruder. Addition of EG

resulted in significant improvements in the properties of the polyester. Tensile

modulus increased more than three-fold, with addition of 20 weight % EG (from

2.35 to 7.27GPa). The enhanced stiffness in the material also led to a reduction

in coefficient of thermal expansion from 7.74×10^5 to $2.9 \times 10^5/°C$. Surface

resistivity decreased from 1.4×10^{13} to 2.5×10^5 ohms. Finally the authors found

that addition of EG also led to a 50% reduction in the hydrogen permeability of

the polyester at 10 weight % EG loading.

Zhao and co-workers studied the effects of the addition of EG (sonicated

and non-sonicated) on the electrical conductivity and flexural strength and

modulus of poly(phenylene sulfide), PPS [122]. Nanocomposites were prepared by

melt blending up to 10 weight % EG in PPS at 300°C in a twin-screw extruder.

Sonication of the EG resulted in the production of smaller aggregates which were

easier to disperse in the PPS matrix. This led to nanocomposites with better

properties than corresponding nanocomposites made with EG which had not

been sonicated. The percolation threshold for electrical conductivity was

achieved at an EG loading of about 4 weight %, with a maximum electrical

conductivity of 0.01 Scm^{-1} observed at 10 weight % EG. The flexural strength of

the EG/PPS nanocomposites initially decreased at EG loading levels up to 4

weight %, but then increased at higher loading levels. In all cases, however, the

flexural strengths of the nanocomposites at any of the EG loadings were always

lower than that of the neat resin. Flexural moduli for the sonicated EG nanocomposites were all higher than that of the neat PPS and increased with increasing EG content. Nanocomposites prepared from EG which had not been sonicated showed flexural moduli that initially decreased with increasing EG content (below 4 weight %) and then increased to about the same value as that of the neat PPS.

Du and co-workers [123] studied the effects of EG addition on the properties of pol(4,4'-oxybis(benzene)disulfide), POBDS, synthesized by *in situ* ring-opening polymerization of cyclo(4,4'-oxybis(benzene)disulfide). EG used in this study was prepared by heating acid-intercalated graphite in a microwave and then sonicating the resulting graphite worms in ethanol. The percolation threshold for electrical conductivity was achieved at an EG loading level of about 4 weight %. The electrical conductivity levelled off at a maximum value of 0.1 Scm^{-1} at an EG loading of about 10 weight %. While flexural modulus increased nearly linearly with increasing EG content, flexural strength was insensitive to EG loading. The authors suggest that these nanocomposites could be useful in a variety of aerospace applications including EMI shielding and lightning strike protection and bipolar plates for proton exchange membrane (PEM) fuel cells.

The development of EG nanocomposites for PEM fuel cell bipolar plates has also been reported by Dhakate [124]. The authors prepared EG by heating-acid intercalated graphite in a muffle furnace at 900°C for 10-20 s. This was dispersed in a novolac phenolic resin by mechanically mixing the EG and resin and then grinding the mixture to give moulding powders with uniform EG and resin particles. These mixtures were then compression moulded to give the desired nanocomposites. Nanocomposites were prepared at high EG loading levels (up to 80 weight %) and the effect of nanocomposite properties (electrical conductivity, air permeability, bending strength and modulus and Shore hardness) was studied as a function of EG loading levels. In addition, the synergistic effects of adding up to 10 weight % carbon black, CB, on the nanocomposite properties were also investigated. A comparison of the properties of two formulations with United States Department of Energy (DOE) 2010 target values for composite bipolar plate materials is given in Table 13.2. With the exception of Shore hardness, the properties of both the 55 weight % EG nanocomposite and a 55 weight % + 5 weight % CB nanocomposite exceeded DOE goals.

13.3.4.2   Graphite oxides and nanocomposites

Graphite oxides have also found interest in polymeric nanocomposites. These are typically prepared by oxidation of flake graphite with sulfuric acid and permanganate following a method described by Hummers in 1958 [125]. This process oxidizes not only the edges of the graphite basal planes but also within the basal planes, thereby introducing epoxide and hydroxyl groups and interrupting the π-conjugation within the graphite. As a result, graphite oxide (GO) is yellow in appearance and has poorer electrical conductivity than graphite or other graphene materials. However, the added functionality on the surface and edges of the basal planes makes GO much easier to exfoliate and disperse into a variety of solvents, including water.

There are a limited number of examples where GO has been used directly in polymer nanocomposites. Zhang and co-workers investigated the use of GO as an additive to improve the thermal stability and reduce the flammability of a styrene-butyl acrylate copolymer. The authors found that addition of only 1 weight % GO led to a 5°C increase in the onset of decomposition of the polymer (measured by thermogravimetric analysis, TGA), a 45% reduction in the peak heat release rate (measured by cone calorimetry) as well as significantly lower total smoke production than the base polymer [126]. These nanocomposites could find use in the interiors of aircraft and spacecraft. Graphite oxide has also been dispersed in poly(ethylene oxide) [127], poly(vinyl alcohol) [128]. and conjugated polymers [129].

More recently, graphene nanosheets have been prepared by dispersing GO in a suitable solvent, typically with some type of organic binder, and then reducing the GO *in situ* to graphene. Stankovich and co-workers [130] found that GO could be suspended in water and reduced with hydrazine monohydrate to generate graphene nanosheets. These nanosheets tended to agglomerate as the chemical reduction process progressed, forming porous films with high surface areas (in the order of 446 m^2g^{-1}) and electrical conductivities around 200 Scm^{-1}, about an order of magnitude lower than pristine graphite. The authors suggested that these nanosheets could be used as conductive fillers in nanocomposites and for hydrogen storage.

<table>
<tr><td>

Table 13.2

A comparison of properties of EG nanocomposite formulations with US

</td></tr>
</table>

Department of Energy 2010 goals for composite bipolar plate materials			
Properties	DOE 2010 Target Values	Nanocomposite with 55 weight % EG	Nanocomposite with 55 weight % EG + 5 weight % CB
Bulk density, g/cc		1.55	1.50
Flexural strength, MPa	25	56	52
Flexural modulus, GPa		8	6
Electrical conductivity, Scm^{-1}	100	250	285
Air permeability, MPa	0.5	No leak, 0.78	No leak, 0.78
Shore hardness	50	46	45

Reproduced with permission from reference 124, American Chemical Society, Dhakate, S.R.; Sharma, S.; Borah, M.; Mathur, R.B.; Dhami, T.L. Development and characterization of expanded graphite-based nanocomposite as bipolar plate for polymer electrolyte membrane fuel cells (PEMFCs). *Energy and Fuels* **2008**, 22, 3329-34.

Si and Samulski have reported the synthesis of water soluble graphene by pre-reducing GO with sodium borohydride, followed by sulfonation with an aryldiazionium salt of sulfanilic acid, and treatment with hydrazine to reduce any unreacted oxygen functionality [131]. The authors measured the electrical conductivity of the sulfonated graphene product and found it to be 12.50 Scm^{-1}. These materials could be useful in the development of multifunctional composites and in flat panel displays.

Xu and co-workers have prepared flexible graphene films by *in situ* reduction of an aqueous dispersion of GO and 1-pyrene butyrate, a surfactant [132]. Filtering the resulting graphene complex gave flexible films whose thickness could be regulated by adjusting the volume and concentration of the GO/surfactant mixture. A 30μm film had a modulus of 4.2 GPa and tensile strength of 8.4 MPa, which the authors claim is comparable to that of flexible graphite foils prepared from expanded graphite. Such films could find use in a variety of aerospace applications, including sensors, electromechanical actuators, and field-effect transistors.

Du and co-workers have used a similar approach to prepare graphene nanocomposites [133]. In this work, the authors prepared graphene nanparticles by the reduction of GO in aqueous media with polysulfide ions generated by the reaction of sulfur, sodium sulfide nonahydrate and potassium hydroxide. The resulting graphene nanosheets have sulfur nanoparticles attached to their surface. X-ray analysis of the nanosheets reveals that the interlayer spacings in GO are lost and that the graphenes are exfoliated. The authors suggest that the sulfur nanoparticles on the graphene surfaces create a steric barrier to agglomeration. The electrical conductivity of the sulfur/graphene nanosheets was measured as 0.69 Scm^{-1} suggesting that these materials would be useful as battery electrode materials and additives in electrically conductive nanocomposites. Poly(arylene sulfide) nanocomposites were made with sulfur/graphene nanoparticles by *in situ* polymerization of a cyclic arylene sulfide oligomer. Examination of these nanocomposites by TEM shows that the sulfur/graphene nanosheets are well dispersed and X-ray diffraction reveals a broad peak, suggesting complete exfoliation of the nanosheets within the matrix.

Graphene/polystyrene nanocomposites were prepared by reduction of isocyanate-functionalized GO in a solution of polystyrene in N,N-dimethyl formamide (DMF) [134]. As the reduction of the GO proceeded, the resulting graphene nanparticles became coated with polystyrene which prevented their agglomeration and kept them well dispersed in solution. After the reduction was complete, the solvent was removed under vacuum, and the resulting material was ground up and then melt processed. Nanocomposites produced by this procedure showed a percolation threshold at 0.1 volume % graphene and a maximum electrical conductivity of 0.1 Scm^{-1}.

13.3.4.3   Functionalized graphene sheets and nanocomposites

The oxygen functionality within the basal planes of GO is thermally labile and can be decomposed by rapid heating (>2000°C/min) to 1050°C to liberate CO_2 and form a graphene-type material having defects and vacancies within the basal plane [135]. The resulting material, known as functionalized graphene sheets (FGS), has a wrinkled appearance compared to the planar, platelet-like structure of graphite. The surface area of FGS, measured by the Brunnauer-Emmett-

Teller method (BET), is in the order of 600-800 m^2g^{-1} compared with 100 m^2g^{-1}

for EG[120] and about 450 m^2g^{-1} for graphene nanosheets prepared by reduction of

GO.[130] XPS analysis (x-ray photoelectron spectroscopy) of FGS reveals that

these materials still retain some oxygen functionality.

Since the development of FGS is fairly recent, there are only a few

examples of its use as a nanoscale filler in polymers. Kim and Macosko [120]

compared the effects of adding graphite and FGS on the properties of a

polyester, poly(ethylene 2,6-naphthalate), PEN. Comparison of TEM

photomicrographs of graphite and FGS nanocomposites showed that the FGS

was much better dispersed in the matrix resin. Electrical conductivity

measurements revealed that the percolation threshold for the FGS

nanocomposites was achieved an FGS loading level roughly one-tenth of that in

the graphite composites. The tensile moduli of FGS nanocomposites were

higher than those of the graphite composites at comparable loading levels due to

the significantly higher aspect ratio and better exfoliation of FGS. Similarly, the

gas barrier performances of FGS nanocomposites were superior to those of the

graphite-based materials. Ramanathan and co-workers [136] studied the effects of

addition of SWCNTs, EG and FGS on the properties of a PMMA matrix. At a

comparable loading level of 1 weight %, FGS produced the nanocomposites with

the highest Young's modulus, Tg, and ultimate tensile strength, and the highest

onset of decomposition.

13.4   Conclusions

Addition of a variety of nanoscale additives to several polymer systems has been shown to lead to enhancements in mechanical, thermal and electrical properties. It should be pointed out that these additives are not a panacea, since they do not always provide across-the-board enhancements in the properties of a given matrix resin.   Changes in properties of nanocomposites are highly dependent upon the nature of the nanoscale additive and polymer, as well as the degree of interaction between the additive and polymer. Thus, functionalization and processing are critical parameters that must be properly addressed to produce nanocomposites with optimal properties. NASA's National Nanotechnology Initiative Grand Challenge Workshop held in 2004 identified the ability to produce nanostructured materials "with controlled morphology and structure over a variety of length scales" as one of the key technical challenges to the use of these materials in aerospace applications [137].

As the field of nanocomposites has matured, a greater emphasis is being placed on controlling nanoparticle dispersion and alignment in an attempt to tailor or optimize material properties for a given application [138]. Several approaches are being explored to develop nanocomposites with controlled morphology.

The use of externally applied magnetic or electric fields to align nanoparticles is beginning to receive attention. For example, Koerner [139] has reported the use of a strong magnetic field to align montmorillonite nanoparticles within an epoxy matrix. The resulting materials show a directionality in coefficient of thermal expansion, where the CTE reduction is the greatest in the direction of the magnetic field.

The use of mechanical forces such as shear to enhance dispersion and/or alignment of nanoparticles has also been studied. Wakabayshi and co-workers have reported the production of polypropylene nanocomposites with highly dispersed graphite nanofiller by solid state shear pulverization followed by compression moulding [140]. Nanocomposites prepared with 2.5 weight % graphite had twice the Young's modulus of polypropylene, and a 60% higher yield strength with only a 30% reduction in elongation at break.

Novel fabrication methods, such as layer-by-layer assembly (LBL), are also being employed in the fabrication of nanocomposites with impressive results. Podsiadlo and co-workers [141] used LBL assembly to fabricate a synthetic mimic of nacre with exceptional mechanical properties. In this work, poly(vinyl alcohol)/Na^+-montmorillonite nanocomposites prepared by LBL assembly were crosslinked by addition of various metal ions. Despite the fact that these films had high clay loadings, as much as 70 weight %, they were highly transparent due to the nanometre scale thicknesses of individual nanocomposite layers within the films. SEM photomicrographs of these films indicated a high degree of alignment of montmorillonite galleries. The tensile behaviour of these films had a saw-tooth failure pattern similar to that of nacre, with ionic crosslinks formed by addition of metal ions acting as sacrificial bonds. Nanocomposite films crosslinked with Cu^{2+} ions had an ultimate tensile strength of 320 MPa, more than twice that of natural nacre, and a modulus of 58 GPa. The ultimate strain of this film was lower than that of nacre – 0.28% compared to 0.8%. LBL assembly has also been used by Shim and Kotov in the preparation of SWCNT/poly(vinyl alcohol) nanocomposites with a high degree of nanotube alignment [142].

Self-assembly and other molecularly directed processes are also being exploited as a means of controlling nanoparticle dispersion and alignment. For example, Orefice and co-workers have employed phase-separation/self-assembly in segmented polyurethanes as a means of templating nanocomposites with structurally tailored silica networks [143]. In this approach, aliphatic polyurethanes end-capped with triethoxysilane groups were prepared with varying ratios of hard and soft blocks. These silane end-caps were then hydrolyzed to silica by addition of water. Small angle X-ray studies revealed that the silica moieties enhanced the microphase separation within the polymer films and tended to organize in the hard segment regions of the films. Desmukh and co-workers have utilized a symmetrical poly(styrene)/PMMA block copolymer to prepare highly aligned gold nanorod composites in which the gold nanorods are confined in the lamellar PMMA domains [144].

Studies such as those discussed above will enable the development of nanocomposites with improved properties and reliability. However, for nanocomposites to find widespread use in aircraft and spacecraft, issues concerning their reliability, durability and inspectability must be addressed. More robust manufacturing techniques must be developed. Inspection methods must be developed to enable real time measurements of nanocomposite quality during manufacturing, as well as means to inspect for damage and degradation during use. In addition, analytical tools must be developed that can reliably predict the behaviour of nanocomposites. While much work remains to be done to enable the use of nanocomposites in aerospace systems, the potential payoffs in weight savings and performance justify continued investment in research in these materials.

References

1.	Hileman, J.I.; Spakovsky, Z.S.; Drela, M.; Sargeant, M.A. Airframe design for "Silent Aircraft". *Proceedings of the 45th Association for Aeronautics and Astronautics Meeting and Exhibition* **2007**, Paper AIAA 2007-0453.

2.	Blumstein, A. Polymerization of adsorbed monolayers. I. Preparation of the clay-polymer complex. *Journal of Polymer Science A* **1965**, *3*, 2653-64.

3.	Kojima, Y.; Usuki, A,; Kawasumi, M.; Okada, A.; Fukushima, Y.; Karauchi, T.; Kamigaito, O. Mechanical properties of nylon 6-clay hybrid. *Journal of Materials Research* **1993**, *8*, 1185-89.

4.	Tjong, S.C. Structural and mechanical properties of polymer nanocomposites. *Materials Science and Engineering R: Reports* **2006**, *53*, 73-197.

5.	Balakrishnan, S.; Start, P.R.; Raghavan, D.; Hudson, S.D. The influence of clay and elastomer concentration on the morphology and fracture energy of preformed acrylic rubber dispersed clay filled epoxy nanocomposites. *Polymer* **2005**, *46*, 11255-11262.

6.	Meng, J.; Hu, X. Synthesis and exfoliation of bismaleimide-organoclay nanocomposites. *Polymer* **2004**, *45*, 9011-18.

7.	Yano, K.; Usuki, A.; Okada, A.; Kurauchi, T. Kamigaito, O. Synthesis and properties of polyimide-clay hybrid. *Journal of Polymer Science, Polymer Chemistry Edition*, **1993**, *31*, 2493-98.

8.	Yasmin, A.; Luo, J.J.; Abot, J.L.; Daniel, I.M. Mechanical and thermal behavior of clay/epoxy nanocomposites. *Composites Science and Technology* **2006**, *66*, 2415-22

9.	Giannelis, E. P. Polymer layered silicate nanocomposites. *Advanced Materials* **1996**, *8*, 29-35.

10. Ray, S.S.; Okamoto, M. Polymer/layered silicate nanocomposites: a review from preparation to processing, *Progress in Polymer Science* **2003**, *28*, 1539-1641.

11. Heydenreich, R. Cyrotanks in future vehicles. *Cryogenics* **1998**, *38*, 125-30.

12. Miller, S.G.; Meador, M.A. , *Proc. of the 48th AIAA/ASME/ASCE/AHS/ACS Structures, Structural Dynamics, and Materials Conference*, April 23-26, 2007, Honolulu, HI.

13. Nielson, L.E. Models for permeability of filled polymer systems. *Journal of Macromolecular Science, Part A: Pure and Applied Chemistry* **1967**, *1*, 929-42.

14. Triantafyllidis, K.S.; LeBaron, P.C.; Park, I.; Pinnavaia, T.J. Epoxy-clay fabric film composites with unprecedented oxygen barrier properties. *Chemistry of Materials* **2006**, *18*, 4393-98.

15. Yang, J-P; Yang, G.; Xu, G.; Fu, S-Y Cyrogenic mechanical behaviors of MMT/epoxy nanocomposites. *Composites Science and Technology* **2007**, *67*, 2934-40.

16. Zhang, Y-H; Wu, J-T; Fu, S-Y; Li, Y.; Fan, L.; Li, R. K-Y; Li, L-F; Yan, Q. Studies on characterization and cryogenic mechanical properties of polyimide-layered silicate nanocomposite films. *Polymer* **2004**, 45, 7579-87

17. Johnson, L.L.; Eby, R.K.; Meador, M.A.B., Investigation of oxidation profile in PMR-15 polyimide using atomic force microscope (AFM). *Polymer* **2003**, *44*, 187-197.

18. Campbell, S.G.; Scheiman, D.A., Orientation of aromatic ion exchange diamines and the effect on the melt viscosity and thermal stability of PMR-15/Silicate nanocomposites. *High Performance Polymers* **2002**, *14*, 17-30.

19. Xie, W.; Xie, R.; Pan, W-P; Hunter, D.; Koene, B.; Tan, L-S, Vaia, R., Thermal stability of quaternary phosphonium modified montmorillonites. *Chemistry of Materials* **2002**, *14*, 4837–45.

20. Kurose, T.; Yudin, V.E.; Otaigbe, J.U.; Svetlichnyi, V.M. Compatibilized polyimide (R-BAPS)/BAPS-modified clay nanocomposites with improved dispersion and properties. *Polymer* **2007**, *48*, 7130-38.

21.   Tyan, H.-L.; Leu, C.-M.; Wei, K.-H. Effect of reactivity of organics-modified montmorillonite on the thermal and mechanical properties of montmorillonite/polyimide nanocomposites. *Chemistry of Materials* **2001**, *13*, 222-26.

22.   Delozier, D.M.; Orwoll, R.A.; Cahoon, J.F.; Ladislaw, J.S.; Smith, J.G., Jr.; Connell, J.W. Polyimde nanocomposites prepared from high-temperature reduced charge organoclays. *Polymer* **2003**, *44*, 2231-41.

23.   Campbell, S.G.; Liang, M.I. High temperature thermosetting polyimide nanocomposites prepared with reduced charge organoclay. *High Performance Polymers* **2006**, 18, 71-82.

24.   Gilman, J.W.; Jackson, C.L.; Morgan, A.B.; Harris, R., Jr.; Manias, E.; Giannelis, E.P.; Wuthenow, M.; Hilton, D.; Phillips, S.H. Flammability properties of polymer-layered silicate nanocomposites. Polypropylene and polystyrene nanocomposites. *Chemistry of Materials* **2000**, *12*, 1866-73.

25.   Ijima, S. Helical microtubules of carbon. *Nature* **1991**, *354*, 56-58.

26.   Ajayan, P.M. Nanotubes from carbon. *Chemical Reviews* **1999**, *99*, 1787- 99.

27.   Harris, C.E.; Shuart, M.J.; Gray, H.R., 2002, A Survey of Emerging Materials   for Revolutionary Aerospace Vehicle Structures and Propulsion Systems, NASA TM-2002-21164.

28.   Dresselhaus, M.S.; Dresselhaus, G. and Saito, R., 1998, "Nanotechnology in Carbon Materials" in Nanotechnology, Timp, G.L.; ed., Springer, 285-329.

29.   Qian, D.; Wagner, G.J.; Liu, W.K.; Yu, M-F; Ruoff, R.S. Mechanics of carbon nanotubes. *Applied Mechanics Reviews* **2002**, *55*, 495-533.

30.   Srivastava, D.; Wei, C.; Cho, K. Nanomechanics of carbon nanotubes and composites. *Applied Mechanics Reviews* **2003**, *56*, 215-230.

31.   Troiani, H.E.; Miki-Yoshida, M.; Camacho-Bragado, G.A.; Marques, M.A.L.; Rubio, A.; Ascencio, J.A.; Jose-Yacaman, M. Direct observation of the mechanical properties of single-walled carbon nanotubes and their junctions at the atomic level. *Nano Letters* **2003**, *3*, 751-55

32. Yu, M.; Lourie, O.; Dyer, M.J.; Moloni, K., Kelly, T.F.; Ruoff, R.S. Strength and breaking mechanism of multiwalled carbon nanotubes under tenisile loading. *Science* **2000**, *287*, 637-40.

33. Yu, M.F.; Files, B.S.; Arepalli, S.; Ruoff, R.S. Tensile loading of ropes of single wall carbon nanotubes and their mechanical properties. *Physical Reviews Letters* **2000**, *84*, 5552-5.

34. Xiao, T.; Ren, Y.; Liao, K.; Wu, P.; Li, F.; Cheng, H.M. Determination of tensile strength distribution of nanotubes from testing of nanotube bundles. *Composites Science and Technology* **2008**, 68, 2937-2942.

35. Popov, V.N. Carbon nanotubes: properties and application. *Materials Science and Engineering R: Reports* **2004**, *43*, 61-102

36. Ebbesen, T.W; Ajayan, P.M. Large-scale synthesis of carbon nanotubes. *Nature* **1992**, *358*, 220 -222.

37. Ijima, S.; Ichahashi, T. Single shell carbon nanotubes of 1 nm diameter. *Nature* **1993**, *363*, 603-605.

38. Bethune, D.S.; Kiang, C.H.; de Vries, M.S.; Gorman, G.; Savoy, R.; Vazquez, J.; Beyers, R. Cobalt-catalysed growth of carbon nanotubes with single atomic layer walls. *Nature* **1993**, *363*, 605 -07

39. Journet, C.; Maser, W.K.; Bernier, P.; Loiseau, A.; Lamy de la Chapelle, M.; Lefrant, S.; Deniard, P.; Lee, R.; Fisher, J.E. Large scale production of single-walled carbon nanotubes by the electric-arc technique. *Nature* **1997**, *388*, 756

40. Thess, A.; Lee, R.; Nikolaev, P.; Dai, H.; Petit, P.; Robert, J.; Xu, C.; Lee, Y.H.; Kim, S.G.; Rinzler, A.G.; Colbert, D.T.; Scuseria, G.E.; Tomanek, D.; Fischer, D.E.; Smalley, R.E. Crystalline ropes of metallic carbon nanotubes. *Science* **1996**, *273*, 483

41. Vander Wal, R.L.; Ticich, T.M. Flame and furnace synthesis of single-walled and multi-walled carbon nanotubes and nanofibers. *Journal of Physical Chemistry B.* **2001**, *105*, 10249-256.

42. Nikolaev, P.; Bronikowski, M.J.; Bradley, R.K; Rohmund, F.; Colbert, D.T.; Smith, K.A.; Smalley, R.E. Gas-phase catalytic growth of single-walled carbon nanotubes from carbon monoxide. *Chemical Physics Letters* **1999**, *313*, 91-97

43.   Resasco, D.E.; Alvarez, W.E.; Pompeo, F.; Balzano, L.; Herrera, J.E.; Kitiyanan, B.; and Borgna, A. A scalable process for production of single-walled carbon nanotubes (SWNTs) by catalytic disproportionation of CO on a solid catalyst. *Journal of Nanoparticle Research* **2002**, *4*, 131-36

44.   Bachilo, S.M.; Balzano, L.; Herrera, J.E.; Pompeo, F.; Resasco, D.E.; Weisman, R.B. Narrow (n,m) distribution of single-walled carbon nanotubes grown using a solid supported catalyst. *Journal of the American Chemical Society* **2003**, *125*, 11186-11187.

45.   Chiang, I.W.; Brinson, B.E.; Huang, A.Y.; Willis, P.A.; Bronikowski, M.J.; Margrave, J.L.; Smalley, R.E.; Hauge, R.H. Purification and characterization of single-wall carbon nanotubes (SWNTs) obtained from the gas-phase decomposition of CO) (HiPco process). *Journal of Physical Chemistry B* **2001**, *105*, 8297-301.

46.   Lebron-Colon, M., 2004, PhD Thesis, Clark Atlanta University.

47.   Lebron-Colon, M.; Meador, M.A. Modified single-wall carbon nanotubes for reinforced thermoplastic polyimide. *Proceedings of the 2006 SAMPE International Symposium*. 2006

48.   Xu, Y-Q, Peng, H.; Hauge, R.H.; Smalley, R.E. Controlled multistep purification of single-walled carbon nanotubes. *Nano Letters* **2005**, *5*, 163-68

49.   Kim, Y.; Luzzi, D.E. Purification of pulsed laser synthesized single wall carbon nanotubes by magnetic filtration. *Journal of Physical Chemistry B* **2005**, *109*, 16636-43

50.   Min, Y-S; Bae, E.J.; Park, W. Sulfidative purification of single wall carbon nanotubes integrated in transistors. *Journal of the American Chemical Society* **2005**, *127*, 8300-8301

51.   Vivekchand, S.R.C.; Govindaraj, A.; Motin Seikh, Md.; Rao, C.N.R. New method of purification of carbon nanotubes based on hydrogen treatment. *Journal of Physical Chemistry B* **2004**, *108*, 6935-37.

52.   Miyata, Y.; Maniwa, Y.; Kataura, H. Selective oxidation of semiconducting single-wall carbon nanotubes by hydrogen peroxide. *Journal of Physical Chemistry B* **2006**, *110*, 25-29.

53. Ramesh, S.; Shan, H.; Haroz, E.; Billups, W.E.; Hauge, R.; Adams, W.W.; Smalley, R.E. Diameter selection of single-walled carbon nanotubes through programmable solvation in binary sulfonic acid mixtures. *Journal of Physical Chemistry C* **2007**, *111*, 17827-34

54. Zheng, M.; Semke, E.D. Enrichment of single chirality carbon nanotubes. *Journal of the American Chemical Society* **2007**, *129*, 6084-85.

55. Chen, F.; Wang, B.; Chen, Y.; Li, L-J. Toward the extraction of single-walled carbon nanotubes using fluorine-based polymers. *Nano Letters* **2007**, *7*, 3013-3017

56. Wei, L.; Wang, B.; Goh, T.H.; Li, L-J, Yang, Y.; Chan-Park, M.B.; Chen, Y. Selective enrichment of (6,5) and (8,3) single-walled carbon nanotubes via cosurfactant extraction from narrow (n,m) distribution samples. *Journal of Physical Chemistry B* **2008**, 112, 2771-4.

57. Tasis, D.; Tagamatarchis, N.; Bianco, A.; Prato, M. Chemistry of carbon nanotubes. *Chemical Reviews* **2006**, *106*, 1105-36

58. Liu, J.; Rinzler, A.G.; Dai, H.J.; Hafner, J.H.; Bradley, R.K.; Boul, P.J.; Lu, A.; Iverson, T.; Shelimov, K.; Huffman, C.B.; Rodriguez-Macias, F.; Shon, Y.S.; Lee, T.R.; Colbert, D.T.; Smalley, R.E. Fullerene pipes. *Science* **1998**, *280*, 1253-1256.

59. Hamon, M.A.; Hui, H.; Bhowmik, P.; Itkis, H.M.E.; Haddon, R.C. Ester-functionalized soluble single-walled carbon nanotubes. *Applied Physics A* **2002**, *74*, 333-38.

60. Gao, J.; Itkis, M.E.; Yu, A.; Bekyarova, E.; Zhao, B.; Haddon, R.C. Continuous spinning of a single-walled carbon nanotube-nylon composite fiber. *Journal of the American Chemical Society* **2005**, *127*, 3847-54

61. Qu, L.; Lin, Y.; Hill, D.E.; Zhou, B.; Wang, W.; Sun, X.; Kitaygorodsky, A.; Suarez, M.; Connell, J.W.; Allard, L.F.; Sun, Y.-P. Polyimide-functionalized carbon nanotubes: synthesis and dispersion in nanocomposite films. *Macromolecules* **2004**, *37*, 6055-60.

62. Bahr, J.L.; Yang, J.; Kosynkin, D.V.; Bronikowski, M.J.; Smalley, R.E.; Tour, J.M. Functionalization of carbon nanotubes by electrochemical reduction of aryl diazonium salts: a bucky paper

electrode. *Journal of the American Chemical Society* **2001**, *123*, 6536-42

63.   Dyke, C.A.; Tour, J.M. Solvent-free functionalization of carbon nanotubes. *Journal of the American Chemical Society* **2003**, *125*, 1156-57

64.   Hu, H.; Zhao, B.; Hamon, M.A.; Kamaras, K.; Itkis, M.E.; Haddon, R.C. Sidewall functionalization of single-walled carbon nanotubes by addition of dichlorocarbene. *Journal of the American Chemical Society* **2003**, *125*, 14893-14900

65.   Holzinger, M.; Vostrowksy, O.; Hirsch, A.; Hennrich, F.; Kappes, M.; Weiss, R.; Jellen, F. Sidewall functionalization of carbon nanotubes. *Angewandte Chemie, International Edition* **2001**, *40*, 4002-06.

66.   Delgado, J.L.; de la Cruz, P.; Langa, F.; Urbina, A.; Casado, J.; Navarrete, J.T.L. Microwave assisted sidewall functionalization of single-wall carbon nanotubes by Diels-Alder cycloaddition. *Chemical Communications* **2004**, *15*, 1734-35.

67.   Sadana, A,; Liang, L.; Brinson, B.; Arepalli, S.; Farhat, S.; Hauge, R.H.; Smalley, R.E.; Billups, W.E. Functionalization and extraction of large fullerenes and carbon-coated metal formed during the synthesis of single wall carbon nanotubes by laser oven, direct current arc, and high-pressure carbon monoxide production methods. *Journal of Physical Chemistry B* **2005**, *109*, 4416-18

68.   Mickelson, E.T.; Huffman, C.B.; Rinzler, A.G.; Smalley, R.E.; Hauge, R.H.; Margrave, J.L. Fluorination of single-wall carbon nanotubes. *Chemical Physics Letters* **1998**, *296*, 188-194

69.   Saini, R.K.; Chiang, I.W.; Peng, H.; Smalley, R.E.; Billups, W.E.; Hauge, R.H.; Margrave, J.L. Covalent sidewall functionalization of single wall carbon nanotubes. *Journal of the American Chemical Society* **2003**, *125*, 3617-21

70.   Zhang, L.; Kiny, V.U.; Peng, H.; Zhu, J.; Lobo, R.F.M.; Margrave, J.L.; Khabashesku, V.N. Sidewall functionalization of single-walled carbon nanotubes with hydroxyl group-terminated moieties. *Chemistry of Materials* **2004**, *16*, 2055-61

71.   Stevens, J.L.; Huang, A.Y.; Peng, H.; Chiang, I.W.; Khabashesku, V.N.; Margrave, J.L. Sidewall amino-functionalization of single-walled carbon nanotubes through

fluorination and subsequent reactions with terminal diamines. *Nano Letters* **2003**, *3*, 331-36

72. Khabashesku, V.N.; Margrave, J.L.; Barrera, E.V. Functionalized carbon nanotubes and nanodiamonds for engineering and biomedical applications. *Diamond & Related Materials* **2005**, *14*, 859-866

73. Liang, F.; Sadana, A.K.; Peera, A.; Chattopadhyay, J.; Gu, Z.; Hauge, R.H.; Billups, W.E. A convenient route to functionalized carbon nanotubes. *Nano Letters* **2004**, *4*, 1257-60

74. Liang, F.; Beach, J.M.; Kobashi, K.; Sadana, A.K.; Vega-Cantu, Y.I.; Tour, J.M.; Billups, W.E. In situ polymerization initiated by single-walled carbon nanotube salts. *Chemistry of Materials* **2006**, *18*, 4764-67

75. Guldi, D.M.; Rahman, G.M.A.; Zerbetto, F.; Prato, M. Carbon nanotubes in electron donor-acceptor nanocomposites. *Accounts of Chemical Research* **2005**, *38*, 871-78

76. Chen, R.J.; Zhang, Y.; Wang, D.; Dai, H. Noncovalent sidewall functionalization of single-walled carbon nanotubes for protein immobilization. *Journal of the American Chemical Society* **2001**, *123*, 3838-3839

77. Lou, X.; Daussin, R.; Cuenot, S.; Duwez, A.-S.; Pagnoulle, C.; Detrembleur, C.; Bailly, C.; Jerome, R. Synthesis of pyrene-containing polymers and noncovalent sidewall functionalization of multiwalled carbon nanotubes. *Chemistry of Materials* **2004**, *16*, 4005-4011

78. Wang, D.; Ji, W.-X.; Li, Z.-C.; Chen, L. A biomimetic "polysoap" for single-walled carbon nanotube dispersion. *Journal of the American Chemical Society* **2006**, *128*, 6556-57

79. Zhang, J.; Lee, J.-K.; Wu, Y.; Murray, R.W. Photoluminescence and electronic interaction of anthracene derivatives adsorbed on sidewalls of single-walled carbon nanotubes . *Nano Letters* **2003**, *3*, 403-407.

80. Li, H.; Zhou B.; Lin, Y.; Gu, L.; Wang, W.; Fernando, K.A.S.; Kumar, S.; Allard, L.F.; Sun, Y. Selective interactions of porphyrins with semiconducting single-walled carbon nanotubes. *Journal of the American Chemical Society* **2004**, *126*, 1014-5.

81.   Satake, A.; Miyajima, Y.; Kobuke, Y. Porphyrin-carbon nanotube composites formed by noncovalent polymer wrapping. *Chemistry of Materials* **2005**, *17*, 716-724

82.   Star, A.; Stoddart, J.F.; Steuerman, D.; Diehl, M.; Boukai, A.; Wang, E.W.; Yang, X.; Chung, S.-W.; Choi, H.; Heath, H.R. Preparation and properties of polymer-wrapped single-walled carbon nanotubes. *Angewandte Chemie, International Edition* **2001**, *40*, 1721-1725

83.   Tang, B. Z.; Xu, H. Preparation, alignment and optical properties of soluble poly(phenylacetylene)-wrapped carbon nanotubes. *Macromolecules* **1999**, *32*, 2569-76.

84.   Star, A.; Stoddart, J.F. Dispersion and solubilization of single-walled carbon nanotubes with a hyperbranched polymer. *Macromolecules* **2002**, *35*, 7516-20

85.   Star, A.; Liu, Y.; Grant, K.; Ridvan, L.; Stoddart, J.F.; Steuerman, D.W.; Diehl, M.R.; Boukai, A.; Heath, J.R. Noncovalent side-wall functionalization of single-walled carbon nanotubes. *Macromolecules* **2003**, *36*, 553-60

86.   O'Connell, M.J.; Boul, P.; Ericson, L.M.; Huffman, C.; Wang, Y.; Haroz, E.; Kuper, C.; Tour, J.; Ausman, K.D.; Smalley, R.E. Reversible water solubilization of single-walled carbon nanotubes by polymer wrapping. *Chemical Physics Letters* **2001**, *342*, 265-71

87.   Delozier, D.L. ; Tigelaar, D.M.; Watson, K.A.; Smith Jr., J.G.; Klein, D.J.; Lillehei, P.T.; Connell, J.W. Investigation of ionomers as dispersants for single wall carbon nanotubes. *Polymer* **2005**, *46*, 2506-21.

88.   Sinani, V.A.; Gheith, M.K.; Yaroslavov, A.A.; Rakhnyanskaya, A.A.; Sun, K.; Mamedov, A.A.; Wicksted, J.P.; Kotov, N.A. Aqueous dispersion of single-wall and multiwall nanotubes with designed amphiphilic polycations. *Journal of the American Chemical Society* **2005**, *127*, 3463-72

89.   Thostenson, R.T.; Chou, T.W. Processing-structure-multi-functional property relationship in carbon nanotube/expoy composites. *Carbon* **2006**, *44*, 3022-29

90.   Moisala, A.; Li, Q.; Kinloch, I.A.; Windle, A.H. Thermal and electrical conductivity of single- and multi-walled carbon

nanotube-epoxy composites. *Composites Science and Technology* **2006**, *66*, 1285-88.

91.   Wang, Q.; Dai, J.; Li, W.; Wei, Z.; Jiang, J. The effects of CNT alignment on electrical conductivity and mechanical properties of SWNT/epoxy nanocomposites. *Composites Science and Technology* **2008**, *68*, 1644-48

92.   Ghose, S.; Watson, K.A.; Working, D.C.; Connell, J.W.; Smith Jr., J.G.; Sun, Y.P. Thermal conductivity of ethylene vinyl acetate copolymer/nanofiller blends. *Composites Science and Technology* **2008**, *68*, 1843-53

93.   Yu, A.; Hu, H.; Bekyarova, E.; Itkis, M.E.; Gao, J.; Zhao, B.; Haddon, R.C. Incorporation of highly dispersed single-walled carbon nanotubes in a polyimide matrix. *Composites Science and Technology* **2006**, *66*, 1190-97

94.   Ounaies, Z.; Park, C.; Wise, K.E.; Siochi, E.J.; Harrison, J.S. Electrical properties of single wall carbon nanotube reinforced polyimide composites. *Composites Science and Technology* **2003**, *63*, 1637-46

95.   Watson, K.A.; Ghose, S.; Delozier, D.M.; Smith Jr., J.G.; Connell, J.W. Transparent, flexible, conductive carbon nanotube coatings for electrostatic charge mitigation. *Polymer* **2005**, *46*, 2076-85

96.   Kang, I.; Heung, Y.Y.; Kim, J.H.; Lee, J.W.; Gollapudi, R.; Subramaniam, S.; Narasimhadevara, S.; Hurd, D.; Kirikera, G.R.; Shanov, V.; Schulz, M.J.; Shi, D.; Boerio, J.; Mall, S.; Ruggles-Wren, M. Introduction to carbon nanotube and nanofiber smart materials. *Composites, B* **2006**, *37*, 382-94.

97.   Park, J-M.; Kim, D.-S.; Kim, S.-J.; Kim P.-G.; Yoon, D.-J.; DeVries, K.L. Inherent sensing and interfacial evaluation of carbon nanofiber and nanotube/epoxy composites using electrical resistance measurement and micromechanical technique. *Composites, B* **2007**, *38*, 847-61.

98.   Boger, L.; Wichmann, M.H.G.; Meyer, L.O., Schulte, K. Load and health monitoring in glass fibre reinforced composites with an electrically conductive nanocomposite epoxy matrix. *Composites Science and Technology* **2008**, *68*, 1886-94.

99.   Yun, y-H; Shanov, V.; Schulz, M.J.; Narasimhadevara, S.; Subramaniam, S.; Hurd, D.; Boerio, F.J. Development of novel

single-wall carbon nanotube-epoxy composite ply actuators. *Smart Materials and Structures* **2005**, *14*, 1526-32.

100. Li, C.; Thostenson, E.T.; Chou, T-W. Sensors and actuators based on carbon nanotubes and their composites: a review. *Composites Science and Technology* **2008**, *68*, 1227-49.

101. Lee, D.Y.; Lee, M-H; Kim, K.J.; Heo, S.; Kim, B-Y; Lee, S-J. Effect of multiwalled carbon nanotube (M-CNT) loading on M-CNT distribution behavior and the related electromechanical properties of the M-CNT dispersed ionomeric nanocomposites. *Surface and Coating Technology* **2005**, *200*, 1920-25

102. Sahoo, N.G.; Jung, Y.C.; Yoo, H.J.; Cho, J.W. Influence of carbon nanotubes and polypyrrole on the thermal, mechanical and electroactive shape-memory properties of polyurethane nanocomposites. *Composites Science and Technology* **2007**, *67*, 1920-29.

103. Coleman, J.N.; Khan, U.; Blau, W.J.; Gun'ko, Y.K. Small but strong: A review of the mechanical properties of carbon nanotube-polymer composites. *Carbon* **2006**, *44*, 1624-52.

104.   Eitan, A.; Fisher, F.T.; Andrews, R.; Brinson, L.C.; Schadler, L.S. Reinforcement mechanisms in MWCNT-filled polycarbonate. *Composites Science and Technology* **2006**, *66*, 1162-73

105. Kim, M-G; Hong, J-S; Kang, S-G; Kim, C-G. Enhancement of the crack growth resistance of a carbon/epoxy composite by adding multi-walled carbon nanotubes at a cryogenic temperature. *Composites A* **2008**, *39*, 647-54.

106.   Hu, N.; Zhou, H.; Dang, G.; Rao, X.; Chen, C.; Zhang, W. Efficient dispersion of multi-walled carbon nanotubes by in situ polymerization. *Polymer International* **2007**, 56, 655-59.

107.   Park, C.; Ounaies, Z.; Watson, K.A.; Crooks, R.E.; Smith Jr, J.; Lowther, S.E.; Connell, J.W.; Siochi, E.J.; Harrison, J.S.; St. Clair, T.L. Dispersion of single wall carbon nanotubes by in situ polymerization under sonication. *Chemical Physics Letters* **2002**, *364*, 303-308.

108.   Zhu, J.; Imam, A.; Crane, R.; Lozano, K.; Khabashesku, V.N.; Barrera, E.V. Processing a glass fiber reinforced vinyl ester composite with nanotube enhancement of interlaminar shear

strength. *Composites Science and Technology* **2007**, *67*, 1509-17.

109. Liu, L.; Gu, A.; Fang, Z.; Tong, L.; Xu, Z. The effects of the variations of carbon nanotubes on the micro-tribological behavior of carbon nanotubes/bismaleimide nanocomposite. *Composites A* **2007**, *38*, 1957-64.

110. Zhu, H.W.; Xu, C.L.; Wu, D.H.; Wei, B.Q.; Vajtai, R., Ajayan, P.M. Direct Synthesis of Long Single-Walled Carbon Nanotube Strands. *Science* **2002**, *296*, 884-86.

111. Zhang, M.; Atkinson, K.R.; Baughman, R.H. Multifunctional Carbon Nanotube Yarns by Downsizing an Ancient Technology. *Science* **2004**, *306*, 1358-61.

112. Motta, M.; Li, Y-L; Windle, A. Mechanical Properties of Continuously Spun Fibers of Carbon Nanutobes. *Nano Letters* **2005**, *5*, 1529-33.

113. Peng, H.; Jain, M.; Li, Q.; Peterson, D.E.; Zhu, Y.; Jai, Q. Vertically aligned pearl-like carbon nanotube arrays for fiber spinning. *Journal of the American Chemical Society* **2008**, *130*, 1130-31.

114. Dalton, A.B.; Collins, S.; Munoz, E.; Razal, J.M.; Ebron, V.H.; Ferraris, J.P.; Coleman, J.M.; Kim, B.G.; Baughman, R.H. Super-tough carbon-nanotube fibers. *Nature* **2003**, *423*, 703.

115. Kumar, S.; Dang, T.; Arnold, F.E.; Bhattacharyya, A.R.; Min, B.G.; Zhang, X.; Vaia, R.A.; Park, C.; Adams, W.W.; Hauge, R.H.; Smalley, R.E.; Ramesh, S.; Willis, P.A. Synthesis, structure and properties of PBO/SWNT nanocomposites. *Macromolecules* **2002**, *35*, 9039-43.

116. Zhou, C.; Wang, S.; Zhang, Y.; Zhuang, Q.; Han, Z. In situ preparation and continuous fiber spinning of poly(p-phenylene benzobisoxazole) composites with oligo-hydroxyamide-functionalized multi-walled carbon nanotubes. *Polymer* **2008**, *49*, 2520-30.

117. Qian, H.; Bismarck, A.; Greenhalgh, E.S.; Kalinka, G.; Shaffer, M.S.P. Heirarchical composites reinforced with carbon nanotube grafted fibers: The potential assessed at the single fiber level. *Chemistry of Materials* **2008**, *20*, 1862-69.

118. Mathur, R.B.; Chatterjee, S.; Singh, B.P. Growth of carbon nanotubes on carbon fibre substrates to produce hybrid/phenolic composites with improved mechanical properties. *Composites Science and Technology* **2008**, *68*, 1608-15.

119. Garcia, E.J.; Wardle, B.L.; Hart, A.J.; Yamamoto, N. Fabrication and multifunctional properties of a hybrid laminate with aligned carbon nanotubes grown in situ. *Composites Science and Technology* **2008**, 68, 2034-41.

120. Kalaitzidou, K.; Fukushima, H.; Drzal, L.T. Mechanical properties and morphological characterization of exfoliated graphite-polypropylene nanocomposites. *Composites: Part A* **2007**, *38*, 1675-*82.*

121. Kim, H.; Macosko, C.W. Morphology and properties of polyester/exfoliated graphite nanocomposites. *Macromolecules* **2008**, *41*, 3317-27.

122. Zhao, Y.F.; Xiao, M.; Wang, S.J.; Ge, X.C.; Meng, Y.Z. Preparation and properties of electrically conductive PPS/expanded graphite nanocomposites. *Composites Science and Technology* **2007**, *67*, 2528-34.

123. Du, X.S.; Xiao, M.; Meng, Y.Z.; Hay, A.S. Synthesis and properties of poly(4,4'-oxybis(benzene)disfulfide)/graphite nanocomposites via in-situ ring-opening polymerization of macrocyclic oligomers. *Polymer* **2004**, *45*, 6713-18.

124. Dhakate, S.R.; Sharma, S.; Borah, M.; Mathur, R.B.; Dhami, T.L. Development and characterization of expanded graphite-based nanocomposite as bipolar plate for polymer electrolyte membrane fuel cells (PEMFCs). *Energy and Fuels* **2008**, 22, 3329-34.

125. Hummers, W.S.; Offeman, R.E. Preparation of graphitic oxide. *Journal of the American Chemical Society* **1958**, *80*, 1339.

126. Zhang, R.; Hu, Y.; Xu, J.; Fan, W.; Chen, Z. Flammability and thermal stability studies of styrene-butyl acrylate copolymer/graphite oxide nanocomposites. *Polymer Degradation and Stability* **2004**, *85*, 583-88.

127. Matsuo, Y.; Tahara, K.; Sugie, Y. Structure and thermal properties of poly(ethylene oxide) intercalated graphite oxide. *Carbon* **1997**, *35*, 113-20.

128. Xu, J.; Hu, Y.; Song, L.; Wang, Q.; Fan, W.; Liao, G.; Chen, Z. Thermal analysis of poly(vinyl alcohol)/graphite oxide intercalated composites. *Polymer Degradation and Stability* **2001**, *73*, 29-31.

129. Han, Y.; Lu, Y. Preparation and characterization of graphite oxide/polypyrrole composites. *Carbon* **2007**, *45*, 2394-99.

130. Stankovich, S.; Dikin, D.A.; Piner, R.D.; Kohlhaas, K.A.; Kleinhammes, A.; Jia, Y.; Wu, Y.; Nguyen, SB. T.; Ruoff, R.S. Synthesis of graphene-based nanosheets via chemical reduction of exfoliated graphite oxide. *Carbon* **2007**, *45*, 1558-65.

131. Si, Y.; Samulski, E.T. Synthesis of water soluble graphene. *Nano Letters* **2008**, *8*, 1679-82.

132. Xu, Y.; Bai, H.; Lu, G.; Shi, G. Flexible graphene films via the filtration of water-soluble noncovalent functionalized graphene sheets. *Journal of the American Chemical Society* **2008**, *130*, 5856-57.

133. Du, X.; Yu, Z-Z; Dasari, A.; Ma, J.; Mo, M.; Meng, Y.; Mai, Y-W. New method to prepare graphite nanocomposites. *Chemistry of Materials* **2008**, *20*, 2066-68.

134. Stankovich, S.; Dikin, D.A.; Dommett, G.H.B.; Kohlhaas, K.M.; Zimney, E.J.; Stach, E.A.; Piner, R.D.; Nguyen, S.B.; Ruoff, R.S. Graphene-based composite materials. *Nature* **2006**, *442*, 282-286.

135. Schniepp, H.C.; Li, J-L, McAllister, M.J.; Sai, H.; Herrera-Alonso, M.; Adamson, D.H.; Prud'homme, R.K.; Car, R.; Saville, D.A.; Aksay, I.A. Functionalized single graphene sheets derived from splitting graphite oxide. *Journal of Physical Chemistry B* **2006**, *110*, 8535-39.

136. Ramanthan, T.; Abdala, A.A.; Stankovich, S.; Dikin, D.A.; Herrera-Alonso, M.; Piner, R.D.; Adamson, D.H.; Schniepp, H.C.; Chen, X.; Ruoff, R.S.; Nguyen, S.T.; Aksay, I.A.; Priud'homme, R.K.; Brinson, L.C. Functionalized graphene sheets for polymer nanocomposites. *Nature Nanotechnology* **2008**, *3*, 327-31.

137. *Nanotechnology in Space Exploration – Report of the National Nanotechnology Initiative Workshop* August 24-26 **2004,** U.S. National Science and Technology Council Committee on

Technology, Subcommitee on Nanoscale Science, Engineering and Technology and the National Aeronautics and Space Administration.

138. Vaia, R.A.; Maguire, J.F. Polymer nanocomposites with prescribed morphology: going beyond nanoparticle-filled polymers. *Chemistry of Materials* **2007**, *19*, 2736-51.

139. Koerner, H.; Hampton, E.; Dean, D.; Turgut, Z.; Drummy, L.; Mirau, P.; Vaia, R. Generating triaxial reinforced epoxy/montmorillonite nanocomposites with uniaxial magnetic fields. *Chemistry of Materials* **2005**, *17*, 1990-96.

140. Wakabayashi, K.; Pierre, C.; Dikin, D.A.; Ruoff, R.S.; Ramanathan, T.; Brinson, L.C.; Torkelson, J.M. Polymer-graphite nanocomposites: effective dispersion and major property enhancement via solid-state shear pulverization. *Macromolecules* **2008**, *41*, 1905-08.

141. Podsiadlo, P.; Kauchik, A.K.; Shim, B.S.; Agarawal, A.; Tang, Z.; Waas, A.M.; Aruda, E.M.; Ktovo, N.A. Can Nature's design be improved upon? High strength, transparent nacre-like nanocomposites with double network of sacrificial cross-links. *Journal of Physical Chemistry B* **2008**, 112, 14359-14363.

142. Shim, B.S.; Kotov, N.A. Single-walled carbon nanotube combing during layer-by-layer assembly: from random adsorption to aligned composites. *Langmuir* **2005**, *21*, 9381-85.

143. Orefice, R.L.; Ayres, E.; Pereira, M.M.; Mansur, H.S. Using the nanostructure of segmented polyurethanes as a template in the fabrication of nanocomposites. *Macromolecules* **2005**, *38*, 4058-60.

144. Deshmukh, R.D.; Liu, Y.; Composto, R.J. Two-dimensional confinement of nanorods in block copolymer domains. *Nano Letters* **2007**, *7*, 3662-68.

Glossary of materials and techniques referred to within this chapter

6-FDA: 2,2-bis(3,4-dicarboxyphenyl)-1,1,1,3,3,3-hexafluoropropane dianhydride, a monomer typically used in the synthesis of high temperature stable polyimides, has the following chemical structure:

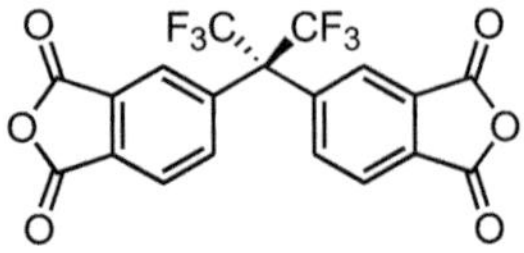

6-FDA/1,3-APB : polyimide prepared from 2,2-bis(3,4-dicarboxyphenyl)-1,1,1,3,3,3-hexafluoropropane dianhydride (6-FDA) and 1,3-bis(4-aminophenoxybenzene) (APB), has the following chemical structure:

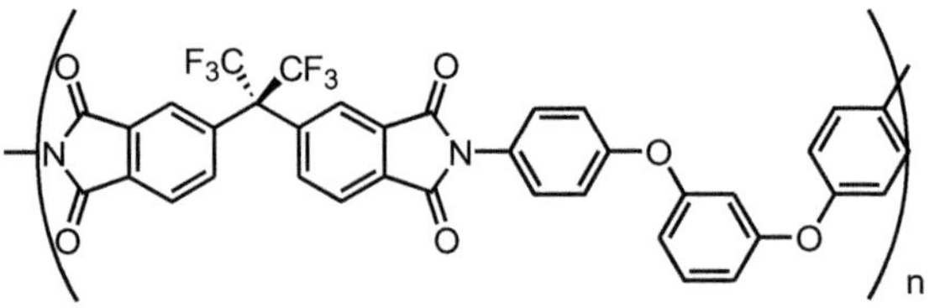

APB-BPDA : polyimide prepared from 1,3-bis(4-aminophenoxy)benzene (APB) and 3,3',4,4'-biphenyltetracarboxylic dianhydride (BPDA), having the following chemical structure:

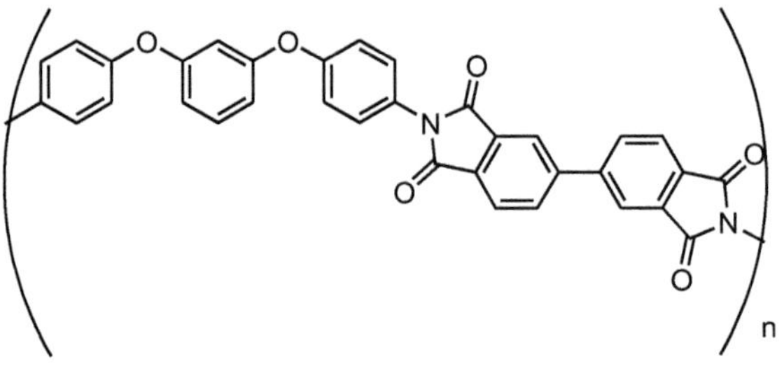

BAPP/BPADA : polyimide prepared from 2,2-bis(4-aminophenoxylphenyl)propane (BAPP) and 2,2'-bis(3,4-dicarboxyphenoxy)biphenyl dianhydride (BPADA), has the following chemical structure:

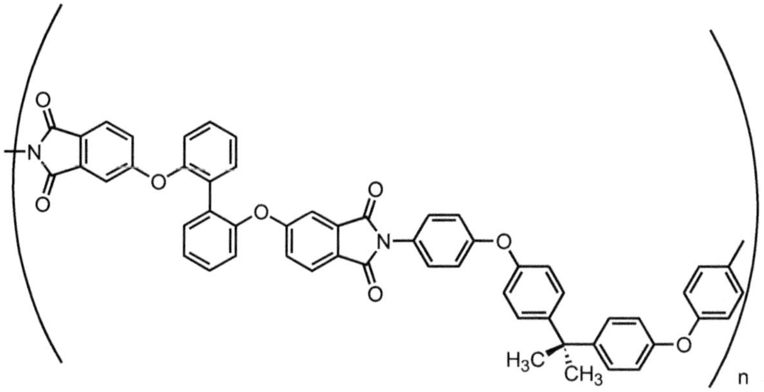

BAPS : bis(4-aminophenoxy)diphenylsulfone, a diamine used in high performance polyimides and epoxy resins, has the following chemical structure:

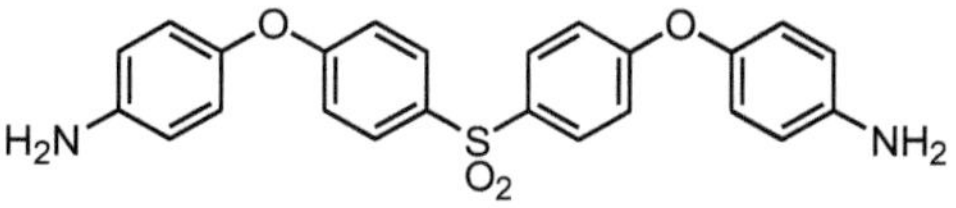

BET : method for determining the surface area of materials based upon gas adsorption measurements, based upon a theory derived by Stephen Brunnauer, Paul Hugh Emmett and Edward Teller (see *Journal of the American Chemical Society* **1938**, *60*, 309.)

BTDA/BAPP : polyimide derived from 3,3',4,4'-benzophenone tetracarboxylic dianhydride (BTDA), and 2,2-bis(4-aminophenoxylphenyl)propane (BAPP), has the following chemical structure:

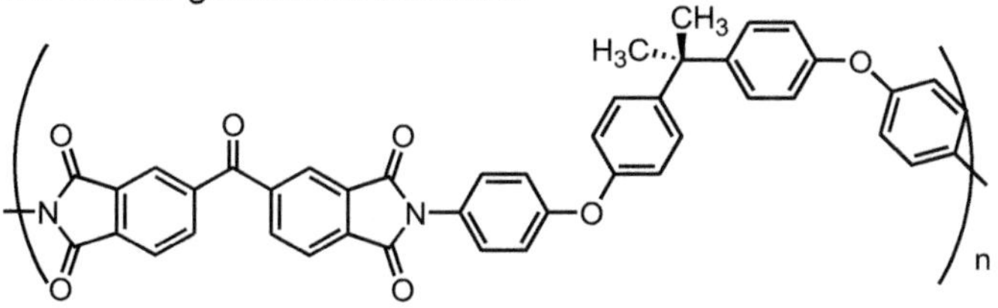

BTDA-ODA : polyimide prepared from 3,3',4,4'- benzophenone tetracarboxylic dianhydride (BTDA) and 4,4'-oxydianiline (ODA), having the following chemical structure:

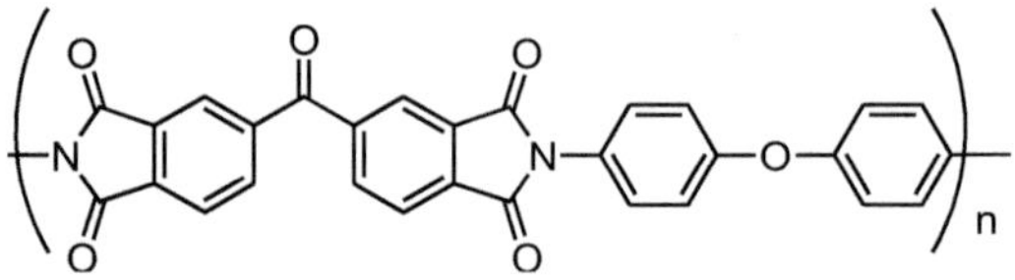

CB : carbon black, an additive used to enhance the thermal and/or electrical conductivity of polymers

Cloisite 30 B : organoclay commonly used in nanocomposites prepared by exchanging Na+ montmorillonite clay with $H_3C-\overset{+}{\underset{CH_2CH_2OH}{\overset{CH_2CH_2OH}{N}}}-T$ where T = Tallow

CNT : carbon nanotube

CTE : coefficient of thermal expansion, the change in length, L, of a material as a function of temperature, T, defined by:

$$\alpha = \left(\frac{1}{Lo}\right)\frac{\partial L}{\partial T}$$ where Lo is the initial length of the specimen.

CVD : chemical vapour deposition, a process for applying coatings to various substrates that involves a gaseous precursor that decomposes onto the surface of a given substrate to provide a coating.

DC : Direct Current

DC volume conductivity/resistivity: electrical (Direct Current) conductivity/resistivity of a bulk material

DGEBA : digylcidyl ether of Bisphenol A, a standard epoxide monomer used in a number of aerospace grade epoxy resins

DMAc : Dimethyl acetamide

DMF : N,N-dimethylformamide, a solvent commonly used in polyimide synthesis

DMBz-15: high temperarature thermosetting polyimide prepared from nadic anhydride, 2,2'-dimethylbenzidine, and 3,3',4,4'-benzophenone tetracarboxylic dianhydride, having the following chemical structure:

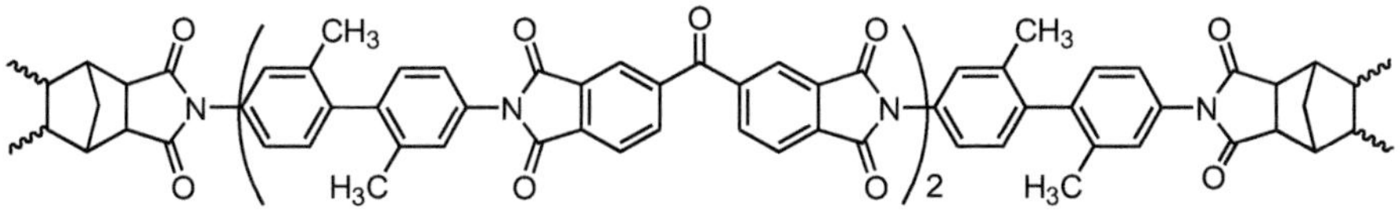

DNA: Deoxyribonucleic acid

DWCNT : double wall carbon nanotube

DOE : United States Department of Energy

EG : expanded graphite, prepared by intercalating flake graphite with sulfuric or nitric acid followed by rapid heating to temperatures above 900°C.

Epon 862 : commercially available epoxy resin prepared from epichlorohydrin and bisphenol F

FGS : Functionalized Graphene Sheets, prepared by rapidly heating graphite oxide to 1050°C

GO : graphite oxide, prepared by oxidation of flake graphite

HiPCO : High Pressure Carbon Monoxide Disproportionation – method for the synthesis of carbon nanotubes developed by the late Nobel Laureate Richard Smalley.

LaRC-CP2 : a colorless polyimide prepared from 6-FDA and APB (see also 6-FDA/APB polyimide.

LBL : layer by layer assembly, a technique for producing materials which typically involves dipping a substrate into a coating solution a series of times to produce a layered structure.

Mode I Interlaminar Fracture Toughness: measure of the interlaminar toughness of a composite under Mode I fracture, in which a laminate is pulled apart perpendicular to a crack, also known as opening mode.

MWCNT : multiwall carbon nanotube

Nafion : sulfonated tetrafluoroethylene copolymer commonly used in fuel cell membranes, has the following chemical structure:

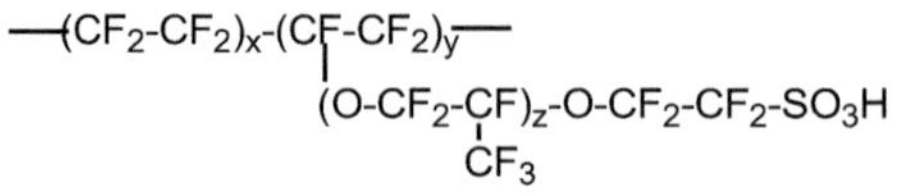

$$—(CF_2\text{-}CF_2)_x\text{-}(CF\text{-}CF_2)_y—$$
$$(O\text{-}CF_2\text{-}CF)_z\text{-}O\text{-}CF_2\text{-}CF_2\text{-}SO_3H$$
$$CF_3$$

ODA-BPDA : polyimide prepared from 4,4'-oxydianiline (ODA) and 3,3',4,4'-benzophenone tetracarboxylic dianhydride (BPDA), has the following chemical structure:

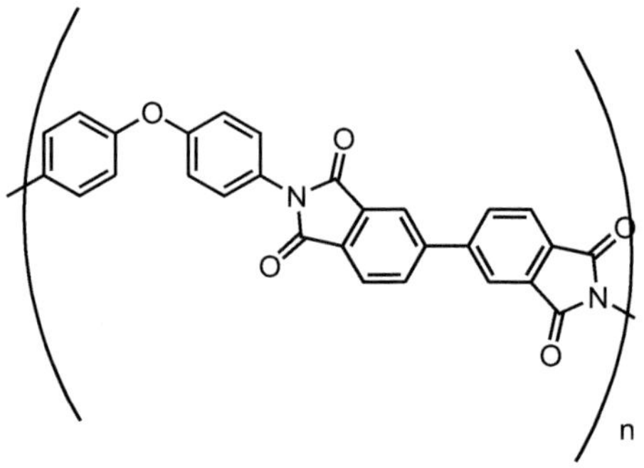

PBO : poly(phenylene benzobisoxazole), a rigid rod polymer with the following chemical structure:

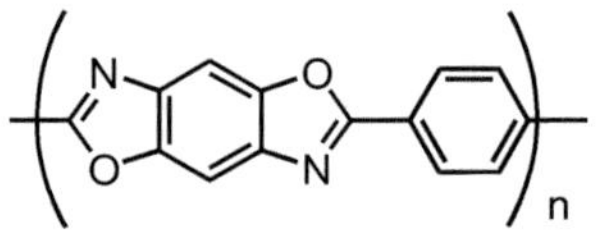

PEM fuel cells: proton exchange membrane fuel cells

PEN : poly(ethylene-2,6-napthalate), a liquid crystalline polyester having the following chemical structure:

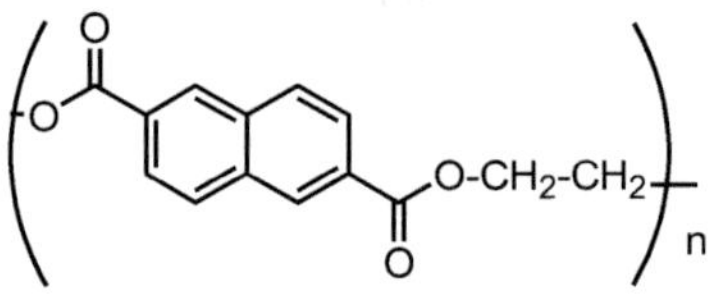

PMDA-ODA : polyimide prepared from pyromellitic dianhydride (PMDA) and 4,4'-oxydianiline (ODA), with the following chemical structure:

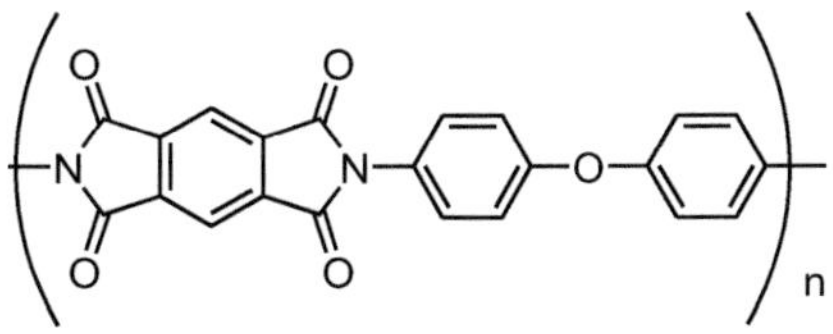

PMMA : poly(methyl methacrylate)

Polyurethane : polymer prepared from an alcohol with an isocyanate via the following scheme:

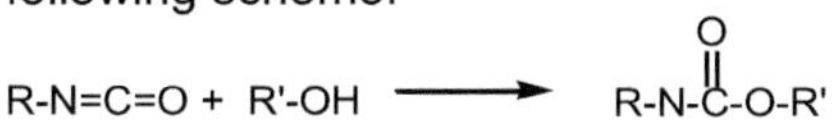

R-N=C=O + R'-OH ⟶ R-N-C-O-R'

POBDS : poly(4,4'-oxybis(benzene) disulfide), a high performance thermoset having the following chemical structure:

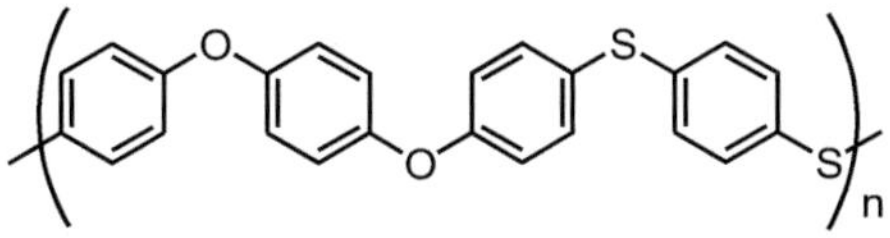

PPS : poly(phenylene sulfide), a high performance thermoplastic polymer having the following chemical structure:

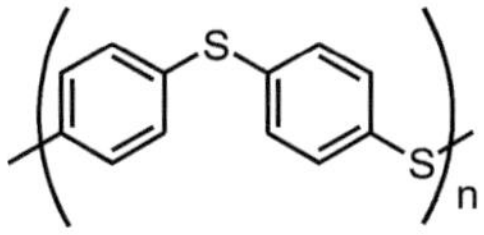

SWCNT : single wall carbon nanotube

T-300 carbon fibres: High modulus carbon fibres (3.75 GPa tensile strength, 231 GPa tensile modulus) prepared by pyrolysis of pitch.

T-650-35 carbon fibres: intermediate modulus (4.28 GPa tensile strength, 255 GPa tensile modulus) PAN (polyacrylonitrile)-based carbon fibre often used in aerospace composites either as continuous fibre reinforcement or in fabric, typically in an 8-HS (8 harness satin) weave.

TEM : Transmission Electron Microscopy

TGA : Thermogravimetric Analysis

THF : tetrahydrofuran

XPS : X-ray Photoelectron Spectroscopy, a quantitative spectroscopic technique that determines elemental composition of materials, commonly used for surface analysis.

AAS	atomic absorption spectrometry (AAS)
ABS	acrylonitrile-butadiene-styrene terpolymer
ACNT	aligned carbon nanotube
ADA	12-aminododecanoic acid
AFM	atomic force microscopy
amu	atomic mass unit
APB-BPDA	polyimide
ASTM	American Society for Testing and Materials
ATH	aluminium trihydrate
AUA	amino undecanoic acid
BAPP/BPADA	polyimide
BAPS	bisaminophenoxy diphenylsulphone
BET	Brunnauer Emmett Teller method for determining surface area
BLT	bondline thickness
BPT	black panel temperature
BTDA/BAPP	polyimide
BTDA/ODA	polyimide
CEC	cation exchange capacity
CB	carbon black
CDNP	combustion derived nanoparticle
CEC	cation exchange capacity
CLTE	coefficient of linear thermal expansion
CNF	carbon nanofibre
CNS	central nervous system
CNT	Carbon nanotube
COC	cyclic olefin copolymer
COPD	chronic obstructive pulmonary disease
CPD	Construction Products Directive (European Commission)
Cperc	percolation concentration limit
CTE	coefficient of thermal expansion
CV	cardiovascular
CVD	chemical vapour deposition
DBDPE	decabromodiphenyl ether
DC	direct current
DGEBA	diglycidyl ether of bisphenol A (epoxy monomer)
DIN EN	Deutsches Institut fur Normung European standard
DMA	dynamic mechanical analysis
DMAc	dimethyl acetamide
DMBz-15	high temperature thermosetting polyimide
DMF	dimethyl formamide
DNA	deoxyribonucleic acid
DOE	United States Department of Energy

DSC	differential scanning calorimetry
DTA	differential thermoanalysis
DTG	differential thermogravimetry
DTUL	heat distortion temperature under load
DWCNT	double-walled carbon nanotube
EG	expanded graphite
EPDM	ethylene-propylene-diene rubber
EVA	ethylene-vinyl acetate copolymer
EVOH	ethylene-vinyl alcohol copolymer
6-FDA	2,2-bis(3,4-dicarboxyphenyl)-1,1,1,3,3,3-hexafluoropropane dianhydride monom
6-FDA/1,3-APB	polyimide
FDA	US Food and Drug Administration
FGS	functionalized graphene sheets
FIGRA	fire index growth rate
FIPEC	Fire Performance of Electrical Cables
FRNH	flame retardant non halogen
FS	flame spread
FTIR	Fourier transform infrared spectroscopy
GI	gastro-intestinal
GM	General Motors Corporation
GO	graphite oxide
GPC	gel permeation chromatography
HALS	hindered amine light stabiliser
HAR	high aspect ratio
HDPE	High density polyethylene
HDT	heat deflection temperature
HHT	high heat treatment
HI-PS	high impact polystyrene
HiPCO	high pressure carbon monoxide disproportionation
ICP-MS	inductively coupled plasma−mass spectrometry
IR	infrared spectroscopy
ITO	indium tin oxide
LaRC-CP2	colourless polyimide
LbL	layer by layer (deposition)
LCA	life cycle assessment
LDH	layered double hydroxides
LDL	low-density lipoprotein
LDPE	low density polyethylene
LLDPE	linear low polyethylene
LHT	low heat treatment
LOI	limiting oxygen index
MAH	maleic anhydride

MD	metal deactivator
MDH	magnesium dihydroxide
MLS	montmorillonite layered silicate
MMVF	man made vitreous fibre
MRE	meal ready-to-eat
MVR	melt volume rate
MWCNT	multi-walled carbon nanotube
Nafion	sulfonated tetrafluoroethylene copolymer
NEXAFS	near edge X-Ray absorption fine structure
NMR	nuclear magnetic resonance spectroscopy
NP	nanoparticles
NVP	N-vinylpyrrolidinone
Octa-TMA	octatetramethylammonium POSS
ODA	octadecylacrylate
ODA-BPDA	polyimide
OIT	oxygen induction time
ORMOCER	organically modified ceramics
ORMOSIL	organically modified silicas
OSHA PEL	U.S. occupational safety & health administration's permissible exposure limit
OTR	oxygen transmittance rates
PA	polyamide
PA-6	polyamide-6
PAH	polycyclic aromatic hydrocarbons
PBAT	poly(butylene adipate)-co-terephthalate
PBO	polyphenylene benzobisoxazole
PBS	polybutylene succinate
PBSA	poly(butylene succinate)-co-adipate
PC	polycarbonate
PCL	polycaprolactone
PCS	gross calorific potential (EN ISO 1716)
PE	polyethylene
PEG	polyethylene glycol
PEM	proton exchange membrane (fuel cells)
PEN	polyethylene naphthalate
PEO	polyethylene oxide
PET	polyethylene terephthalate
PFA	perfluoroalkoxy (fluoropolymer)
PHA	polyhydroxyalkanoate
PHB	polyhydroxybutyrate
PHBV	polyhydroxybutyrate-co-valerate
PHRR	peak of heat release rate
PLA	polylactic acid

PM	particulate matter
PM10	particulate matter smaller than 10 micrometres
PMDA-ODA	polyimide
PMMA	polymethyl methacrylate
POBDS	polyoxybisbenzene disulfide
POSS	polyhedral oligomeric silsesquioxane
PP	polypropylene
PPE	polyphenylene ether
PP-g-MAH	polypropylene-g-maleic anhydride
PPS	polyphenylene sulphide
PS	polystyrene
PSP	poorly soluble particles
PVA	polyvinyl acetate
PVC	polyvinyl chloride
PVDC	polyvinylidene chloride
PVNO	polyvinyl–pyridine-N-oxide
RH	relative humidity
RHR	rate of heat release
RNS	reactive nitrogen species
ROS	reactive oxygen species
rpm	revolutions per minute
RTD	resistance temperature detector
SAN	styrene-acrylonitrile copolymer
SEM	scanning electron microscopy
SPR	smoke production rate
SWCNT	single-walled carbon nanotube
TEM	transmission electron microscopy
TEOS	tetraethoxysilane
TGA	thermogravimetric analysis
THF	tetrahydrofuran
THR	total heat released
TIM	thermal interface materials
TKD	twist kneading rotor
TMOS	tetramethoxysilane
TPO	thermoplastic olefin
TPU	thermoplastic polyurethane
TSP	total smoke production
UFP	ultrafine particles
VGCF	vapour grown carbon fibre
WVTR	water vapour transmittance rates
XPS	x-ray photoelectron spectroscopy
XRD	X-ray diffraction